PLASMIDS

CURRENT RESEARCH AND FUTURE TRENDS

Edited by:

Georg Lipps
University of Bayreuth
Department of Biochemistry II
Universitätsstr. 30
D-95440 Bayreuth
Germany

Caister Academic Press

www.caister.com

British Library Cataloguing-in-Publication Data

A catalogue record for this book is available from the British Library

ISBN: 978-1-904455-35-6

Description or mention of instrumentation, software, or other products in this book does not imply endorsement by the author or publisher. The author and publisher do not assume responsibility for the validity of any products or procedures mentioned or described in this book or for the consequences of their use.

Printed and bound in Great Britain

Contents

Contributors

Wolfgang Schumann*
Institute of Genetics
University of Bayreuth
D-95440 Bayreuth
Germany

Georg Lipps*
University of Bayreuth
Department of Biochemistry II
Universitätsstr. 30
D-95440 Bayreuth
Germany

Armin Baiker
Max von Pettenkofer-Institute
Pettenkoferstrasse 9a
80336 Munich
Germany

Rudolf Haase
Max von Pettenkofer-Institute
Pettenkoferstrasse 9a
80336 Munich
Germany

Hans Joachim Lipps*
University of Witten/Herdecke
Institute for Cell Biology
Stockumer Strasse 10
58448 Witten
Germany

Jason R. Grant
Department of Agricultural
Food and Nutritional Science
University of Alberta
T6G 2P5
Canada

Paul Stothard*
Department of Agricultural
Food and Nutritional Science
University of Alberta
T6G 2P5
Canada

Masahiro Sota
Department of Biological Sciences
University of Idaho
Moscow ID 83844-3051
USA

Eva M. Top*
Department of Biological Sciences
University of Idaho
PO Box 443051
Moscow ID 83844-3051
USA

Silvia Russi
Institute for Research in Biomedicine
Institut de Biologia Molecular de
Barcelona (CSIC)
Barcelona Science Park
Baldiri Reixac 10-12
08028 Barcelona
Spain

Roeland Boer
Institute for Research in Biomedicine
Institut de Biologia Molecular de
Barcelona (CSIC)
Barcelona Science Park
Baldiri Reixac 10-12
08028 Barcelona
Spain

Miquel Coll*
Institute for Research in Biomedicine
Institut de Biologia Molecular de Barcelona (CSIC)
Barcelona Science Park
Baldiri Reixac 10-12
08028 Barcelona
Spain

Karthikeyan Kandavelou
Department of Environmental Health Sciences
Bloomberg School of Public Health
Johns Hopkins University
615 North Wolfe Street
Baltimore, MD 21205
USA

Srinivasan Chandrasegaran*
Department of Environmental Health Sciences
Bloomberg School of Public Health
Johns Hopkins University
615 North Wolfe Street
Baltimore, MD 21205
USA

Devin B. Lowe
Department of Microbiology and Immunology
Texas Tech University Health Sciences Center
3601 4th Street, MS 6591
Lubbock, TX 79430
USA

Michael H. Shearer
Department of Microbiology and Immunology
Texas Tech University Health Sciences Center
3601 4th Street, MS 6591
Lubbock, TX 79430
USA

Cynthia A. Jumper
Department of Internal Medicine
Texas Tech University Health Sciences Center
3601 4th Street, MS 9410
Lubbock, TX 79430
USA

Ronald C. Kennedy*
Department of Microbiology and Immunology
Texas Tech University Health Sciences Center
3601 4th Street, MS 6591
Lubbock, TX 79430
USA

En-Min Zhou
Department of Preventive Veterinary Medicine
College of Animal Science and Veterinary Medicine
Shandong Agricultural University
Shandong
China

*** Corresponding author**

Preface

Plasmids are fascinating entities which can replicate autonomously in bacterial, archaeal and eukaryotic cells. Plasmids profit from the cellular environment of the host and cellular enzymes ensure their replication. However many plasmids are not selfish. Plasmids can carry a rich diversity of genes which might be beneficial for the host: Plasmids provide bacteria with the ability to degrade calcitrant organic compounds or to fix elemental nitrogen. Plasmids can also carry antibiotic resistance genes and their spread in pathogenic bacteria is of great medical importance. In addition, plasmids were and are continuing to be of primordial importance for studying the molecular biology of various organisms.

The first three chapters of the book are focused on the occurrence of plasmids in the three domains of life, the bacteria, archaea and eukarya. W. Schumann gives a historical perspective on *E. coli* plasmids and provides some scientific background of the plasmids now used routinely in the laboratories to clone genes, to express proteins and to study various biological phenomena by genetic methods. G. Lipps gives an overview on the plasmids of archaea. Our knowledge on archaeal plasmids is still quite limited but more and more plasmids are discovered and shed light onto the molecular biology and ecology of these novel genetic elements. In the last chapter of this section the plasmids of eukarya are described by H.J. Lipps. Approaches to construct novel plasmids which replicate faithfully in mammalian cells and could eventually be used for gene therapy are summarized.

The following chapter by P. Stothard gives an account on bioinformatic tools which can be used to analyse plasmid sequences. The chapter –although focussed on plasmids- is also a perfect primer for everybody interested in nucleotide sequence analysis. In this chapter the reader will also find a detailed list of useful web resources for plasmid biology.

Plasmids have devised a number of strategies to be efficiently transferred from cells to cells. The impact of horizontal gene transfer on evolution and the adaptability of microbial consortia is well recognized. Two chapters are dedicated to horizontal gene transfer: E. Tops reviews in the first chapter the properties of plasmids and mobile elements and their interesting molecular biology and ecology. In the second chapter by M. Coll the structures of the molecular machines of the conjugational DNA transfer are presented and our knowledge on the process of DNA transfer is reviewed.

The last two chapters are focussed on two important medical applications of plasmids. S. Chandrasegaran describes the use of plasmids to correct gene defects. In the last chapter R.C. Kennedy summarizes the use of plasmids as vaccines to treat cancer or infectious diseases.

To conclude, plasmids continue to be a fascinating field of research with ramification in synthetic biology, medicine, ecology and evolution as well as basic research in molecular and structural biology.

Georg Lipps (April 2008)

Other Books of Interest

Chapter 1

Escherichia coli Cloning and Expression Vectors

Wolfgang Schumann

ABSTRACT

The introduction of gene technology methods together with the DNA sequencing and PCR methodologies have revoluzionarized basic and applied research in molecular biology involving all kinds of organisms from eubacteria up to humans. Since almost all of the cloning experiments start by insertion of DNA fragments into *Escherichia coli* plasmid vectors, this chapter describes important cloning vectors from the very beginning up to the present days. Furthermore, it deals in detail with expression vectors and the promoter elements needed for high-level production of recombinant proteins. The last part describes special vectors allowing for the analysis of promoters and terminators, addition of localization and epitope tags.

INTRODUCTION

In the past three decades, we have witnessed a revolution in the fields of molecular genetics and biochemistry. This revolution is the result of three major technical advancements termed "Recombinant DNA Technology", "DNA Sequencing" and "PCR". While the first methods allows cloning of any DNA sequence and mainly genes, the second technique not only allows sequencing of single genes but since more than ten years complete bacterial genomes and even eukaryotic genomes including our own. The last technique, PCR, complements the first, also known as genetic engineering, by allowing amplification of any DNA sequence followed by subsequent cloning of these amplicons. Two different vector types have been developed based on naturally occurring plasmids and phages. But plasmid vectors are the first choice because of their easy preparation and handling. Plasmids are extrachromosomal genetic elements found in almost all bacterial species. They are double-stranded, closed circular DNA molecules (a few linear plasmids have also been described) that range in size from 1 kb to greater than 200 kb. Among the phenotypes conferred by different plasmids are:

- resistance to antibiotics
- production of antibiotics
- production of bacteriocins (proteins that can kill cells of the same or closely related species)

- production of enterotoxins
- degradation of complex organic compounds
- production of virulent factors

Based on such naturally occurring plasmids, a variety of vector plasmids have been developed mainly for *E. coli*, the first-choice organism in cloning experiments. To be able of autonomous replication, each plasmid carries an origin of replication abbreviated *ori* or *oriV* and codes for one or more proteins involved in their replication. Plasmids can be classified by several criteria:

- conjugative or mobilizable
- incompatibility groups
- copy number
- host range

Plasmids of Gram-negative bacteria are either conjugative or mobilizable depending on the presence or absence of three different elements. These three elements are present on all conjugative plasmids and include a *cis*-element called *oriT* (origin of transfer), one or more *mob* (mobilization) genes where the Mob proteins constitute the relaxosome and the *tra* (transfer) genes which code for the pilus anchored in the two membranes and responsible for making contact with the recipient cells. Mobilizable plasmids do not code for the *tra* genes and need the presence of a conjugative plasmid to be able to transfer (mobilize) their single-stranded DNA into the recipient cell. Transfer of the conjugative and the mobilizable plasmid occur independent of each other. Most plasmid vectors are mobilizable; a few are devoid of the *oriT* sequence. Conjugation and horizontal gene transfer are dealt in more details in the chapters "Molecular Machinery for DNA Translocation in Bacterial Conjugation" and "Horizontal Gene Transfer Mediated by Plasmids".

Another important characteristic to classify plasmids is their incompatibility status. Two different plasmids are said to be compatible with each other, if they can stably coexist without selective pressure. If one plasmid is lost, both plasmids are incompatible with each other, and all known plasmids have been arranged in different incompatibility groups (Bouet et al., 2007). The knockledge of the incompatibility group of plasmids becomes important when two or more plasmids are introduced into the same cell.

The third characteristic of a given plasmid is its copy number per bacterial chromosome. Based on the copy number, plasmids can be arranged into four different groups: low-copy-number plasmids (1-2 copies per chromosome; example: the mini-F plasmids), medium-copy-number plasmids (5-10; pSC101, pACYC177/184), high-copy-number plasmids (20-50; pBR3222 and its derivatives) and very high-copy-number plasmids (100-500; the pUC plasmids).

The fourth characteristic is the host-range of the plasmids. Most plasmids have a narrow host range that means they can replicate only in a few bacterial species. All ColE1-derived plasmids such as pBR322 are able to replicate only in the *Enterobacteriaceae*. Plasmids with a broad host-range are able to replicate in many different species. One of the most prominent examples is RP4 (also known as RP1, RP8 and RK2), which replicates in all Gram-negatives (Thomas, 1981). Another example is pC194 which has been reported to replicate in low G+C Gram-positives and in *E. coli* (Goursot et al., 1982).

VERSATILE CLONING VECTORS

The first cloning experiments using restriction enzymes has been published by Stanley Cohen and coworkers (Cohen et al., 1973; Morrow et al., 1974). In the first experiment, they cloned DNA fragments of the R factor R6-5 into pSC101 and fused pSC101 and the plasmid RSF1010 using their unique *Eco*RI sites (Cohen et al., 1973). In the second experiment, they cloned rRNA genes from *Xenopus laevis* into the plasmid pSC101 and demonstrated stable replication and transcription of eukaryotic DNA within *E. coli* cells (Morrow et al., 1974). The cloning vector pSC101 is the first plasmid used in cloning experiments carrying an antibiotic resistance marker (tetracycline); it is a deletion derivative of the natural R factor R6 and has been derived by sheering (Cohen and Chang, 1973). Later, a temperature-sensitive replicative derivate was isolated and termed pHSG1 (Hashimoto and Sekiguchi, 1976). This plasmid replicates stably at around 30°C, but fails to be replicated at temperatures ≥37°C. Derivatives with three antibiotic resistance markers and no *oriT* have been created later (Hashimoto-Gotoh et al., 1981).

A second vector used in the early days of cloning was the natural plasmid ColE1 (Hershfield et al., 1974). Since this plasmid does not code for any antibiotic resistance gene, the recombinant genes were cloned into the gene coding for colicin E1 and selection was carried out by plating on rich medium spread with a crude preparation of colicin E1. Only cells carrying a plasmid could grow under these conditions since they express the immunity gene which codes for a protein able to neutralize the colicin. In a second step, cells carrying the empty vector are distinguished from recombinant plasmids by the failure to produce the colicin. This was tested by toothpicking candidates on a law of sensitive bacteria. Cells carrying the empty vector yielded colonies surrounded by a halo originating from the sensitive cells killed by the colicin.

The next step in the development of versatile cloning vectors resulted in the construction of plasmids carrying two different antibiotic resistance markers allowing the identification of recombinant plasmids by the principle of insertional inactivation. Here, the foreign DNA fragment is inserted into one of the two antibiotic resistant genes. Transformants are selected on plates containing the appropriate antibiotic encoded by the intact gene and screened for the loss of resistance by the inactivated gene. The first cloning vectors containing two antibiotic resistance genes are the pBR series starting with pBR312 and pBR313 (Bolivar et al., 1977b) followed by pBR322 and pBR325 (Bolivar et al., 1977a) and some others. All these vectors have been derived from the natural plasmid pMB1, closely related to ColE1. pBR322 codes for two antibiotic resistance markers, ampicillin (derived from the transposon Tn*3*) and tetracycline resistance (derived from pSC101). Addition of a chloramphenicol resistance marker to pBR322 resulted in pBR325; pBR327 and pBR328 are deletion derivates of pBR322 and pBR325, respectively, from which the *oriT*s have been removed (Soberon et al., 1980).

Other important cloning vectors constructed in the early days are pACYC177 and pACYC184 (Chang and Cohen, 1978). Both vectors are based on the naturally occurring plasmid P15A to which two different antibiotic resistance markers have been added, ampicillin and kanamycin in the case of pACYC177 and tetracycline and chloramphenicol in the case of pACYC184. Both vectors are medium-copy number plasmids and fully compatible with the pBR series. These are useful vectors in cloning genes coding for proteins which are harmful for the cell upon production from a high-copy number vector such as integral membrane proteins.

Another important series of cloning vectors are the pUC plasmids. They are derivatives of pBR322 containing only the ampicillin resistance gene and part of the *lacZ* gene of the *E. coli lac* operon. The *lacZ* gene codes for the enzyme β-galactosidase which exerts two activities. The minor activity is to convert the sugar lactose into its isomer allolactose and its major activity is to cleave lactose (and later the allolactose) into glucose and galactose. The enzyme consists of 1024 amino acid residues, and its active form is a homotetramer. One of the many *lacZ* mutations with the allele number M15 carries an in-frame internal deletion removing the codons 11 to 41 and leading to an inactive enzyme. The M15 deletion can be complemented by a truncated form of the β-galactosidase where only the first 146 amino acid residues are present. This deletion derivative has been designated *lacZ*'. Here, two monomers of LacZM15 and two of LacZ' form an active tetramer. This kind of complementation has been called α-complementation. Furthermore, it could be shown that the first eight amino acid residues encoded by *lacZ*' are dispensable and that even additional amino acids can be inserted between the seventh and the eighth amino acid of *lacZ*' without destroying its complementing activity. All pUC vectors contain the *lacZ*' allele and the different vectors are distinguished by the number of restriction sites introduced between the seventh and eighth codon without destroying the reading frame. Furthermore, the pUC vectors have been constructed as pairs, where one pair is distinguished by the sequence of restriction sites. While pUC8 carries unique sites for *Eco*RI-*Sma*I-*Bam*HI-*Sal*I-*Pst*I-*Hin*dIII, the sequence of sites is inverted in pUC9 (Vieira and Messing, 1982). If one of the pUC vectors is transformed into an *E. coli* strain carrying the M15 deletion, the transformants are able to degrade lactose or a chromogenic compound termed X-gal (stands for 5-bromo-4-chloro-3-indolyl-β-D-galactopyranoside). Cleavage of X-gal will turn the colour of the transformants from white to blue. If the pUC vectors carry an insert within the multiple cloning site (MCS) correct expression of the *lacZ'* allele will be prevented abolishing α-complementation. After transformation of a ligation mixture into an *E. coli* M15 strain, blue colonies indicate empty vectors while white colonies represent candidates with a DNA insert. If this *E. coli* strain carries an intact copy of *lacI*, isopropyl-β-D-thiogalactopyranoside (IPTG) has to be added to the medium. The *lacI* gene codes for the Lac-repressor, a homotetramer, which prevents expression of the *lacZ* and two additional genes of the *lac* operon. While the above mentioned allolactose is the natural inducer of the *lac* operon, IPTG is a gratuitous inducer which does not need an uptake system, but enters the cytoplasm via diffusion.

Another group of vectors is derived from the F factor. Since this naturally occurring plasmid is too large to be used as a vector (100 kb) and does not code for an antibiotic resistance gene, smaller derivatives have been constructed called mini-F (Eichenlaub et al., 1977; Eichenlaub, 1979). Later, the mini-F plasmids have been renamed BAC (bacterial artificial chromosome) which can accept and stably replicate DNA inserts up the 300 kb (Shizuya et al., 1992).

Vectors with specialized features are cosmids, phagemids and fosmids. Cosmids (Fig. 1A) are plasmids that contain the cohesive end sites (*cos*) of bacteriophage

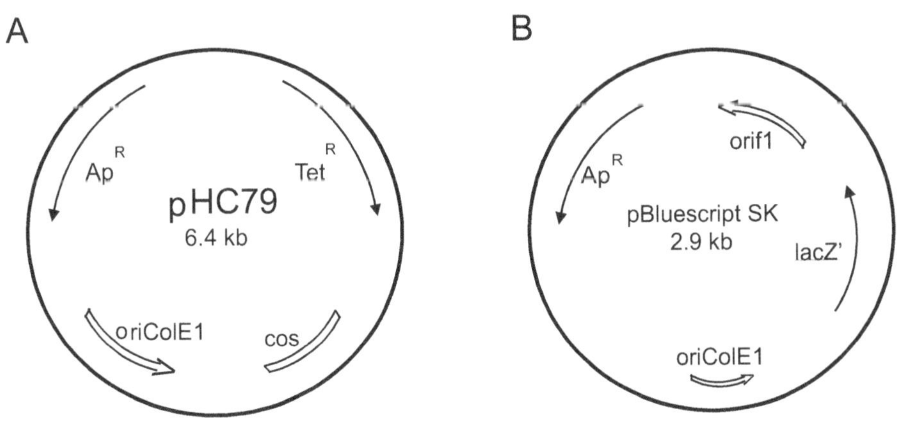

Fig. 1. Genetic maps of the cosmid pHC79 (A) and the phagemid pBluescript SK (B). Indicated are the antibiotic resistance marker (Ap^R, ampicillin resistance; Tet^R, tetracycline resistance), the truncated *lacZ* gene (*lacZ'*), the region of the plasmid (oriColE1) and phage f1 origin of replication (oriF1) and the *cos* region allowing *in vitro* packaging.

λ (Hohn and Collins, 1980). A *cos* site is a DNA sequence derived from phage λ which serves as a packaging signal (Skalka, 1977). Packing requires that two *cos* sites are located in parallel on the same DNA molecule. The packaging enzyme (the λ terminase) will carry out staggered cuts within the two *cos* sequences leaving 12 nucleotides 5' overhang at each end. The DNA between two *cos* sites will be stably packed provided its size is not smaller than 37 kb and not larger than 52 kb. Cloning in cosmids requires the insertion of DNA fragments about 35 kb in length. If these fragments are ligated into cosmid molecules in such a way that they are flanked by two *cos* sequences within the same orientation and they are 37 to 52 kb apart, the DNA between these *cos* regions will be packaged in λ phage particles. Upon infection of sensitive *E. coli* host strains, the linear DNA will be injected, circularized via the cohesive ends and further replicated using the origin of replication of the plasmid. Since cosmids are derivatives of pBR322, they are present in a high copy number which can lead to instability depending on the DNA sequences cloned. Therefore, fosmids have been developed (Kim et al., 1992). These are based on mini-F vectors and are therefore present in 1-2 copies per chromosome.

Phagemids (Fig. 1B) are defined as cloning vectors carrying two different origins of replication, one derived from a plasmid and one from a phage (Melnikov et al., 1984; Mead and Kemper, 1988). These two origins allow replication of the phagemids either as plasmids or as phages. Most phagemids contain the origin of replication of a filamentous phage such as f1 allowing the production and packaging of single-stranded DNA upon overinfection of *E. coli* cells carrying recombinant phagemids. This will lead to the production of phage particles, and their single-stranded DNA can be directly used, *e.g.* in mutagenesis experiments.

In summary, a versatile cloning vector should fulfil the following characteristics:

- small size (2 – 5 kb)
- two selectable antibiotic resistance markers
- stable replication
- several unique restriction sites organized in a MCS

VERSATILE EXPRESSION VECTORS

With the advent of gene cloning, overexpression of genes has replaced purification of proteins from large amounts of cells. Recombinant proteins can be produced either in their authentic form or fused to a purification tag. High level production of recombinant proteins needs strong expression signals including efficient promoters. A promoter is defined as a stretch of DNA sequence that encodes information for RNA polymerase (RNAP) binding and initiation of transcription. At the beginning, constitutive promoters were used, but it turned out quickly that continuous high-level production of even a normal non-toxic protein may be disadvantageous and lead to decreased stability of the particular host-vector system. These observations led to the development of regulated promoters which are induced in most cases by a chemical compound such as IPTG or by a shift in the growth conditions, *e.g.*, a temperature downshift (reviewed in (Makrides, 1996; Hannig and Makrides, 1998; Baneyx, 1999; Baneyx, 1999; Schumann and Ferreira, 2004)). But besides an efficient promoter, two additional parameters influence the expression level of genes coding for recombinant proteins: an efficient SD sequence and a stabilizer of the mRNA.

Bacteria code for a multisubunit RNAP with a conserved subunit composition. The core enzyme is composed of one β, β', ω subunit each and of two α subunits. Binding of a σ subunit with the core enzyme constitutes the holoenzyme and determines the specificity of promoter utilization. All bacterial species code for one primary or housekeeping σ factor and one or more secondary or alternative σ factors. While the housekeeping σ factor (σ^{70} in *E. coli*) is needed mainly during the vegetative growth phase under physiological conditions to transcribe all the genes involved in replication, cell division, transcription, translation etc., alternative σ factors are employed mainly under stressful conditions. One example is σ^{S} which largely replaces σ^{70} when *E. coli* cells enter the stationary growth phase. Another example is σ^{32} which becomes activated under heat shock conditions. A heat shock is defined as a drastic increase in the growth temperature. To induce a heat shock in *E. coli* cells, the growth temperature is increased from 30 to 42°C, and σ^{32} is responsible for enhanced expression of approximately 60 genes to combat the consequences of the temperature increase leading to the formation of denatured polypeptide chains (Yura and Nakahigashi, 1999). Interestingly, all available *E. coli* expression systems make use of σ^{70}. This σ factor recognizes the canonical hexameric sequences TATAAT and TTGACA, centred ~10 and 35 bp upstream of the transcription start site, respectively (Dombroski et al., 1992; Busby and Ebright, 1994). The similarity of the -35 and -10 elements to the consensus recognition hexamers and the spacing between them determine to a great extent the strength of a promoter. In addition, the presence of a third A+T-rich recognition element, located upstream of the -35 hexamer of some promoters can enhance the promoter strength significantly (Gourse et al., 2000). Due to its localization, this A+T-rich sequence has been designated UP element. A full UP element actually consists of two subsites, each of which can bind one α subunit (Estrem et al., 1998; Estrem et al., 1999). A second element can increase transcription called "extended -10" promoter. This element contains an extra 2-bp TG sequence located 1 bp 5' to the -10 element (consensus sequence TGNTATAAT) (Bown et al., 1997). The TG element also binds to the RNAP by contacting the σ^{70} subunit (Barne et al., 1997). Furthermore, a third element has been described and referred as -15 sequence (Liu et al., 2004). It consists of a specific 10-bp-long AT-rich DNA sequence within the spacer, which must be juxtaposed to the upstream end of the -10 region and enhances both RNAP binding and open complex formation.

In summary, a useful promoter (Fig. 2) must fulfil the following criteria:

- consisting of an UP element from -70 to -36, a consensus promoter sequence including, besides the -35 and the -10 region, a TnTG motif from -17 to -14, a six nucleotide spacer region between the -10 region and the transcription start site +1, which should be a purine
- an efficient transcriptional terminator to ensure termination of transcription; in most cases, two different terminators are arranged in tandem
- a low basal activity in the absence of inducer (tightly regulated)
- high level of expression after induction
- must be easily transferable to other *E. coli* strains
- induction must be simple and cost-effective
- should be independent of the commonly used ingredients of culturing media

The presence of secondary structures within the 5' untranslated regions of prokaryotic mRNA can provide significant mRNA stabilization and thereby increase the amount of the gene product (Carrier and Keasling, 1997). This stabilization is thought to result mainly from protection of the mRNA 5' end from binding by the decay-initiating ribonuclease, RNase E (Cohen and McDowall, 1997). One of the best analyzed 5' stabilizers is encoded by the *ompA* gene of *E. coli* where the OmpA protein is a major outer membrane protein. In rapidly growing cells at 30°C, the *ompA* transcript decays with a half-life of ~17 min, making it one of the most stable mRNAs in *E. coli* (von Gabain et al., 1983). It could be shown that the presence of a stem-loop no more than 2-4 nucleotides from the extreme 5' terminus of the transcript is crucial to its stabilizing influence (Emory et al., 1992). Addition of this stem-loop structure to the 5' end of *ompA* is able to enhance the lifetime of many normally labile mRNAs.

Besides a strong promoter and a mRNA stabilizer, unique structural features at the 5' end of the mRNA transcript are the major determinants of the efficiency of mRNA translation initiation.

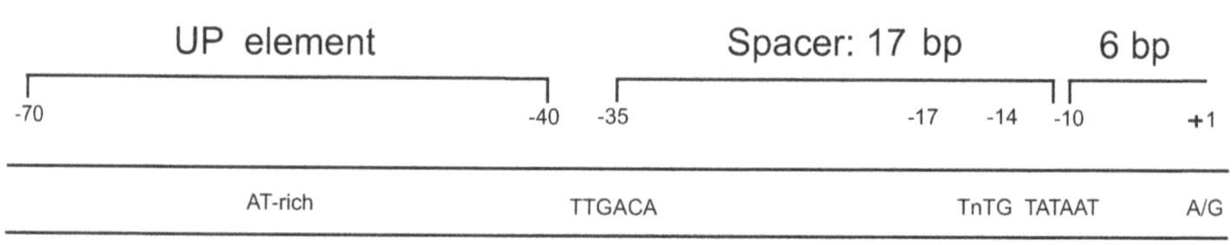

Fig. 2. The optimal promoter region. The optimal promoter consists of the consensus sequence (TTGACA for the -35 element and TATAAT for the -10 element separated by a 17 bp spacer) and in addition by an AT-rich UP element to which the αCTD will bind. Furthermore, it contains the TnTG motif, the transcriptional start codon is a purine, which is 6 bp away from the -10 element.

DNA sequences involved in translation strongly influence the amount of recombinant proteins produced. These are: (1) The ribosome binding site (RBS), (2) the start codon and (3) the stop codon. The RBS consists of the Shine-Dalgarno sequence followed by an AT-rich translational spacer that has an optimal length of approximately eight nucleotides. The major start codon is AUG used in 91% of the genes followed by GUG with 8% in *E. coli*. Therefore, AUG is the preferred start codon. The presence of a stop codon is an indispensable signal for termination of mRNA translation, and *E. coli* expression vectors frequently contain all three stop codons to prevent ribosome skipping. *E. coli* exhibits a strong bias towards the UAA codon, but translation-termination is further enhanced in the presence of the tetranucleotide UAAU, also called the extended stop codon (Poole et al., 1995).

Eight major inducible promoters are used for the production of recombinant proteins (Table 1):

1. IPTG-inducible promoters (*lac*, *lacUV5*)
2. the autoinducible *trp* promoter
3. the hybrid promoters *tac* and *trc*
4. the heat-inducible phage λ promoters P_L and P_R
5. the tetracycline-inducible promoter
6. the arabinose-inducible P_{BAD} promoter
7. the cold-inducible *cspA* promoter
8. the T7 promoter

Based on the *lac* operon and its well-know regulation mechanism, several expression systems have been designed. The *lac* system consists of the *lac* promoter (*lacO*$_1$) and its downstream operator to which the Lac repressor encoded by the *lacI* gene

Table 1. Important *E. coli* promoters used for recombinant protein production

Promoter	Inducer range	Level of expression
lac	IPTG: 0.05-2.0 mM	low level
lacUV5	IPTG: 0.05-2.0 mM	moderate level
trp	depletion of tryptophan	moderately level
trc and *tac*	IPTG: 0.05-2.0 mM	moderately high
phage T7	IPTG: 0.05-2.0 mM	very high
phage λ	temperature shift from 30 to 42°C	moderately high
tetA	andydrotetracycline: 200 μg/l	middle to high level
araBAD	L-arabinose: 0.001-1.0%	low to high level

binds as a tetramer. There are two additional operators, *lacO*$_2$ and *lacO*$_3$, located at the beginning of the *lacZ* gene and at the end of the *lacI* gene, respectively. By binding to *lacO*$_1$ and either *lacO*$_2$ or *lacO*$_3$, the DNA in between these two operators is looped out resulting in a 10-fold stronger repression. The natural inducer of the Lac repressor is allolactose. Lactose is taken up by the LacY permease, an integral inner membrane protein. The β-galactosidase coded for by the *lacZ* gene converts a few molecules of lactose into allolactose. But in combination with *lac*-based expression systems, IPTG is used as inducer. IPTG offers two important advantages over lactose: First, its uptake is not dependent on the LacY permease, but it diffuses passively through the inner membrane, and second, it is not cleaved by the β-galactosidase. The very first promoter used was the wild-type *lac* promoter (Gronenborn, 1976). Since this promoter is rather weak, the first improvement to be applied was its *lacUV5* derivative (Fuller, 1982). This promoter is distinguished from the wild-type promoter by two important characteristics: First, its -10 region corresponds to the consensus sequence due to two point mutations and second, more important, this promoter is less sensitive to catabolite repression by glucose (Silverstone et al., 1970). To obtain strong repression in the absence of inducer, a sufficient amount of Lac repressor has to be synthesized. Since the wild-type promoter of the *lacI* gene is rather weak (there are only 10-20 molecules of Lac repressor per wild-type cell (Gilbert and Muller-Hill, 1966), a promoter-up mutation is used designated *lacI*^{q} where "q" stands for quantity (Müller-Hill et al., 1998). Here, about 100 repressor molecules are present per cell if the *lacI*^{q} allele is present in the chromosome. Another promoter-up mutation has been isolated later which leads to the synthesis of approximately 1700 molecules per cell termed *lacI*^{q1} (Calos and Miller, 1981).

The second promoter system that has been introduced is that of the *trp* operon (Hallewell and Emtage, 1980; Tacon et al., 1983; De Boer et al., 1983b; Itoh et al., 1984). The *trp* promoter is a strong, regulated *E. coli* promoter that has been extensively studied (Yanofsky, 2007). It is repressed by tryptophan and derepressed by the addition of a tryptophan analog such as 3-β-indole-acrylic acid (IAA) or by removal of tryptophan. The *trp* operon is under dual control by the Trp repressor and by a transcription attenuation mechanism. The activity of the Trp repressor (encoded by *trpR*) is regulated by tryptophan which acts as a corepressor. As long as the tryptophan concentration within the cell is high, the Trp repressor-corepressor complex binds its operator to prevent expression of the genes involved in the biosynthesis of tryptophan. If the amount of tryptophan within the cell drops beyond a threshold level, the corepressor fails to saturate the Trp repressor, it will be unable to bind to its promoter and the *trp* operon will be transcribed. Such a system is called to be auto-inducible. To derepress cloned gene expression, tryptophan can be removed from the medium by the induction of intracellular tryptophanase through changing the glucose feed to glycerol (Kawai et al., 1986), or with cross-filtration

(Iijima et al., 1987). But there is also the possibility to inactivate the Trp repressor by the addition of IAA (Doolittle and Yanofsky, 1968). But his compound exerts a disadvantage which has to be taken into consideration. It inhibits charging of $tRNA^{trp}$ by tryptophanyl-tRNA synthetase which results in starvation for tryptophan and a subsequent translation arrest. The first vectors based on the Trp-promoter are the ptrpED series where the amount of recombinant protein accounted for up to 30% of the total protein in the cell (Hallewell and Emtage, 1980).

While the *trp* promoter contains a consensus -35 element and the *lacUV5* promoter a consensus -10 region, two different groups decided independently of each other to combine both elements into a consensus promoter. These novel promoters have been termed *tac* and *trc* (De Boer et al., 1983a; Brosius et al., 1985). The difference between these two promoters is that the spacer length in the *tac* promoter is 16 bp (suboptimal), while that in *trc* is 17 bp (optimal). Nevertheless, unexpectedly, the *tac* promoter is stronger than the *trc* promoter (about 11 times stronger than the *lacUV5* promoter). Thus the hybrid promoter is far more efficient than either one of the parental promoters.

Initiation of transcription at the leftward (P_L) and rightward promoter (P_R) of phage λ is repressed by the product the phage *cI* repressor gene to levels below detection limit (Joyner et al., 1966). Mutants of the *cI* gene are available that code for a temperature-sensitive repressor (Lieb, 1966). The *cIts857* allele is the most well-known temperature-inducible promoter. While this repressor is active at 30°C, it becomes inactive after a shift to 42°C (Elvin et al., 1990). A family of plasmid cloning vectors based on P_L provide efficient expression of cloned genes in *E. coli* (Remaut et al., 1981). Under optimal conditions, between 30 and 40% of the total protein synthesis is directed by the cloned gene under P_L control. Improved expression vectors have been constructed based on the runaway-replication mutants of plasmid R1 and the λ P_L promoter (Remaut et al., 1983). Miniplasmids derived from R1 are present in 20-30 copies per cell at 30°C. At temperatures above 35°C, these plasmids replicate without copy number control during 2-3 h to more than 2000 copies (Uhlin et al., 1979). At the end of this period plasmid DNA amounts to about 75% of the total DNA. During the increase in copy number, growth and protein synthesis continues at normal rate leading to a drastic amplification of plasmid-encoded gene products. Upon insertion of the λ expression system (vectors pCP3, pCP39 and pCP40), both the plasmid copy number and the P_L promoter can be temperature-induced at the same time (Remaut et al., 1983).

A fifth system is based on *tetA* promoter/operator sequence which is under the negative control of the Tet repressor (Skerra, 1994). In Gram-negative bacteria, resistance to tetracycline is mediated by the TetA protein, a proton-$[tetracycline{\cdot}Mg]^+$ antiporter embedded in the cytoplasmic membrane (Yamaguchi et al., 1990). In enteric bacteria, the efflux-encoding *tetA* genes are strictly regulated at the level of transcription by the tetracycline-responsive Tet repressor (TetR). In the absence of inducer, TetR dimers bind to the operators *tet*O_1 and *tet*O_2, abolishing transcription of its own gene, *tetR*, and *tetA*. When tetracycline has entered the cytoplasm, it binds TetR with high affinity, inducing a conformational change in TetR (Orth et al., 2000) resulting in dissociation from the *tet* operators (Lederer et al., 1996). One of the *tet* promoter/operator-based expression systems is the vector pASK75, and induction is carried out by the addition of anhydrotetracycline (Skerra, 1994). This compound binds more strongly to the Tet-repressor and is less antibiotic.

The sixth system is arabinose-inducible. This expression system consists of the P_{BAD} promoter of the arabinose operon and its regulatory gene *araC*. The AraC protein is both a positive and a negative regulator (Schleif, 2000). In the presence of arabinose, transcription from the P_{BAD} promoter is turned on and in the absence of arabinose, transcription occurs at very low levels (Lee et al., 1987). The uninduced levels can be reduced even further by growth in the presence of glucose. The P_{BAD} series of expression vectors allows modulated expression over a wide range of inducer concentrations (Guzman et al., 1995).

After a rapid decrease in temperature called cold-shock, bacteria express a well-defined set of proteins to cope with the stressful conditions (Weber and Marahiel, 2003). These cold-shock proteins include helicases, nucleases and ribosome-associated proteins. The major cold-shock protein in *E. coli* is CspA which is virtually undetectable at 37°C, but more than 10% of the total protein synthesis is devoted to its production 60 min after a temperature downshift to 15°C (Goldstein et al., 1990). This induction is based on the increase in transcript stability by almost two orders of magnitude (Jiang et al., 1993). Fusion of heterologous genes to the *cspA* transcriptional control (plasmids pCS22 and pCS24) resulted in production of proteolytically sensitive recombinant proteins after a temperature downshock to 23 or 15°C (Mujacic et al., 1999). Similar vectors have been constructed by another group (Qing et al., 2004).

The last and most widely used expression system is based on the phage T7-expression system which is the first choice for protein production in *E. coli*. One important reason for the success of this system is that recombinant genes are fused to the very strong T7-promoter, which is not recognized by the *E. coli* RNAP (Rosenberg et al., 1987). Therefore, virtually no expression is observed until the gene coding for the T7 RNAP is expressed. The T7 RNAP is distinguished from bacterial ones by three different characteristics: (1) It consists of only one single polypeptide chain of about 100 kDa; (2) it possesses a high processivity that means it initiates RNA chains very efficiently and elongates them about five times faster than the *E. coli* enzyme; (3) it is rifampicin resistant. The gene coding for the T7

RNAP is normally present together with its promoter on the expression vector fused to an IPTG-inducible promoter. To induce for the production of recombinant proteins, IPTG is added to the growing cells resulting in the production of T7 RNAP. After about 10 min, rifampicin can be added to inactivate the *E. coli* RNAP and thereby further increase the amount of recombinant protein. To reduce low-level recombinant gene expression in the absence of IPTG due to the leakiness of the promoter, T7 lysozyme is expressed from a compatible plasmid (Studier, 1991). Several T7-expression systems have been published (Studier and Moffat, 1986; Tabor and Richardson, 1985; Liang et al., 2007; Kang et al., 2007).

Besides allowing overexpression of authentic proteins, many of the above mentioned expression systems allow cotranslational addition of a purification tag either to the N- or to the C-terminus of the recombinant protein. Purification tags allow single step purification of the proteins by affinity chromatography and sometimes keep them soluble (Terpe, 2003). These tags can provide 100- or even 1000-fold purification from crude extracts without prior steps to remove nucleic acids or other cellular material. Among the purification tags widely used are:

- the His and HAT tag
- the Strep tag
- the maltose-binding protein (MBP) tag
- the glutathione S-transferase (GST) tag
- the chitin-binding domain
- the calmodulin-binding peptide

The available protein and affinity tags can be categorized into three classes depending on the nature of the affinity tag. The first class of tags uses a peptide or protein fusion that binds to small molecule ligands immobilized on a solid support (*e.g.*, His- and GST-tag). The second class of affinity tags consists of a peptide tag interacting with a protein-binding partner attached on a chromatography resin (calmodulin). The third class of epitope affinity tag uses as a binding partner an antibody attached to the resin (FLAG).

The most widely employed method utilizes metal-affinity chromatography for the purification of recombinant proteins containing a polyhistidine tag. Immobilized metal-affinity chromatography (IMAC) is based on the interaction between a transition metal ion such as Ni^{2+} or Cu^{2+} immobilized on a matrix and specific amino acid side chains. Here, histidine exhibits the strongest interaction with immobilized metal ion matrices, as electron donor groups on the histidine imidazole ring readily form coordination bonds with the immobilized transition metal. Recombinant proteins containing sequences of consecutive histidine residues (6-10) are efficiently retained on IMAC and can be eluted by either adjusting the pH of the column buffer or by adding free imidazole. This purification method has been first described in 1987 (Hochuli et al., 1987). These authors developed a nitrilotriacetic acid (NTA) adsorbent for metal-chelate chromatography. Another material that has been developed for the purification of His-tagged proteins is TALON. This material consists of Co^{2+}-carboxymethylaspartate (Co^{2+}-CMA), which is coupled to a solid-support resin. TALON allows the elution of tagged proteins under mild conditions and exhibits less non-specific protein binding than Ni^{2+}-NTA resin (Chaga et al., 1999b; Chaga et al., 1999a). Purification with Co^{2+}-CMA allowed the development of a natural 19-amino-acid poly-histidine affinity tag (HAT-tag). His-tags are commonly placed on either the N- or C-terminus of the recombinant protein, and optimal placement of the tag is protein-specific. The general characteristics of the His-tag are as follows: (1) It is poorly immunogenic; (2) it is small and uncharged at pH 8.0; (3) it does not generally affect secretion, compartmentalization or folding of the fusion protein within the cells; and (4) in most cases it does not interfere with the structure or function of the purified protein. There are a few *E. coli* histidine-rich proteins which could contaminate the recombinant protein (Roche and Sauer, 2001).

The 40-kDa maltose-binding protein (MBP) of *E. coli* is located in the periplasm where it involved in binding maltose and in delivering at the maltose-specific ATP-transporter complex inserted in the inner membrane (Boos and Bohm, 2000). Vectors that facilitate the expression and purification of recombinant proteins fused to MBP were first published in 1988 (Guan et al., 1988). Fusion proteins can be purified by single step affinity chromatography on cross-linked amylose. Bound proteins can be eluted with 10 mM maltose in physiological buffer. MBP can increase the solubility of overexpressed fusion proteins, especially eukaryotic proteins (Sachdev and Chirgwin, 1999). A spacer sequence coding for ten asparagine residues between the MBP and the passenger protein increase the chances that a particular fusion protein will bind tightly to the amylose resin. Some fusions proteins cannot be efficiently absorbed onto amylose resin. A solution to this problem can be the construction of a His-MBP-fusion protein (Nallamsetty et al., 2005).

The Strep-tag is an amino acid peptide that has been developed as an affinity tool for the purification of recombinant proteins on streptavidin columns (Schmidt and Skerra, 1993). Strep-tagged proteins are bound under physiological buffer conditions in the biotin binding pocket and are eluted gently with biotin derivatives such as 2.5 mM desthiobiotin. Steptavidin mutants with a specific mutation at position 44, 45 and 47 have a higher affinity for the octapeptide Strep-tag II than for the native form (Schmidt et al., 1996) (Korndorfer and Skerra, 2002; Voss and Skerra, 1997). This mutant streptavidin has been termed Strep-Tactin. A variant of the Strep-tag is the 38-amino-acid streptavidin-binding peptide- (SBP) tag with a dissociation constant of 2.5 nM (Wilson et al., 2001). The elution occurs under mild conditions, using 2 mM biotin. SBP-tagged proteins can be reversibly immobilized on a surface plasmon resonance biosensor chip (Li et al., 2006). Other variants of the Strep-tag are the nano-tags (Lamla and Erdmann, 2004). The Nano-tag(15) is 15 amino acids long and binds to streptavidin with a dissociation constant of 4 nM, while the Nano-tag(9) is a 9-mer peptide with a dissociation constant of 17 nM. The Nano-tag allows the detection of recombinant proteins on Western blots by a streptavidin-alkaline phosphatase conjugate.

Single-step purification of recombinant proteins with glutathione S-transferase (GST) was first described in 1988 (Smith and Johnson, 1988). GST is a 26-kDa protein from *Schistosoma japonicum*, and fusion proteins are purified from crude lysates by affinity chromatography on immobilized glutathione. The bound recombinant proteins can be eluted with 10 mM reduced glutathione under nondenaturing conditions. In the majority of the cases, the fusion proteins stay soluble in aqueous solutions. Even integral membrane proteins such as the ATP-dependent FtsH metalloprotease stay soluble when fused to the GST-tag (Teff et al., 2000). The GST-tag can be easily detected using specific antibodies.

The chitin-binding domain (CBD) from *Bacillus circulans* consists of 51 amino acids (Watanabe et al., 1994). The purification tag is commonly available in combination with self-splicing inteins (Banki and Wood, 2005). Here, the intein from the *Saccharomyces cerevisiae* VMA1 gene (454 amino acids) (Chong et al., 1997) and shorter inteins have also been employed (Xu et al., 2000). Self-cleavage of the thioester bond between one end of the intein and the recombinant protein can be induced by thiol reagents, such as 1,4-dithiothreitol or β-mercaptoethanol. The uncleaved fusion precursor and the intein tag remain bound to the chitin residue during passenger protein elution and can be removed from the chitin resin by 1% SDS or 6 M guanidine HCl.

The calmodulin-binding peptide (CBP) binds specifically to calmodulin and purification of recombinant proteins with a calmodulin-tag have been first described in 1992 (Stofko-Hahn et al., 1992). The affinity peptide is 26 amino acids long (4 kDa) and is derived from the skeletal-muscle myosin light-chain kinase, which binds calmodulin with nanomolar affinity in the presence of 0.2 mM $CaCl_2$ (Krueger et al., 1998). The binding of CBP to calmodulin is very tight ($K_d = 10^{-9}$) and needs the presence of low concentrations of calcium (0.2-2 mM). This tight binding allows more stringent washing conditions largely reducing the copurification of contaminating proteins. Elution of the bound fusion protein is accomplished with virtually any buffer containing 2 mM EGTA. A special advantage of the CBP tag is that it acts as a good substrate for a cAMP-dependent protein kinase. This allows direct labelling of the fusion protein with [γ-^{32}P]ATP and its use as a probe for protein-protein interactions and other applications (Zheng et al., 1997). The system is highly specific for *E. coli* cells since there are no endogenous proteins that interact with calmodulin.

The presence of the purification tag may affect important characteristics or functions of the protein during further studies. Removal of the affinity tag from the recombinant protein can be accomplished with the aid of a site-specific protease. If the protease is equipped with a tag, the purification tag and the protease can be removed at the same time by affinity chromatography after cleavage. Cleavage of the tag without using a protease is also possible by introducing a self-splicing intein is mentioned above. The most commonly used proteases are: enterokinase, tobacco etch virus (TEV), thrombin and factor Xa (see Table 2 for the amino acid residues recognized and the cleavage sites).

MISCELLANEOUS VECTORS

Besides cloning and expression vectors, there are additional vectors serving several purposes:

- promoter- and terminator-test vectors
- bending vectors
- epitope-tag vectors
- localization-tag vectors
- riboprobe vectors
- multifunctional vectors
- shuttle vectors

Genetic fusions have from the earliest times of molecular genetics provided an important means of analyzing basic biological problems. The ability to fuse different regulatory DNA sequences to easily assayed or selectable genes allowed for detailed analysis of these genetic regions. Three different kinds of vectors have been developed allowing for the analysis of promoters, terminators and translational signals.

Table 2. Recognition and cleavage sites for some site-specific endoproteases

Endoprotease	Recognition and cleavage site
Thrombin	Leu-Val-Pro-Arg ↓ Gly-Ser
Factor Xa	Ile-Glu-Gly-Arg ↓
Enterokinase	Asp-Asp-Asp-Asp-Lys ↓
Tobacco Etch Virus	Glu-X-X-Tyr-X-Glu ↓ Ser

Promoter-test vectors contain a reporter gene coding for an easily measurable enzyme without its own promoter. DNA fragments to be tested for promoter activity are cloned in front of the reporter gene followed by measuring its activity. If the inserted DNA fragment carries a promoter pointing into the direction of the reporter gene, it will be expressed. The reporter gene most widely used is the *lacZ* gene of *E. coli* which codes for β-galactosidase activity reviewed in (Silhavy and Beckwith, 1985). These kinds of fusions are also called transcriptional fusions or operon fusions. They are not only used to demonstrate promoter activity, but to study the regulation of a given promoter. In contrast, a terminator test vector contains a constitutive promoter fused to a reporter gene and separated by one or more unique restriction sites. Upon insertion of a DNA fragment carrying a transcriptional terminator between the promoter and the reporter gene, the enzymatic activity of the enzyme will be reduced. The vector pKG1800 is one of these terminator test vectors (Bernardi and Bernardi, 1990; McKenney et al., 1981).

Promoters which direct transcription bidirectionally from closely spaced sites are widespread in bacteria, phages and plasmids (Yamada et al., 2003). There are three types of divergent promoters: a back-to-back arrangement with an intervening sequence between two promoters, an overlapping arrangement and a face-to-face arrangement that produces partially overlapping transcripts. The widespread occurrence of promoters transcribing in opposite directions, frequently regulated from common sites, indicates that this type of control region represents a general principle evolved for the regulation of gene expression. Analysis of the regulatory characteristic of such control regions can be greatly facilitated by fusing the promoters and regulatory sites to promoterless indicator genes such that gene expression in both directions can be monitored simultaneously. The construction of promoter-probe vectors which contain the gene combinations *lacZ/galK* and *lacZ/phoA* has been described (Schneider and Beck, 1986).

Bent DNA structures have been identified in a wide variety of cellular and viral genomes from bacteria to man. We distinguish between intrinsic bends and protein-induced bends. Intrinsic bends form when special base sequences, usually short (3-6 bp) runs of adenines termed A-tracts occur repeatedly in phase with the DNA helical repeat of about 10.5 bp. Bent DNA appears to be implicated in both activation and repression of transcription. While in activation, it plays a direct role, whereas in repression, a silencer protein binding to the bend plays a crucial role. During transcription activation, bent DNA often occurs in the region from about -100 to about -35, relative to the transcription initiation site. Such an intrinsic bend facilitates the binding of RNAP to promoters and/or the formation of open complexes. Is it possible to demonstrate that a DNA sequence contains a bend? It has been shown that bent DNA exhibits an unusual electrophoretic mobility. The location of the bend within DNA fragments of identical size determines its mobility.

A DNA fragment where the bend is located within the middle will migrate more slowly than a fragment where the bend is present near one end. This behaviour led to the development of so-called bending vectors. Such a bending vector carries a series of restriction sites in a tandem dimer. The desired DNA fragment to be tested will be cloned at a unique restriction site placed at the joint of the repeat. With such a construct one can easily clone a fragment of interest between the two copies of the repeat. When the recombinant vector is cut by the restriction enzymes cleaving within the tandem dimers, permutated monomers of identical size are liberated which will be subjected to gel electrophoresis (Robertson and Nash, 1988).

Another important class of vectors allow the addition of a tag to the protein of interest. Besides the purification tag already described above, epitope and localization tags are used. Epitope-tagging vectors allow the covalent addition of a short amino acid sequence (around 10 residues) to the protein of interest which is recognized by commercially available antibodies. This tagging procedure avoids overproduction and purification of the protein, followed by immunization of an animal, normally rabbits or mice. Different epitopes have been described (see Table 3 for their amino acids), c-Myc (10 amino acid residues) and HA (9 amino acid residues) were derived from the human *c-myc* proto-oncogene and the hemagglutinin of the influenza virus, respectively (Evan et al., 1985; Wilson et al., 1984). FLAG is an artificial tag. Localization tags allow the detection of proteins within the cell, *e.g.* as part of the cytoplasmic membrane or near one of the poles of a rod-shaped bacterial cell. In most cases, the green fluorescence protein (GFP) or one of its variants such as the cyan fluorescence protein (CFP) or the yellow fluorescence protein (YFP) are used (Margolin, 2000). Vectors have been constructed which allow the cotranslational fusion of one of these and other tags to the protein of interest (Kaltwasser et al., 2001).

Riboprobe vectors are designed for the *in vitro* preparation of single-stranded RNA probes for hybridization experiments or microgram quantities of defined RNA transcripts from cloned inserts. This vector type contains a multiple cloning site (MCS) to accept the insert DNA. The MCS is typically flanked by two different promoters which are recognized by phage RNAP (T3, T7 or SP6). These polymerases are extremely promoter-specific (there is almost no transcriptional

Table 3. Epitope tags used

Epitope	Amino acid sequence
FLAG	Asp-Tyr-Lys-Asp-Asp-Asp-Asp-Lys
cMyc	Glu-Gln-Lys-Leu-Ile-Ser-Glu-Glu-Asp-Leu
HA	Tyr-Pro-Tyr-Asp-Val-Pro-Asp-Tyr-Ala

cross-talk), virtually homogenous RNA can be obtained using plasmid DNA as the template in a transcription reaction. When it is desirable to transcribe only the insert DNA sequences, the recombinant plasmid is linearized at an appropriate restriction site prior to the transcription reaction leading to discrete run-off transcripts, virtually free of vector sequences. The choice of promoter used to initiate transcription determines which strand will be transcribed resulting in the production of sense or antisense RNA. The RNA probes can be labelled either radioactively or non-radioactively during their *in vitro* synthesis and can be used to generate probes for hybridization to Northern and Southern blots, plaque and colony hybridization. These RNA transcripts are also useful for S1 nuclease mapping, mRNA synthesis for *in vitro* translation and antisense RNAs to block translation.

Multifunctional vectors are distinguished from simple cloning vectors by the following additional features:

- A multiple cloning site (MCS) allowing directed cloning of restriction or PCR fragments. Often, the MCS is flanked by sites for a homing endonuclease (Guhan and Muniyappa, 2003). This group of enzymes recognizes DNA sequences between 15 and 35 bp in length and cleave within these sequences. Since the occurrence of such a recognition sequences in almost any DNA fragment is extremely unlikely, the cloned fragment independent of its length can be released by treatment with the appropriate homing endonuclease.
- An origin of replication from a filamentous phage allowing the generation of single-stranded DNA upon overinfection with the appropriate phage.
- The *lacZ'* gene allowing blue-white screening after transformation.
- The MCS is often flanked by two promoters recognized by two different phage-encoded RNAPs such as those coded for by the *E. coli* phages T3 or T7 or by the *S. typhimurium* phage SP6. After cloning of a DNA fragment into the MCS, its DNA sequence can be converted into RNA by *in vitro* transcription using the purified plasmid and the appropriate RNAP. In principle, sense and antisense RNA can be prepared in two different reactions and labelled either radioactively or non-radioactively.

Shuttle vectors are defined as vectors able to replicate in *E. coli* (which is transformed with the ligation products) and in another species such as *B. subtilis*, *Streptomyces*, *Saccharomyces cerevisiae* or animal cells. These vectors contain two different origins of replication, one active in *E. coli* and the second in the other species. The advantage of shuttle vectors is that they allow cloning and analysis of the insert within *E. coli*, and the corrcct recombinant plasmids is then transferred (shuttled) into the appropriate host where the final analysis is carried out.

REFERENCES

Baneyx, F. (1999). Recombinant protein expression in *Escherichia coli*. Curr. Opin. Biotechnol. *10*, 411-421.

Banki, M.R. and Wood, D.W. (2005). Inteins and affinity resin substitutes for protein purification and scale up. Microb. Cell Fact. *4*, 32.

Barne, K.A., Bown, J.A., Busby, S.J.W., and Minchin, S.D. (1997). Region 2.5 of the *Escherichia coli* RNA polymerase σ^{70} subunit is responsible for the recognition of the 'extended -10' motif at promoters. EMBO J. *16*, 4034-4040.

Bernardi, F. and Bernardi, A. (1990). Completed sequence of pKG1800, a vector for determination of transcription terminators. DNA Seq. *1*, 147-150.

Bolivar, F., Rodriguez, R.L., Greene, P.J., Betlach, M.C., Heyneker, H.L., and Boyer, H.W. (1977a). Construction and characterization of new cloning vehicles. II. A multipurpose cloning system. Gene *2*, 95-113.

Bolivar, F., Rodriquez,R., Betlach, M., and Boyer, H. (1977b). Construction and characterization of new cloning vehicles I: Ampicillin-resistant derivatives of the plasmid pMB9. Gene *2*, 75-93.

Boos, W. and Böhm, A. (2000). Learning new tricks from an old dog: MalT of the *Escherichia coli* maltose system is part of a complex regulatory network. Trends Genet. *16*, 404-409.

Bouet, J.Y., Nordstrom, K., and Lane, D. (2007). Plasmid partition and incompatibility--the focus shifts. Mol. Microbiol. *65*, 1405-1414.

Bown, J., Barne, K.A., Minchin, S., and Busby, S. (1997). Extended -10 promoters. Nucleic Acids Mol. Biol. *11*, 41-52.

Brosius, J., Erfle, M., and Storella,J . (1985). Spacing of the -10 and -35 regions in the tac promoter. Effect on its *in vivo* activity. J. Biol. Chem. *260*, 3539-3541.

Busby, S. and Ebright, R.H. (1994). Promoter structure, promoter recognition, and transcription activation in prokaryotes. Cell *79*, 743-746.

Calos, M.P. and Miller, J.H. (1981). The DNA sequence change resulting from the IQ1 mutation, which greatly increases promoter strength. Mol. Gen. Genet. *183*, 559-560.

Carrier, T.A. and Keasling, J.D. (1997). Controlling messenger RNA stability in bacteria: strategies for engineering gene expression. Biotechnol. Prog. *13*, 699-708.

Chaga, G., Bochkariov, D.E., Jokhadze, G.G., Hopp, J., and Nelson, P. (1999a). Natural poly-histidine affinity tag for purification of recombinant proteins on cobalt(II)-carboxymethylaspartate crosslinked agarose. J. Chromatogr. A *864*, 247-256.

Chaga, G., Hopp, J., and Nelson, P. (1999b). Immobilized metal ion affinity chromatography on Co2+-carboxymethylaspartate-agarose Superflow, as demonstrated by one-step purification of lactate dehydrogenase from chicken breast muscle. Biotechnol. Appl. Biochem. *29 (Pt 1)*, 19-24.

Chang, A.C.Y. and Cohen, S.N. (1978). Construction and characterization of amplifyable multicopy DNA cloning vehicles derived from the P15A cryptic miniplasmid. J. Bacteriol. *134*, 1141-1156.

Chong, S., Mersha, F.B., Comb, D.G., Scott, M.E., Landry, D., Vence, L.M., Perler, F.B., Benner, J., Kucera, R.B., Hirvonen, C.A., Pelletier, J.J., Paulus, H., and Xu, M.Q. (1997). Single-column purification of free recombinant proteins using a self-cleavable affinity tag derived from a protein splicing element. Gene *192*, 271-281.

Cohen, S.N. and Chang, A.C. (1973). Recircularization and autonomous replication of a sheared R-factor DNA segment in *Escherichia coli* transformants. Proc. Natl. Acad. Sci. U. S. A *70*, 1293-1297.

Cohen, S.N., Chang, A.C., Boyer, H.W., and Helling, R.B. (1973). Construction of biologically functional bacterial plasmids *in vitro*. Proc. Natl. Acad. Sci. U. S. A *70*, 3240-3244.

Cohen, S.N. and McDowall, K.J. (1997). RNase E: still a wonderfully mysterious enzyme. Mol. Microbiol. *23*, 1099-1106.

De Boer, H.A., Comstock, L.J., and Vasser, M. (1983a). The tac promoter: a functional hybrid derived from the trp and lac promoters. Proc. Natl. Acad. Sci. U. S. A *80*, 21-25.

De Boer, H.A., Hui, A., Comstock, L.J., Wong, E., and Vasser, M. (1983b). Portable Shine-Dalgarno regions: a system for a systematic study of defined alterations of nucleotide sequences within *E. coli* ribosome binding sites. DNA *2*, 231-235.

Dombroski, A.J., Walter, W.A., Record, M.T., Jr., Siegele, D.A., and Gross, C.A. (1992). Polypeptides containing highly conserved regions of transcription initiation factor σ^{70} exhibit specificity of binding to promoter DNA. Cell *70*, 501-512.

Doolittle, W.F. and Yanofsky, C. (1968). Mutants of *Escherichia coli* with an altered tryptophanyl-transfer ribonucleic acid synthetase. J. Bacteriol. *95*, 1283-1294.

Eichenlaub, R. (1979). Mutants of the mini-F plasmid pML31 thermosensitive in replication. J. Bacteriol. *138*, 559-566.

Eichenlaub, R., Figurski, D., and Helinski, D.R. (1977). Bidirection replication from a unique origin in a mini-F plasmid. Proc. Natl. Acad. Sci. U. S. A *74*, 1138-1141.

Elvin, C.M., Thompson, P.R., Argall, M.E., Hendry, P., Stamford, N.P.J., Lilley, E., and Dixon, N.E. (1990). Modified bacteriophage lambda promoter vectors for overproduction of proteins in *Escherichia coli*. Gene *87*, 123-126.

Emory, S.A., Bouvet, P., and Belasco, J.G. (1992). A 5'-terminal stem-loop structure can stabilize mRNA in *Escherichia coli*. Genes Dev. *6*, 135-148.

Estrem, S.T., Gaal, T., Ross, W., and Gourse, R.L. (1998). Identification of an UP element consensus sequence for bacterial promoters. Proc. Natl. Acad. Sci. USA *95*, 9761-9766.

Estrem, S.T., Ross, W., Gaal, T., Chen, Z.W., Niu, W., Ebright, R.H., and Gourse, R.L. (1999). Bacterial promoter architecture: subsite structure of UP elements and interactions with the carboxy-terminal domain of the RNA polymerase α subunit. Genes Dev. *13*, 2134-2147.

Evan, G.I., Lewis, G.K., Ramsay, G., and Bishop, J.M. (1985). Isolation of monoclonal antibodies specific for human *c-myc* proto-oncogene product. Mol. Cell. Biol. *5*, 3610-3616.

Fuller, F. (1982). A family of cloning vectors containing the *lacUV5* promoter. Gene *19*, 43-54.

Gilbert, W. and Muller-Hill, B. (1966). Isolation of the Lac repressor. Proc. Natl. Acad. Sci. U. S. A *56*, 1891-1898.

Goldstein, J., Pollitt, N.S., and Inouye, M. (1990). Major cold shock protein of *Escherichia coli*. Proc. Natl. Acad. Sci. USA *87*, 283-287.

Gourse, R.L., Ross, W., and Gaal, T. (2000). UPs and downs in bacterial transcription initiation: the role of the alpha subunit of RNA polymerase in promoter recognition. Mol. Microbiol. *37*, 687-695.

Goursot, R., Goze, A., Niaudet, B., and Ehrlich, S.D. (1982). Plasmids from *Staphylococcus aureus* replicate in yeast *Saccharomyces cerevisiae*. Nature *298*, 488-490.

Gronenborn, B. (1976). Overproduction of phage lambda repressor under control of the lac promotor of *Escherichia coli*. Mol. Gen. Genet. *148*, 243-250.

Guan, C.-D., Li, P., Riggs, P.D., and Inouye, H. (1988). Vectors that facilitate the expression and purification of foreign peptides in *Eschrichia coli* by fusion to maltose-binding protein. Gene *67*, 21-30.

Guhan, N. and Muniyappa, K. (2003). Structural and functional characteristics of homing endonucleases. Crit. Rev. Biochem. Mol. Biol. *38*, 199-248.

Guzman, L.M., Belin, D., Carson, M.J., and Beckwith, J. (1995). Tight regulation, modulation, and high-level expression by vectors containing the arabinose P_{BAD} promoter. J. Bacteriol. *177*, 4121-4130.

Hallewell, R.A. and Emtage, S. (1980). Plasmid vectors containing the tryptophan operon promoter suitable for efficient regulated expression of foreign genes. Gene *9*, 27-47.

Hannig, G. and Makrides, S.C. (1998). Strategies for optimizing heterologous protein expression in *Escherichia coli*. Trends Biotechnol. *16*, 54-60.

Hashimoto, T. and Sekiguchi, M. (1976). Isolation of temperature-sensitive mutants of R plasmid by *in vitro* mutagenesis with hydroxylamine. J. Bacteriol. *127*, 1561-1563.

Hashimoto-Gotoh, T., Franklin, F.C., Nordheim, A., and Timmis, K.N. (1981). Specific-purpose plasmid cloning vectors. I. Low copy number, temperature-sensitive, mobilization-defective pSC101-derived containment vectors. Gene *16*, 227-235.

Hershfield, V., Boyer, H.W., Yanofsky, C., Lovett, M.A., and Helinski, D.R. (1974). Plasmid ColEl as a molecular vehicle for cloning and amplification of DNA. Proc. Natl. Acad. Sci. U. S. A *71*, 3455-3459.

Hochuli, E., Döbeli, H., and Schacher, A. (1987). New metal chelate adsorbent selective for proteins and peptides containing neighboring histidine residues. J. Chromatogr. *411*, 177-184.

Hohn, B. and Collins, J. (1980). A small cosmid for efficient cloning of large DNA fragments. Gene *11*, 291-298.

Iijima, S., Kawai, S., Mizutani, S., Taniguchi, M., and Kobayashi, T. (1987). On-off regulation of gene expression from tryptophan promoter with cross-flow filtration. Appl. Microbiol. Biotechnol. *26*, 542-545.

Itoh, S., Mizukami, T., Matsumoto, T., Nishi, T., Saito, A., Oka, T., Furuya, A., Takaoka, C., and Taniguchi, T. (1984). Efficient expression in *Escherichia coli* of a mature and a modified human interferon-beta 1. DNA *3*, 157-165.

Jiang, W., Jones, P., and Inouye, M. (1993). Chloramphenicol induces the transcription of the major cold shock gene of *Escherichia coli, cspA*. J. Bacteriol. *175*, 5824-5828.

Joyner, A., Isaacs, L.N., Echols, H., and Sly, W.S. (1966). DNA replication and messenger RNA production after induction of wild-type lambda bacteriophage and lambda mutants. J. Mol. Biol. *19*, 174-186.

Kaltwasser, M., Wiegert, T., and Schumann, W. (2001). Construction and application of epitope- and GFT-tagging integration vectors for *Bacillus subtilis*. Appl. Environ. Microbiol. *68*, 2624-2628.

Kang, Y., Son, M.S., and Hoang, T.T. (2007). One step engineering of T7-expression strains for protein production: increasing the host-range of the T7-expression system. Protein Expr. Purif. *55*, 325-333.

Kawai, S., Mizutani, S., Iijima, S., and Kobayashi, T. (1986). On-off regulation of the tryptophan promoter in fed-batch culture. J. Fermen. Technol. *64*, 503-510.

Kim, U.J., Shizuya, H., de Jong, P.J., Birren, B., and Simon, M.I. (1992). Stable propagation of cosmid sized human DNA inserts in an F factor based vector. Nucleic Acids Res. *20*, 1083-1085.

Korndorfer, I.P. and Skerra, A. (2002). Improved affinity of engineered streptavidin for the Strep-tag II peptide is due to a fixed open conformation of the lid-like loop at the binding site. Protein Sci. *11*, 883-893.

Krueger, J.K., Bishop, N.A., Blumenthal, D.K., Zhi, G., Beckingham, K., Stull, J.T., and Trewhella, J. (1998). Calmodulin binding to myosin light chain kinase begins at substoichiometric Ca2+ concentrations: a small-angle scattering study of binding and conformational transitions. Biochemistry *37*, 17810-17817.

Lamla, T. and Erdmann, V.A. (2004). The Nano-tag, a streptavidin-binding peptide for the purification and detection of recombinant proteins. Protein Expr. Purif. *33*, 39-47.

Lederer, T., Kintrup, M., Takahashi, M., Sum, P.E., Ellestad, G.A., and Hillen, W. (1996). Tetracycline analogs affecting binding to Tn*10*-Encoded Tet repressor trigger the same mechanism of induction. Biochemistry *35*, 7439-7446.

Lee, N., Francklyn, C., and Hamilton, E.P. (1987). Arabinose-induced binding of AraC protein to *araI2* activates the *araBAD* operon promoter. Proc. Natl. Acad. Sci. U. S. A *84*, 8814-8818.

Li, Y.J., Bi, L.J., Zhang, X.E., Zhou, Y.F., Zhang, J.B., Chen, Y.Y., Li, W., and Zhang, Z.P. (2006). Reversible immobilization of proteins with streptavidin affinity tags on a surface plasmon resonance biosensor chip. Anal. Bioanal. Chem. *386*, 1321-1326.

Liang, R., Liu, X., Liu, J., Ren, Q., Liang, P., Lin, Z., and Xie, X. (2007). A T7-expression system under temperature control could create temperature-sensitive phenotype of target gene in *Escherichia coli*. J. Microbiol. Methods *68*, 497-506.

Lieb, M. (1966). Studies of heat-inducible lambda bacteriophage. I. Order of genetic sites and properties of mutant prophages. J. Mol. Biol. *16*, 149-163.

Liu, M.F., Tolstorukov, M., Zhurkin, V., Garges, S., and Adhya, S. (2004). A mutant spacer sequence between -35 and-10 elements makes the P_{lac} promoter hyperactive and cAMP receptor protein-independent. Proc. Natl. Acad. Sci. USA *101*, 6911-6916.

Makrides, S.C. (1996). Strategies for achieving high-level expression of genes in *Escherichia coli*. Microbiol. Rev. *60*, 512-538.

Margolin, W. (2000). Green fluorescent protein as a reporter for macromolecular localization in bacterial cells. Methods *20*, 62-72.

McKenney, K., Shimatake, H., Court, D., Schmeissner, U., Brady, C., and Rosenberg, M. (1981). A system to study promoter and terminator signals recognized by *Escherichia coli* RNA polymerase. Gene Amplif. Anal. *2*, 383-415.

Mead, D.A. and Kemper, B. (1988). Chimeric single-stranded DNA phage-plasmid cloning vectors. Biotechnology *10*, 85-102.

Melnikov, A.A., Tchernov, A.P., Fodor, I., and Bayev, A.A. (1984). Lambda phagemids and their transducing properties. Gene *28*, 29-35.

Morrow, J.F., Cohen, S.N., Chang, A.C., Boyer, H.W., Goodman, H.M., and Helling, R.B. (1974). Replication and transcription of eukaryotic DNA in *Escherichia coli*. Proc. Natl. Acad. Sci. U. S. A *71*, 1743-1747.

Mujacic, M., Cooper, K.W., and Baneyx, F. (1999). Cold-inducible cloning vectors for low-temperature protein expression in *Escherichia coli*: application to the production of a toxic and proteolytically sensitive fusion protein. Gene *238*, 325-332.

Müller-Hill, B., Crapo, L., and Gilbert, W. (1998). Mutants that make more *lac* repressor. Proc. Natl. Acad. Sci. USA *59*, 1259-1264.

Nallamsetty, S., Austin, B.P., Penrose, K.J., and Waugh, D.S. (2005). Gateway vectors for the production of combinatorially-tagged His6-MBP fusion proteins in the cytoplasm and periplasm of *Escherichia coli*. Protein Sci. *14*, 2964-2971.

Orth, P., Schnappinger, D., Hillen, W., Saenger, W., and Hinrichs, W. (2000). Structural basis of gene regulation by the tetracycline inducible Tet repressor-operator system. Nat. Struct. Biol. *7*, 215-219.

Poole, E.S., Brown, C.M., and Tate, W.P. (1995). The identity of the base following the stop codon determines the efficiency of *in vivo* translational termination in *Escherichia coli*. EMBO J. *14*, 151-158.

Qing, G.L., Ma, L.C., Khorchid, A., Swapna, G.V.T., Mal, T.K., Takayama, M.M., Xia, B., Phadtare, S., Ke, H.P., Acton, T., Montelione, G.T., Ikura, M., and Inouye, M. (2004). Cold-shock induced high-yield protein production in *Escherichia coli*. Nat. Biotechnol. *22*, 877-882.

Remaut, E., Stanssens, P., and Fiers, W. (1981). Plasmid vectors for high-efficiency expression controlled by the P_L promoter of coliphage lambda. Gene *15*, 81-93.

Remaut, E., Tsao, H., and Fiers, W. (1983). Improved plasmid vectors with a thermoinducible expression and temperature-regulated runaway replication. Gene *22*, 103-113.

Robertson, C.A. and Nash, H.A. (1988). Bending of the bacteriophage lambda attachment site by *Escherichia coli* integration host factor. J. Biol. Chem. *263*, 3554-3557.

Roche, E.D. and Sauer, R.T. (2001). Identification of endogenous SsrA-tagged proteins reveals tagging at positions corresponding to stop codons. J. Biol. Chem. *276*, 28509-28515.

Rosenberg, A.H., Lade, B.N., Chui, D.S., Lin, S.W., Dunn, J.J., and Studier, F.W. (1987). Vectors for selective expression of cloned DNAs by T7 RNA polymerase. Gene *56*, 125-135.

Sachdev, D. and Chirgwin, J.M. (1999). Properties of soluble fusions between mammalian aspartic proteinases and bacterial maltose-binding protein. J. Protein Chem. *18*, 127-136.

Schleif, R. (2000). Regulation of the L-arabinose operon of *Escherichia coli*. Trends Genet. *16*, 559-565.

Schmidt, T.G., Koepke, J., Frank, R., and Skerra, A. (1996). Molecular interaction between the Strep-tag affinity peptide and its cognate target, streptavidin. J. Mol. Biol. *255*, 753-766.

Schmidt, T.G. and Skerra, A. (1993). The random peptide library-assisted engineering of a C-terminal affinity peptide, useful for the detection and purification of a functional Ig Fv fragment. Protein Eng *6*, 109-122.

Schneider, K. and Beck, C.F. (1986). Promoter-probe vectors for the analysis of divergently arranged promoters. Gene *42*, 37-48.

Schumann,W. and Ferreira,L.C.S. (2004). Production of recombinant proteins in *Escherichia coli*. Gen. Mol. Biol. *27*, 442-453.

Shizuya, H., Birren, B., Kim, U.J., Mancino, V., Slepak, T., Tachiiri, Y., and Simon, M. (1992). Cloning and stable maintenance of 300-kilobase-pair fragments of human DNA in *Escherichia coli* using an F-factor-based vector. Proc. Natl. Acad. Sci. U. S. A *89*, 8794-8797.

Silhavy,T.J. and Beckwith,J.R. (1985). Uses of *lac* fusions for the study of biological problems. Microbiol. Rev. *49*, 398-418.

Silverstone, A.E., Arditti, R.R., and Magasanik, B. (1970). Catabolite-insensitive revertants of *lac* promoter mutants. Proc. Natl. Acad. Sci. U. S. A *66*, 773-779.

Skalka, A.M. (1977). DNA replication--bacteriophage lambda. Curr. Top. Microbiol. Immunol. *78*, 201-237.

Skerra, A. (1994). Use of the tetracycline promoter for the tightly regulated production of a murine antibody fragment in *Escherichia coli*. Gene *151*, 131-135.

Smith, D.B. and Johnson, K.S. (1988). Single-step purification of polypeptides expressed in *Escherichia coli* as fusions with glutathione S-transferase. Gene *67*, 31-40.

Soberon, X., Covarrubias, L., and Bolivar, F. (1980). Construction and characterization of new cloning vehicles. IV. Deletion derivatives of pBR322 and pBR325. Gene *9*, 287-305.

Stofko-Hahn, R.E., Carr, D.W., and Scott, J.D. (1992). A single step purification for recombinant proteins. Characterization of a microtubule associated protein (MAP 2) fragment which associates with the type II cAMP-dependent protein kinase. FEBS Lett. *302*, 274-278.

Studier, F.W. (1991). Use of bacteriophage T7 lysozyme to improve an inducible T7 expression system. J. Mol. Biol. *219*, 37-44.

Studier, F.W. and Moffat, B.A. (1986). Use of bacteriophage T7 RNA polymerase to direct selective high-level expression of cloned genes. J. Mol. Biol. *189*, 113-130.

Tabor, S. and Richardson, C.C. (1985). A bacteriophage T7 RNA polymerase/promoter system for controlled exclusive expression of specific genes. Proc. Natl. Acad. Sci. U. S. A *82*, 1074-1078.

Tacon, W.C., Bonass, W.A., Jenkins, B., and Emtage, J.S. (1983). Expression plasmid vectors containing *Escherichia coli* tryptophan promoter transcriptional units lacking the attenuator. Gene *23*, 255-265.

Teff, D., Koby, S., Shotland, Y., Ogura, T., and Oppenheim, A.B. (2000). A colicin-tolerant *Escherichia coli* mutant that confers Hfl phenotype carries two mutations in the region coding for the C-terminal domain of FtsH (HflB). FEMS Microbiol. Lett. *183*, 115-117.

Terpe, K. (2003). Overview of tag protein fusions: from molecular and biochemical fundamentals to commercial systems. Appl. Microbiol. Biotechnol. *60*, 523-533.

Thomas, C.M. (1981). Complementation analysis of replication and maintenance functions of broad host range plasmids RK2 and RP1. Plasmid *5*, 277-291.

Uhlin, B.E., Molin, S., Gustafsson, P., and Nordstrom, K. (1979). Plasmids with temperature-dependent copy number for amplification of cloned genes and their products. Gene *6*, 91-106.

Vieira, J. and Messing, J. (1982). The pUC plasmids, an M13mp7-derived system for insertion mutagenesis and sequencing with synthetic universal primers. Gene *19*, 259-268.

von Gabain, A., Belasco, J.G., Schottel, J.L., Chang, A.C., and Cohen, S.N. (1983). Decay of mRNA in *Escherichia coli*: investigation of the fate of specific segments of transcripts. Proc. Natl. Acad. Sci. U. S. A *80*, 653-657.

Voss, S. and Skerra, A. (1997). Mutagenesis of a flexible loop in streptavidin leads to higher affinity for the Strep-tag II peptide and improved performance in recombinant protein purification. Protein Eng *10*, 975-982.

Watanabe, T., Ito, Y., Yamada, T., Hashimoto, M., Sekine, S., and Tanaka, H. (1994). The roles of the C-terminal domain and type III domains of chitinase A1 from *Bacillus circulans* WL-12 in chitin degradation. J. Bacteriol. *176*, 4465-4472.

Weber, M.H.W. and Marahiel, M.A. (2003). Bacterial cold shock responses. Science Progress *86*, 9-75.

Wilson, D.S., Keefe, A.D., and Szostak, J.W. (2001). The use of mRNA display to select high-affinity protein-binding peptides. Proc. Natl. Acad. Sci. U. S. A *98*, 3750-3755.

Wilson, I.A., Niman, H.L., Houghten, R.A., Cherenson, A.R., Connolly, M.L., and Lerner, R.A. (1984). The structure of an antigenic determinant in a protein. Cell *37*, 767-778.

Xu, M.Q., Paulus, H., and Chong, S. (2000). Fusions to self-splicing inteins for protein purification. Methods Enzymol. *326*, 376-418.

Yamada, M., Kabir, M.S., and Tsunedomi, R. (2003). Divergent promoter organization may be a preferred structure for gene control in *Escherichia coli*. J. Mol. Microbiol. Biotechnol. *6*, 206-210.

Yamaguchi, A., Ono, N., Akasaka, T., Noumi, T., and Sawai, T. (1990). Metal-tetracycline/H^+ antiporter of *Escherichia coli* encoded by a transposon, Tn*10*. The role of the conserved dipeptide, Ser65-Asp66, in tetracycline transport. J. Biol. Chem. *265*, 15525-15530.

Yanofsky, C. (2007). RNA-based regulation of genes of tryptophan synthesis and degradation, in bacteria. RNA. *13*, 1141-1154.

Yura, T. and Nakahigashi, K. (1999). Regulation of the heat-shock response. Curr. Opin. Microbiol. *2*, 153-158.

Zheng, C.F., Simcox, T., Xu, L., and Vaillancourt, P. (1997). A new expression vector for high level protein production, one step purification and direct isotopic labeling of calmodulin-binding peptide fusion proteins. Gene *186*, 55-60.

Chapter 2

Archaeal Plasmids

Georg Lipps

ABSTRACT

Although many archaea harbour plasmids our knowledge on the biology of archaeal plasmids has remained quite limited. In most cases it is not known how the plasmids are replicated and whether the host benefits from the extrachromosomal element. Sequence analysis of the archaeal plasmids has identified many protein genes whose function cannot be predicted by sequence similarity to known protein. These orphan genes provide a rich repertoire for rewarding functional studies on the molecular biology of archaea. Despite this lack of knowledge archaeal plasmids are increasingly important as genetic tools and allow to perform genetic studies in archaea.

INTRODUCTION

Plasmids and viruses have been extremely helpful for studying the molecular biology of various organisms and to unravel the molecular mechanism of the cellular information processing pathways in bacteria and eukarya. These genetic elements replicate independently from the host chromosome and are amenable to experimental studies due to their small size and their reduced complexity.

There is a long tradition of studying bacterial plasmids that is also reflected in the great number of known, sequenced and studied bacterial plasmids. In the current Genbank release there are deposited approximately 1000 sequences of bacterial plasmids. In contrast only the sequences of about 50 archaeal plasmids are known today. Although the importance of plasmids for horizontal gene transfer, genome plasticity and microbial adaptation is widely acknowledged for both prokaryotic domains, only a few archaeal plasmids have been studied. For most of the known archaeal plasmids only the nucleotide sequence has been determined, which only allows for a functional prediction of a minor fraction of the encoded proteins. For a number of plasmids many predicted proteins display no clear sequence similarity to characterised proteins, rendering the functional prediction of these open reading frames (ORF) difficult or even impossible. Likewise the benefit of the host carrying the respective plasmid, is to our knowledge, only known for some haloarchaeal plasmids where the genes for the gas vesicle synthesis are encoded on some larger

Table 1. Types of archaeal plasmids. The known archaeal plasmids can be classified in five distinct groups: plasmids with the putative replication gene *repH*, plasmids which replicate with a rolling circle mechanism (RC), plasmids with a *cdc6* homologue, plasmids of the pRN type and conjugative plasmids.

	repH	RC	*cdc6*	pRN	conjugative	unclassified
Euryarchaeota						
Halobacteriaceae	pHV2, …	pHGN1, pGRB1, pHK2, …				pZMX101, pSCM211, pHH205, pL47
Thermococcaceae		**pTN1**, **pGT5**, …				
Thermoplasmataceae			pTA1			
Methanobacteriales			pFV1,pFZ1			pME2200
Methanococcales						pURB500, pMMC501
Methanosarcinales		pML	Plasmid1			pC2A
Crenarchaeota						
Thermoproteales						pTPEN01
Sulfolobaceae				**pRN1**, …	**pNOB8**, …	

Bold: characterized. Underline: used as shuttle.

plasmids. Therefore the majority of the archaeal plasmids appear to be cryptic, but as stated above the function of many plasmidal genes remains currently unknown. Most probably we currently only know a minor fraction of archaeal plasmids (the tip of the iceberg) and also the number of archaeal plasmid families is likely to increase considerable as soon as the diversity of the plasmids is sampled more deeply.

The two major branches of the archaea are the *Crenarchaeota* and the *Euryarchaeota*, although other branches include *Nanoarchaeota* (with only a single representative, *i.e. Nanoarchaeum equitans*) and the *Korarchaeota* (so far only environmental samples with no cultured representative). The *Euryarchaeota* are more diverse, and encompass the halophiles, the extreme thermophiles and the methanogens. The *Crenarchaeota* are best known for their acidophilic thermophiles. But recent metagenomic studies indicate a large proportion of mesophilic *Crenarchaeota* in the environment (Karner et al., 2001; Schleper et al., 2005).

The sequences of the archaeal plasmid genomes allow a broad classification into six groups (Table 1). The last group is heterogeneous and includes all remaining plasmids which could not be grouped in the first five groups. Interestingly the major division of the archaea in *Euryarchaeota* and *Crenarchaeota* is mirrored by the plasmid groups. None of the five distinct well-defined plasmid groups has host organisms in both *Euryarchaeota* and *Crenarchaeota*. It remains to be seen whether this strict separation of the host range will persist when more archaeal plasmids are known.

repH TYPE PLASMIDS

Plasmids with a *repH* replication gene have only been isolated from a few haloarchaeal species (Table 2), in addition a *repH* gene is found in the two viruses ϕCh1 and ϕH infecting the haloarchaea *Natrilba magadii* and *H. salinarum* (Klein et al., 2002; Gropp et al., 1992). Although experimental evidence is lacking, it is believed that the viral *repH* genes are involved in replicating the viral genomes. The plasmid pHV2 (Fig. 1) from *Haloferax volcanii* was one of first known archaeal plasmids (Pfeifer et al., 1981; Charlebois et al., 1987) and derivates of pHV2 have been extensively used as shuttle vectors for decades (Lam and Doolittle, 1989). Astonishingly the plasmid itself has not been well studied and it is unknown how the plasmid is replicated and what the functions of the three to five plasmid genes beside the *repH* gene are. The protein RepH is the only large protein encoded by the plasmid and therefore the prime candidate for a replication protein. Intriguingly, there is no sequence similarity to known proteins and only the approximately 100 last amino acids can be predicted to fold into a winged-helix DNA binding domain. This type of sequence specific DNA-binding protein is common for initiator proteins of replication (Giraldo, 2003). The DNA binding domain is conserved within the RepH proteins, adding support to the notion that the RepH protein could be involved

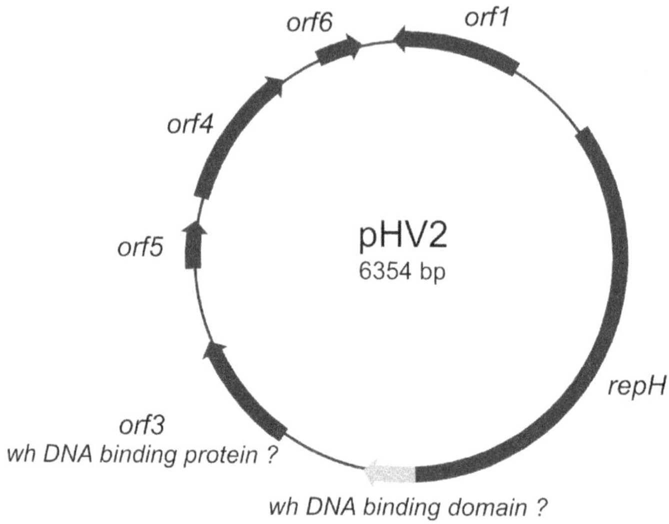

Fig. 1. The repH plasmid pHV2 from *Haloferax volcanii*. The plasmid pHV2 is the smallest plasmid containing a *repH* gene. Additional evidence that the *repH* gene is the replication protein comes from the mapping of the minimal replicon of pNRC100, which also contains the *repH* gene. None of the open reading frames of this plasmid has been studied experimentally. However sequence analysis suggests that the C-terminal 100 amino acids of RepH could fold into a winged helix DNA-binding domain. Likewise there is a weak sequence homology of *orf3* to winged helix DNA binding proteins. *Orf1-4* were predicted when the plasmid was originally sequenced (Charlebois et al., 1987). *Orf5 and orf6* are smaller open reading frames which are supported by a Genemark analysis (Lukashin and Borodovsky, 1998).

in plasmidal replication initiation by binding to the replication origin. The precise functions of the RepH proteins are however completely unclear.

The *repH* genes are found on plasmids form *Haloferax volcanii*, *Haloarcula marismortui* and several *Halobacterium* species. These archaea have fragmented genomes with a number of replicons. Besides one or two main chromosomes, up to seven minichromosomes/megaplasmids are present in these species. Without experimental data, the distinction between minichromosome and megaplasmids is not straightforward: the replication mechanism cannot be determined from the nucleotide sequence alone. The mere presence of the chromosomal replication initiator protein such as a Cdc6/Orc1 protein or of a plasmidal replication protein does not prove that the encoded proteins are actively involved in replicating the respective element.

In case of *Haloferax volcanii* the origins of the replicons pHV1 (86 kb), pHV3 (442 kb) and pHV4 (690 kb) were mapped and could be shown to be a chromosomal type replication origin with typical sequence motifs and a neighbouring *cdc6/orc1*

Table 2. Plasmids and virus with a *repH* replication protein.

plasmid	size	original host	remarks	reference
pHV2	6354 bp	*Haloferax volcanii*		Charlebois et al., 1987
pHH1	143 kbp	*Halobacterium salinarium*	gas vesicle genes, not completely sequenced	Pfeifer and Ghahraman, 1993
pNRC100	191346 bp	*Halobacterium sp.* NRC-1	gas vesicle genes, minimal replicon defined	Ng et al., 2000
pNRC200	365425 bp	*Halobacterium sp.* NRC-1	gas vesicle genes	Ng et al., 2000
pNG200	33452 bp	*Haloarcula marismortui*		Baliga et al., 2004
pNG300	39521 bp	*Haloarcula marismortui*		Baliga et al., 2004
pNG400	50060 bp	*Haloarcula marismortui*		Baliga et al., 2004
φCh1	58498 bp	*Natrilba magadii*		Klein et al., 2002
φH	~ 59 kbp	*Halobacterium salinarium*		Schnabel et al., 1982

gene (Norais et al., 2007). In contrast the plasmid pHV2 (6 kb) lacks these features and appears to be replicated by the RepH protein.

A contrasting situation is observed for *Halobacterium sp.* NRC1 that has a main chromosome of 2 Mb and two additional replicons, pNRC100 (191 kb) and pNRC200 (365 kb). Both plasmids encode the important gas-vesicle proteins, which enable the archaeon to synthesize gas vesicles, which allows the cells to float. Interestingly, pNRC100 encodes two RepH proteins and a Cdc6/Orc1 homologue. The minimal replication origin of pNRC1 was determined and found to encompass a region of 3.9 kb which encodes the RepH protein and has an AT-rich region of about 500 bases upstream of the *repH* gene (Ng and Dassarma, 1993). Therefore it seems that the *repH* gene is the solely essential replication gene of the *repH* plasmids and that the pNRC100 plasmid is also replicated through this mechanism. The replicon pNRC200 encodes two RepH proteins and six Cdc6/Orc1 proteins. In this case the mechanism of replication is unknown, however the deletion of one pNRC200-borne *cdc6/orc1* gene is not possible (Berquist et al., 2007), pointing towards a chromosomal replication type of this replicon.

ARCHAEAL PLASMIDS WHICH REPLICATE THROUGH A ROLLING CIRCLE MECHANISM

A number of euryarchaeal plasmids encode an initiator protein for rolling circle replication (Table 3). The rolling circle replication is a specialized replication mode employed by plasmids and viruses (reviewed by (Khan, 2000)). The initiator protein (usually termed Rep protein) binds to the double-stranded origin of replication (*dso*). One strand of the *dso* is specifically cut by the endonuclease activity of the initiator protein resulting into a liberated 3' hydroxyl group and a covalent linkage of the 5' phosphate group to a conserved tyrosine in the active site of the nicking domain of the initiator protein. The 3' hydroxyl group is then used as primer by the host replication machinery. While the leading strand is replicated the non-template strand with the covalently attached initiator protein is displaced. After a complete round of leading strand replication the *dso* is regenerated and after transesterification reactions catalysed by the initiator protein a double-stranded circular DNA is released while the displaced strand forms a single-stranded circular DNA. To complete replication this ssDNA has to be converted into dsDNA: at the single-stranded origin of replication (*sso*) a short ribonucleotide primer is synthesized by the host RNA polymerase which is then extended by the replicative DNA polymerase of the host to generate the second copy of double-stranded circular DNA. Since replication at the *sso* starts after completion of the leading strand replication, circular ssDNA is an intermediate product and its presence in dividing cells is a good indication of rolling circle replication.

Rolling circle replicating plasmids can have a very compact genome as only one plasmidal gene product, the initiator protein, is required for replication. In fact

Table 3. Archaeal plasmids with rolling-circle replication.

plasmid	size	original host	remarks	reference
pHGN1	1765 bp	*Haloarchaeal coccus*		Hall and Hackett, 1989
pHSB2	1736 bp	*Halobacterium salinarum*		Akhmanova et al., 1993
pGRB1	1781 bp	*Halobacterium* strain GRB		Hackett et al., 1990
pHK2	10,5 kb	*Haloferax volcanii*	only replication region sequenced	Holmes et al., 1994
pML	2158 bp	*Methanohalophilus mahii*		NC_005563
pZMX201	1668 bp	*Natrinema sp.* CX2021		NC_006996
pNB101	2538 bp	*Natronobacterium sp.* AS-7091		Zhou et al., 2004
pGT5	3444 bp	*Pyrococcus abyssi* GE5		Erauso et al., 1996
pTN1	3367 bp	*Thermococcus nautilus* 30-1		Soler et al., 2007
pRT1	3373 bp	*Pyrococcus sp.* JT1	experimental and bioinformatic support for rolling circle replication is low	Ward et al., 2002

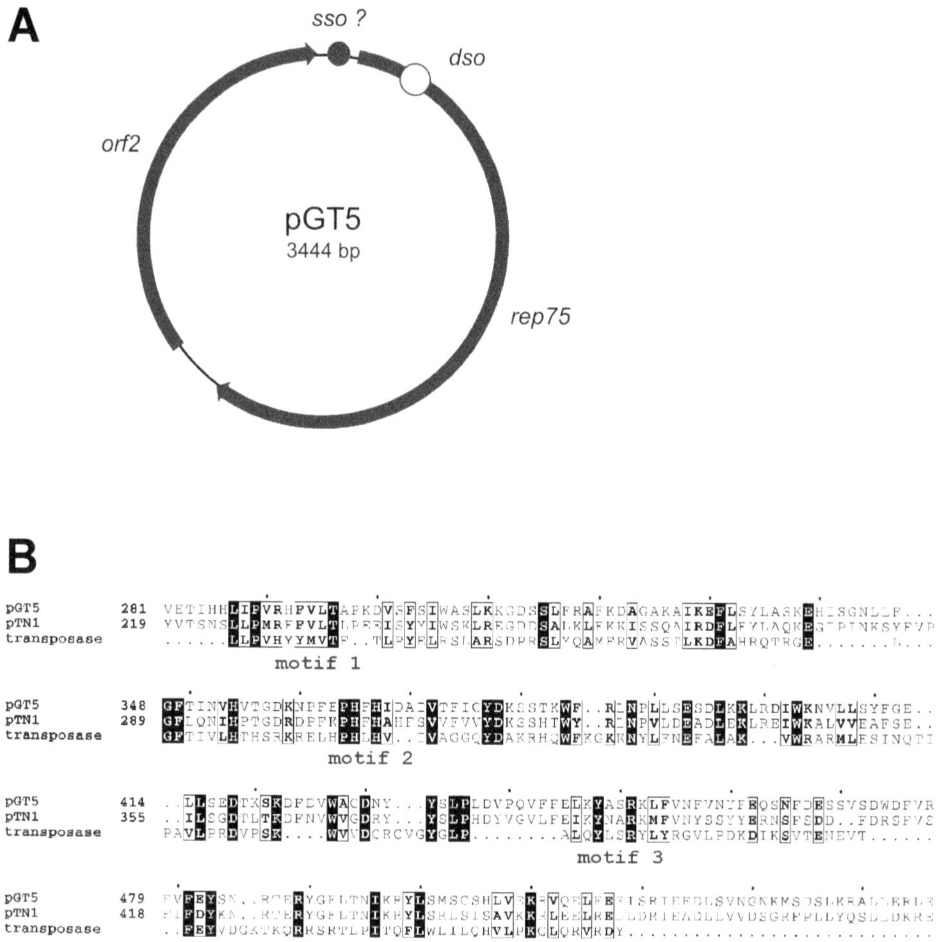

Fig. 2. The plasmid pGT5 from *Pyrococcus abyssi.* A: The plasmid pGT5 is the best-understood of the archaeal plasmids, which replicate through a rolling circle mechanism. The double-stranded origin (*dso*) 5'- TTATCTTGATA-3' has been identified with recombinant replication protein and is cleaved after the guanosine (Marsin and Forterre, 1998). The location of the single-stranded origin (*sso*) is not known, however typical features of a bacterial *sso* has been detected upstream of the *rep75* gene (Erauso et al., 1996). B: Alignment of the initiator proteins of pGT5 and pTN1 with a transposase of *Vibrio vulnificus*. Aligned are the central parts of the initiator proteins containing the conserved motifs 1-3 originally described by (Ilyina and Koonin, 1992). Interestingly these initiator proteins display higher sequence similarity to some bacterial transposases like the transposase from *Vibrio vulnificus* (Transposase_32, pfam PF04986) than to the haloarchaeal plasmidal initiator proteins and to bacterial rolling circle initiator proteins. The two conserved histidine residues of motif 2 could be involved in chelating a catalytic metal ion. When the tyrosine residue in motif 3 (Y448) is changed to phenylalanine, the nicking and closing activity of the initiator protein of pGT5 is abolished, suggesting that this tyrosine residue forms the phosphoester bond with the DNA. The neighbouring arginine residue (R451) is involved in nucleotidyl transferase activity (Marsin and Forterre, 1999).

three haloarchaeal plasmids have a genome size of less than 1800 base pairs and appear to encode only this single protein. These plasmids are similar in sequence and in support of rolling circle replication circular single-stranded intermediates have been detected (Akhmanova et al., 1993).

Among the rolling circle replication plasmids, two archaeal plasmids have been studied in more detail: the plasmid pGT5 from *Pyrococcus abyssi* and recently the plasmid pTN1 from *Thermococcus nautilus*. The initiator proteins of both plasmids are clear homologues (Fig. 2) and the pGT5 initiator protein has been studied in detail. It could be shown that the initiator protein from pGT5 has endonuclease activity on single-stranded oligonucleotides derived from the nicking region of the predicted *dso* (Marsin and Forterre, 1998). Interestingly the *dso* has sequence similarity to a number of bacterial rolling circle replication plasmids of the pC194 group. In addition the catalytic tyrosine of the initiator protein was experimentally confirmed by a point mutational study (Marsin and Forterre, 1999). The recombinant initiator protein was also able to perform a transesterification reaction of the oligonucleotide bound at the catalytic tyrosine back to DNA substrate, the so-called closing activity that is required to terminate the replication of the leading strand (see above). Interestingly the enzyme had also a nucleotidyl transferase activity. This activity could have a regulatory function by blocking the catalytic tyrosine in order to prevent additional rounds of replication (Marsin and Forterre, 1999).

Both plasmids encode a second protein: these are not related to each other and their exact physiological function is unclear. Whereas the second encoded protein of pGT5 has not been studied it could be demonstrated that the second protein of pTN1 is a DNA-binding protein. It binds to single- and double-stranded DNA without apparent sequence specificity. The recombinant protein devoid of its hydrophobic C-terminal tail forms a dimer or tetramer in solution. Electron microscopy and atomic force microscopy revealed that the protein is able to form large nucleoprotein complexes by condensing the DNA. Possibly this protein assists the segregation of the plasmids during cell division (Soler et al., 2007).

PLASMIDS WITH A Cdc6/Orc1 REPLICATION PROTEIN

There are currently four euryarchaeal plasmids known which encode a Cdc6/Orc1 protein (Table 4). These proteins are AAA$^+$-ATPase that are related to both the eukaryal Cdc6 and Orc1 proteins. In eukaryotes the multimeric origin binding complex (ORC) containing the Orc1 subunit binds to replication origins, and later on the Cdc6 protein is required to load the replicative helicase onto the origin. Archaeal genomes encode for a variable number of Cdc6/Orc1 homologues, *e.g. Pyrococcus* has a single Cdc6/Orc1, *Sulfolobus* species have three and *Halobacterium salinarium* has ten homologues. The first archaeal replication origin was mapped in *Pyrococcus abyssi* (Myllykallio et al., 2000) and found to be adjacent to the *cdc6/orc1* gene.

Table 4. Plasmids with a Cdc6/Orc1 replication protein.				
plasmid	size	original host	remarks	reference
pTA1	15723 bp	*Thermoplasma acidophilum*		NC_008318
pFV1	13514 bp	*Methanothermobacter thermautotrophicus*	DNA glycosylase, restriction/ modification, endonuclease III	Nolling et al., 1992
pFZ1	11014 bp	*Methanothermobacter thermautotrophicus*	DNA glycosylase, Restriction/ modification	Nolling et al., 1992
Plasmid 1	36358 bp	*Methanosarcina barkeri*	5 transposases, 4 genes for methanochondroitin synthesis	Maeder et al., 2006

Archaeal Cdc6/Orc1 proteins bind sequence specifically albeit with relatively low affinity to the replication origins identified or predicted in various archaea (Grainge et al., 2006; Robinson et al., 2004; Capaldi and Berger, 2004). Nevertheless, *in vivo* the Cdc6/Orc1 proteins binds highly specifically to the replication origin, as could be demonstrated by a whole-genome microarray chromatin immunoprecipitation experiment performed with *Pyrococcus abyssi* (Matsunaga et al., 2007). Recently the structures of two Cdc6/Orc1 proteins in complex with DNA demonstrated how the binding of the archaeal proteins to the DNA bend and distort the origin DNA (Dueber et al., 2007; Gaudier et al., 2007) and provided structural evidence that these protein are also important in opening the double-stranded DNA at the replication origin.

The presence of Cdc6/Orc1 homologues on plasmids raises the intriguing possibility that these plasmids use the same replication initiation mechanism as the archaeal chromosomes and function like an *oriC* plasmid. Sequence analysis of the plasmidal Cdc6/Orc1 homologues shows that the plasmidal proteins are in general most similar to the homologues encoded on the chromosomes of the native hosts. However the sequence divergence from plasmidal to chromosomal copy has the same magnitude as between chromosomal Cdc6/Orc1 proteins from different species. The replication origin of the plasmids is unknown. In the archaeal chromosomes, characteristic sequence motifs termed origin recognition boxes (ORB) are commonly found in the upstream neighbourhood of the *cdc6/orc1* genes. In the plasmids pFV1

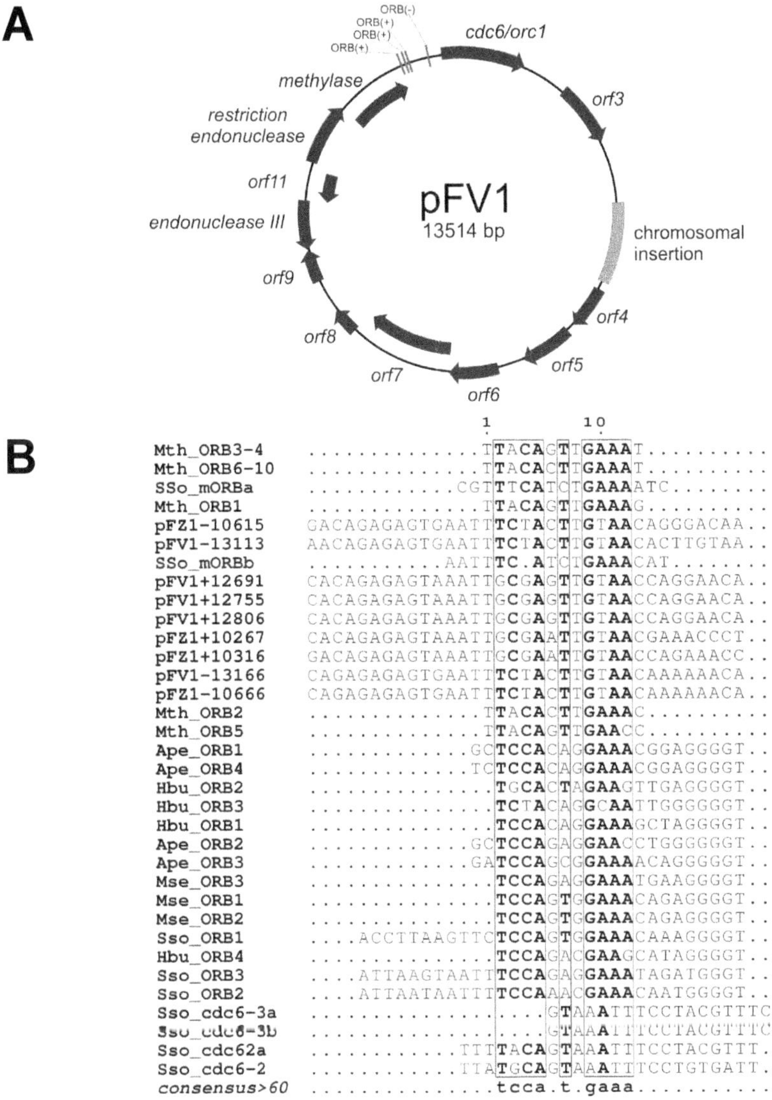

Fig. 3. The plasmid pFV1 from *Methanothermobacter thermautotrophicus*. A: The function of most of the open reading frames of this plasmid cannot be predicted. The plasmid encodes a GGCC-specific type II restriction/modification system (Nolling et al., 1992). In addition, an endonuclease III that could be involved in mismatch repair is present. Most remarkably the plasmid encodes a Cdc6/Orc1 homologue. These proteins initiate chromosomal replication in archaea and eukarya. Upstream of the *cdc6/orc1* gene several ORBs (origin recognition boxes) could be identified, suggesting that this intergenic region is the replication origin of the plasmid. Between ORF3 and ORF4, a region with high similarity to the chromosome of *Methanothermobacter thermautotrophicus* could be identified. B: Alignment of the origin binding boxes (ORB). Aligned are experimentally confirmed ORB (origin binding boxes) from *Methanothermobacter thermoautotrophicus* (Mth) (Capaldi and Berger, 2004), *Sulfolobus solfataricus* (Sso) (Robinson et al., 2004), *Aeropyrum pernix* (Ape) (Grainge et al., 2006) and predicted ORB from the genome of *Metallosphaera sedula* (Mse) and *Hyperthermus butylicus* (Hbu) predicted by (Robinson and Bell, 2007) as well as the putative ORB from the plasmids pFZ1 and pFV1. The sign (+/-) and the numbers indicate strand and position of the ORB.

(Fig. 3) and pFZ1 we identified three ORB sequences with high resemblance to the established chromosomal ORB. However on the plasmid pTA1 we were not able to unambiguously identify the sequence elements characteristic for chromosomal origins. The plasmid 1 from *Methanosarcina barkeri* contains an array of 38 direct repeats with a length of 143 nucleotides upstream of the *cdc6/orc1* gene. Although its function remains unclear, this array seems suggestive of a replication origin.

PLASMIDS OF THE pRN TYPE

Plasmids of the pRN type (Table 5) were isolated from *Sulfolobus* strains and the crenarchaeote *Acidianus ambivalens*. The characteristics of the plasmids of this group are two rather small DNA binding proteins and a large replication protein. Four plasmids isolated from *Sulfolobus* strains from Iceland and the plasmid pDL10 from *Acidianus ambivalens* encode these three types of genes in the same order and orientation and make up the core group of the pRN type plasmids. Four additional plasmids isolated from *Sulfolobus* strains from New Zealand and Italy do not encode all three genes and are only distantly related to the core group of the pRN plasmid family.

The best studied plasmid of the pRN group is the plasmid pRN1, which was isolated 1994 by Zillig and coworkers (Zillig et al., 1994) and was the first plasmid of this family to be sequenced (Keeling et al., 1996). In the original isolate *S. islandicus* REN1H1 the plasmid pRN1 (Fig. 4) and the related plasmid pRN2 coexist. Both plasmids appear to be cryptic and are present in multiple copies. However both plasmids do not depend on the other, since it is possible to obtain clones containing neither plasmid, a single plasmid or both plasmid after prolonged cultivation (Purschke and Schaefer, 2001).

With 5350 bp the plasmid pRN1 is one of the smaller pRN type plasmids. Analysis of the nucleotide sequence of pRN1 revealed that the plasmid could code for five small proteins (< 100 amino acids) and one large protein (904 amino acids). Among these six proteins three are conserved within the plasmid family and were therefore considered to be the essential core of the pRN plasmids (Fig. 4) (Peng et al., 2000; Lipps, 2006). The remaining three smaller ORFs probably do not code for proteins. In support of this notion, an analysis of the transcripts of pRN1 failed to detect transcripts corresponding to these ORFs. Nevertheless two counter-transcripts were identified in this part of the plasmid (Berkner and Lipps, 2007).

The three conserved proteins of pRN1, named ORF56, ORF80 and ORF904, were studied by our group. For this purpose the proteins were overexpressed in *E. coli*, purified and characterized. The amino acid sequence of ORF56 revealed sequence similarity to several DNA binding proteins of the strand-helix-turn-helix fold. These proteins are dimeric and their DNA binding interface consists of an

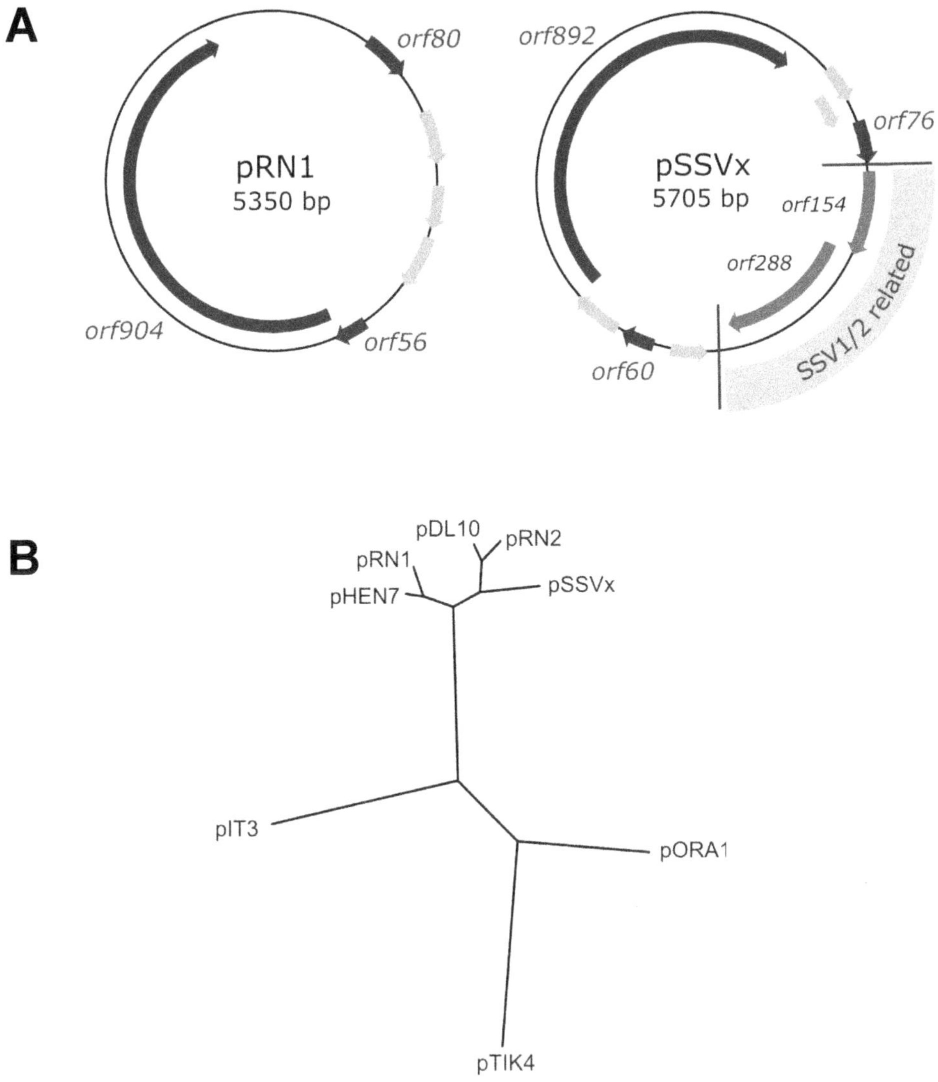

Fig. 4. Comparison of the plasmid pRN1 with the plasmid-virus hybrid pSSVx. Both plasmids code for the three conserved proteins of the pRN plasmid family (*orf56/60*, *orf904/892* and *orf80/orf76*, black arrows). ORF56 and ORF80 are sequence specific DNA-binding proteins, the former protein is involved in plasmidal copy control. The protein ORF904 is a multifunctional replication protein with primase, DNA polymerase and helicase activity. Additional genes not conserved are shown in grey arrows. The virus-plasmid hybrid pSSVx contains a region containing two open reading frames (dark grey) also found in Fuselloviridiae. The Fuselloviruses SSV1 and SSV2 can act as helper viruses for pSSVx (Arnold et al., 1999). B: Phylogenetic tree of the replication protein of the pRN plasmids. The replication proteins from the plasmids pIT3, pORA1 and pTIK4 are only distantly related to the replication proteins from the plasmids of the core group (pRN1, pRN2, pDL10, pHEN7, pSSVx), which were all isolated from strains from Iceland. The plasmid pTAU4 encodes an MCM protein instead of a replication protein of the pRN type and was not included in this analysis.

Table 5. Plasmids of the pRN type.

plasmid	size	original host	ORF56/ CopG	prim/pol	SF3	ORF80/Plr	reference
core group							
pRN1	5350 bp	*S. islandicus*	+	+	+	+	Keeling et al., 1996
pRN2	6959 bp	*S.islandicus*	+	+	+	+	Keeling et al., 1998
pHEN7	7830 bp	*S. islandicus*	+	+	+	+	Peng et al., 2000
pSSVx	5705 bp	*S. islandicus*	+	+	+	+	Arnold et al., 1999
pDL10	7598 bp	*A. ambivalens*	+	+	+	+	Kletzin et al., 1999
related							
pTIK4	13638 bp	*S. neozealandicus*	+	-	+	+	Greve et al., 2005
pTAU4	7192 bp	*S. neozealandicus*	+	-	MCM	+	Greve et al., 2005
pORA1	9689 bp	*S. neozealandicus*	+	+	+	-	Greve et al., 2005
pIT3	4967 bp	*S. solfataricus*	+	+	+	-	Prato et al., 2006

anti-parallel β-sheet made up of from the N-terminal β-strands of both subunits. Well-studied DNA binding proteins of this fold are the *arc* repressor (Raumann et al., 1994) and the CopG protein from the rolling circle plasmid pLS1 (del Solar et al., 1990; Gomis et al., 1998). The latter protein binds as a tetramer to an inverted repeat located upstream of a bicistronic operon of its own gene and the replication gene. A regulatory counter transcript RNA is mainly responsible for the plasmidal copy control, CopG appears to have a fine-tuning role in copy control (del Solar and Espinosa, 1992).

The recombinant protein ORF56 from pRN1 could be shown to be a dimeric DNA binding protein. Its DNA binding site was identified to be an imperfect inverted repeat of twelve base pairs which is located upstream of its own gene. ORF56 binds as a tetramer to the inverted repeat (Lipps et al., 2001b). The binding of ORF56 to the operator could interfere with the recruitment of the RNA polymerase to the promoter since the transcript start site is situated within the inverted repeat. In fact a reporter gene study supported that ORF56 could act as an repressor (Berkner and Lipps, 2007). The genes *orf56* and *orf904* are cotranscribed, suggesting that ORF56 would also down-regulate transcription of the replication protein gene *orf904* that could ultimately lead to reduced replication initiation. In this scenario ORF56 would be central in plasmidal copy control. It is not known whether a regulatory RNA is also involved in copy control. However there is a counter-transcript complementary to the 5' end of the *orf56*/*orf904* cotranscript that could participate in regulation. The homologues of the ORF56 proteins of the pRN plasmid family are also called CopG reflecting the functional similarity to the CopG protein from pLS1 (see above).

The gene *orf904* codes for a 110 kDa protein and is the only large protein of this plasmid family. In fact this gene occupies roughly half of the coding capacity of pRN1. Sequence analysis of ORF904 showed that the C-terminal part of the protein is a helicase domain of superfamily 3. Superfamily 3 helicases are found in the replication proteins of bacteriophages (such as P4) and of viruses, *e.g.* SV40 and Papilloviruses. These viral proteins have been studied extensively and crystal structures have been determined. Superfamily 3 helicases are AAA+-ATPases, which form hexameric rings around single-stranded or double-stranded DNA. These helicases are able to translocate along DNA. Steric exclusion of one strand of the DNA while translocating on the other strand would result in unwinding of double-stranded DNA. A structure of the complex of the helicase from the bovine Papillovirus type 1 with a short single-stranded DNA show that β-hairpins reach into the central pore of the hexameric ring and contact the DNA substrate. The hairpins move along the axis of the pore and transport the DNA substrate concomitant to conformational changes caused by binding and hydrolysis of ATP. Although ORF904 is only distantly related to the viral proteins a homology model of the core helicase domain of ORF904 could be constructed and showed that the critical residues for helicase activity are present in the ORF904 protein (Lipps, unpublished). In addition the helicase activity of ORF904 has the same polarity (3' – 5') as the other characterized helicases of superfamily 3. It is noteworthy that superfamily 3 helicases are related to the MCM proteins, the presumed replicative helicases of the archaea and eukarya (Iyer et al., 2004). One of the pRN plasmids, the plasmid pTAU4, does not encode a protein with a superfamily 3 helicase domain but instead an MCM homologue, suggesting that the MCM proteins and the superfamily 3 helicases perform similar functions in plasmid replication.

The helicase domains of superfamily 3 are located within the C-terminal half of the replication proteins. Although the N-terminal halves of the replication proteins are conserved it was not possible to predict the function of this part of proteins by sequence analysis, due to the lack of similarity to characterized proteins. The study of the recombinant replication protein ORF904 from pRN1 unexpectedly revealed that the N-terminal part of the protein has primase and DNA polymerase activity (Lipps et al., 2003). The corresponding domain was consequently termed prim/pol domain and its crystal structure unexpectedly revealed that the prim/pol domain is structurally similar to the structures of the small catalytic subunits of the archaeal-eukaryal primases (Lipps et al., 2004). The structural similarity is confined to the active site of both enzymes. Interestingly both proteins appear to have a common ancestor (Iyer et al., 2005) which could have been a simple bifunctional DNA replication enzymes with primase and DNA polymerase activity capable of replicating simple DNA genomes (Lipps et al., 2004).

The third conserved protein of the pRN family is a sequence-specific DNA binding protein. This protein is the most conserved protein of the plasmid family and homologues are also found in several conjugative plasmids isolated from *Sulfolobus* strains. The ORF80 protein from pRN1 binds to two neighbouring binding sites upstream of its own gene. Remarkably the binding sites and the distances between the binding sites are conserved. In addition EMSA experiments suggest that a larger nucleoprotein complex is formed (Lipps et al., 2001a). Attempt to visualize this complex by atomic force microscopy failed however (Lysetska et al., 2005) and the physiological function of this protein remains unclear. Previous suggestions that this protein could bind to the replication origin can no longer be maintained, since a systematic study of insertion mutants revealed that the *orf80* gene can be interrupted without compromising its replication in *Sulfolobus*. However the mutant plasmid with the interrupted *orf80* gene was lost at a significantly higher rate when grown under non-selective conditions as the other insertion mutants (Berkner et al., 2007). Possibly ORF80 could contribute to plasmid partitioning. However there is no sequence similarity to the known plasmid-partitioning proteins. Whereas a previous analysis suggested that ORF80 could be a novel type of coiled-coil DNA binding protein with a C-terminal DNA binding domain, a more recent analysis

Table 6. Conjugative plasmids.

plasmid	size	original host	remarks	reference
pNOB8	41229 bp	*Sulfolobus sp. NOB8H2*	ParA, ParB, deletion variant pNOB8-33	Schleper et al., 19955
pING1	24554 bp	*Sulfolobus islandicus*	deletion variants pING2, 3, 4, 6	Stedman et al., 20000
pKEF9	28930 bp	*Sulfolobus islandicus*		Greve et al., 20044
pARN3	26200 bp	*Sulfolobus islandicus*		Greve et al., 20044
pARN4	26476 bp	*Sulfolobus islandicus*		Greve et al., 20044
pHVE14	35422 bp	*Sulfolobus islandicus*		Greve et al., 20044
pSOG1	29000 bp	*Sulfolobus islandicus*		Erauso et al., 20066
pSOG2	26960 bp	*Sulfolobus islandicus*		Erauso et al., 20066
pTC	20417 bp	*Sulfolobus tengchongensis*		NC_005969

hints that the ORF80 proteins fold into a winged helix DNA binding protein (Lipps, unpublished). Although the function of these proteins is still unknown the genes were tentatively termed *plr* for plasmid regulatory.

One of pRN plasmids is very special, since it contains a virus-related fragment (Fig. 4). The plasmid-virus hybrid pSSVx was isolated together with the fusellovirus SSV2 from a *Sulfolobus islandicus* isolate (Arnold et al., 1999). Remarkably the virus SSV2 and also the related virus SSV1 function as helper virus for pSSVx. Cells infected with the supernatant of the original culture produced two types of viral particles, the smaller particle probably contains the pSSVx DNA and the larger particles are SSV2 virions. No viral particles were observed when the cells are transformed with pSSVx DNA alone. In contrast, transformation with pSSVx and viral DNA leads to a viral infection and spreading of the pSSVx through the culture. No such spreading has been observed with the other pRN plasmids. It is therefore believed that the viral fragment of pSSVx contains the signals which are required for the "viral" life-style of pSSVx which ultimately leads to the packaging of the pSSVx DNA in the viral particles and its spreading through the culture.

CONJUGATIVE PLASMIDS

A number of *Sulfolobus* strains have been isolated which harbour conjugative plasmids (Table 6). These plasmids are transferred from one cell to another by a yet uncharacterised transfer involving direct cell-cell contacts. The first example of an archaeal conjugative plasmid was the plasmid pNOB8 which was isolated from a Japanese *Sulfolobus* strain (Schleper et al., 1995). This 45 kb plasmid could be transferred to different *Sulfolobus* strains by mixing with a small amount of donor cells. The plasmid accumulates to a copy number of about 20-40 and causes significant growth retardation. The plasmid pNOB8 and other conjugative plasmids are not genetically stable, and deletions of larger parts of the plasmids are commonly observed (Prangishvili et al., 1998; She et al., 1998; Stedman et al., 2000). It is noteworthy that conjugative archaeal plasmids have so far only been isolated from *Sulfolobus* strains.

Comparing the genomes of several conjugative plasmids (Fig. 5) allowed for the delineation of conserved genes and their putative function (Greve et al., 2004). The genomes of the plasmids can be divided in three sections. The first section encodes several rather long proteins, which might be involved in conjugational transfer of DNA (Fig. 5). The second part of the genomes contains the putative replication origin, which is highly conserved, rich in repeats and has a high A/T content. The third section of the plasmids encodes for a number of relatively short genes including a homologue of the *orf80/plr* gene from the *Sulfolobus islandicus* plasmid pRN1 (see above). The function of this gene is unclear. In addition this section contains a tyrosine integrase, which can probably catalyse the reversible

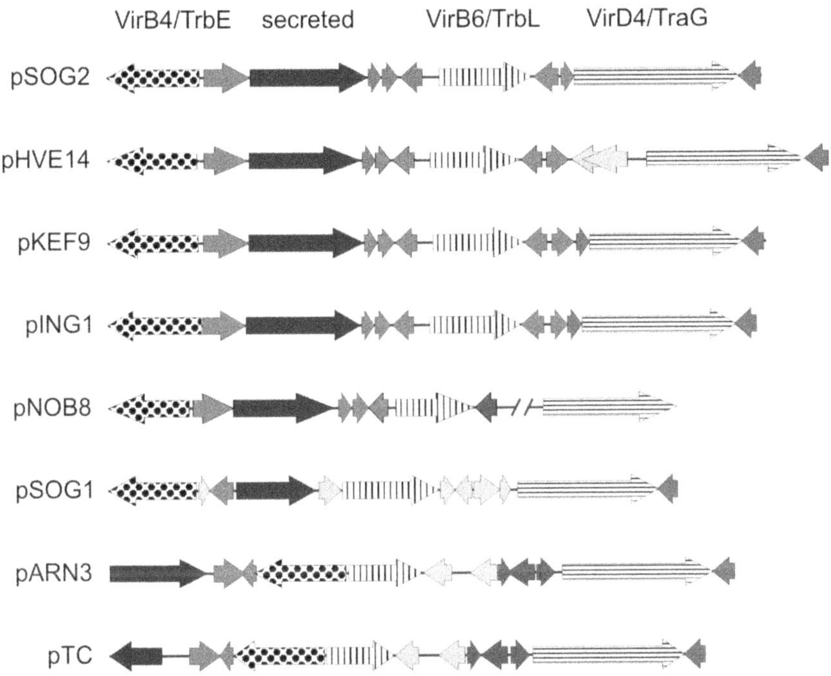

Fig. 5. Organization of the putative transfer region of the archaeal conjugative plasmids. Four proteins are completely conserved in all known archaeal conjugative plasmids. Three of them (VirB4/TrbE, VirB6/TrbL and VirD4/TraG) are related to bacterial type 4 secretion system proteins. In dark grey are shown open reading frames that are partially conserved. Unique open reading frames are shown in light grey.

integration of the plasmid into the host chromosome. The plasmidal attachment sites are sequences of about 40 bases that are identical or nearly identical to parts of the glutamyl-tRNA genes of the host. After integration the tRNA genes remain intact. Since the tRNA genes are highly conserved, the plasmids could potentially integrate into the genomes of a broad number of strains (Erauso et al., 2006). In fact, integrated copies of conjugative plasmids are found in the genomes of *S. acidocaldarius* and *S. tokodaii* (Greve et al., 2004).

It is not known how these plasmids replicate, and a clear candidate for a protein initiating plasmid replication could not be identified. Apart from several putative conjugative proteins in the first section of the genomes and the integrase only rather small proteins are encoded by these plasmids. Possibly, host proteins cooperate with the plasmid encoded proteins to initiate replication.

The first section of the conjugative plasmids contains about a dozen open reading frames (Fig. 5). Some of these putative proteins are conserved within a number of the plasmids but only a core set of four genes is found in all conjugative plasmids.

These proteins are also the only proteins that have low sequence similarity to characterised proteins, further supporting their importance. Three of these core proteins are weakly similar to proteins of the type 4 secretion system which is able to transport proteins and DNA across bacterial membranes. The well-studied type 4 secretion system of *Agrobacterium tumefaciens* consists of 11 VirB proteins that assemble into a pore and a pilus. Additionally five VirD proteins are required to process and to transport the substrate DNA (see also the chapter of M. Coll about the molecular mechanism of DNA transport).

The coupling protein VirD4 (TrbE) is a hexameric protein, which is attached onto the cytoplasmic face of the cell membrane. It interacts with the relaxases, which prepare the DNA to be transferred, and with the membrane spanning VirB components. The VirD4 protein is an ATPase that in cooperation with the VirB11 and VirB4 delivers the energy for the directional transport of the DNA. Relaxases prepare the DNA for transport by cleaving the double-stranded DNA and transforming it into single-stranded DNA. These proteins have not been identified on the archaeal conjugative plasmids, nor could a VirB11 protein be identified on the plasmids. However three VirB11 proteins are found in the three sequenced genomes of *Sulfolobus*, raising the possibility that these proteins might be involved in the conjugational transfer of the plasmids. The VirB6 (TrbL) proteins are membrane proteins that form the pore through the cytoplasmic membrane. Indeed the VirB6 proteins of the *Sulfolobus* conjugative plasmids have 10 to 12 predicted transmembrane helices. The third conserved protein is a VirB4-like protein. VirB4 proteins are also membrane bound ATPases and act together with the coupling protein and VirB11.

The fourth protein appears to be a secreted protein with peptidase activity. Possibly this protein processes or activates another extracellular protein required for conjugation transport. Although conjugation has not been studied in archaea, it appears that the conjugative apparatus of the conjugative *Sulfolobus* plasmids is quite different from the bona fide type 4 secretion system.

ARCHAEAL PLASMIDS AS TOOLS FOR GENETIC STUDIES

A number of archaeal plasmids have successfully been used as basis for the construction of shuttle vectors. These vectors are able to replicate in *E. coli* and the respective archaeal species. The first successful shuttle vectors were constructed for halophiles. Shuttle vectors based on the *Haloferax volcanii* plasmid pHV2 and pHH1 have been used to transform *Halobacterium* species, *Haloferax volcanii* and *Haloarcula marismortui* with a polyethylene-based procedure. Transformants are selected in mevinolin-containing media. The plasmid-borne 3-hydroxy-3-methylglutaryl coenzyme A (HMG-CoA) reductase gene confers resistance to the antibiotic mevinolin (Allers and Mevarech, 2005).

The plasmid pC2A from *Methanosarcina acetivorans* was used to construct shuttle vectors for several *Methanosarcina* species. The transformation is liposome-mediated and selection is performed with puromycin. Resistance to puromycin is provided by the *pac* gene from *Streptomyces alboniger* (Metcalf et al., 1997). Another shuttle vector for methanogens is based on the plasmid pURB500 from *Methanococcus maripaludis* (Tumbula et al., 1997). However this plasmid is unstable in *E. coli*.

Recently progress has been made in constructing shuttle vectors for hyperthermophiles. A shuttle vectors based on pGT5 from *Pyrococcus abyssi* has been used to transform a uracil auxotrophic strain from *Pyrococcus abyssi* (Lucas et al., 2002). Likewise a construct based on the plasmid pTN1 from *Thermococcus nautilus* was successfully applied to transform *Thermococcus kodakaraensis* (Soler et al., 2007).

Finally our group has achieved to construct a shuttle vector based on the plasmid pRN1 from *Sulfolobus islandicus*. The vectors are electroporated into *S. acidocaldarius* and transformants are selected in uracil-free medium. In addition derivates of the shuttle vector can also be employed in β-galactosidase deficient *Sulfolobus solfataricus* strains (Berkner et al., 2007).

FUTURE DIRECTIONS

Our knowledge on archaeal plasmids is still very fragmentary. In contrast to many bacterial plasmids, which provide resistance to antibiotics or other beneficial metabolic capabilities, the benefit for the host organism is unclear for most archaeal plasmids. The currently known archaeal plasmids appear to be selfish or parasitic. Clearly more archaeal plasmids need to be characterized to establish whether this is a general trend. There is also very little knowledge concerning the host range of most archaeal plasmids. In this respect antibiotic resistance genes would be of great experimental utility, but so far there is no case of an archaeal resistance gene encoded on a plasmid. Archaea and bacteria live together in many environments. Are these populations able to share genetic material by horizontal gene transfer mediated by plasmids or conjugation? Notably conjugative plasmids have so far been only isolated from *Sulfolobus* strains, which is very surprising given that conjugation is very common in bacteria.

REFERENCES

Akhmanova, A.S., Kagramanova, V.K., and Mankin, A.S. (1993). Heterogeneity of small plasmids from halophilic archaea. J. Bacteriol. *175*, 1081-1086.

Allers, T. and Mevarech, M. (2005). Archaeal genetics - the third way. Nat. Rev. Genet. *6*, 58-73.

Arnold, H.P. et al. (1999). The genetic element pSSVx of the extremely thermophilic crenarchaeon *Sulfolobus* is a hybrid between a plasmid and a virus. Mol. Microbiol. *34*, 217-226.

Baliga, N.S. et al. (2004). Genome sequence of *Haloarcula marismortui*: a halophilic archaeon from the Dead Sea. Genome Res. *14*, 2221-2234.

Berkner, S. et al. (2007). Small multicopy, non-integrative shuttle vectors based on the plasmid pRN1 for *Sulfolobus acidocaldarius* and *Sulfolobus solfataricus*, model organisms of the (cren-)archaea. Nucleic Acids Res.

Berkner, S. and Lipps, G. (2007). Characterization of the transcriptional activity of the cryptic plasmid pRN1 from *Sulfolobus islandicus* REN1H1 and regulation of its replication operon. J. Bacteriol. *189*, 1711-1721.

Berquist, B.R., DasSarma, P., and Dassarma, S. (2007). Essential and non-essential DNA replication genes in the model halophilic Archaeon, *Halobacterium* sp. NRC-1. BMC. Genet. *8*, 31.

Capaldi, S.A. and Berger, J.M. (2004). Biochemical characterization of Cdc6/Orc1 binding to the replication origin of the euryarchaeon *Methanothermobacter thermoautotrophicus*. Nucleic Acids Res. *32*, 4821-4832.

Charlebois, R.L. et al. (1987). Characterization of pHV2 from *Halobacterium volcanii* and its use in demonstrating transformation of an archaebacterium. Proc. Natl. Acad. Sci. U. S. A *84*, 8530-8534.

del Solar, G. and Espinosa, M. (1992). The copy number of plasmid pLS1 is regulated by two trans-acting plasmid products: the antisense RNA II and the repressor protein, RepA. Mol. Microbiol. *6*, 83-94.

del Solar, G.H., Perez, M.J., and Espinosa, M. (1990). Plasmid pLS1-encoded RepA protein regulates transcription from repAB promoter by binding to a DNA sequence containing a 13-base pair symmetric element. J. Biol. Chem. *265*, 12569-12575.

Dueber, E.L. et al. (2007). Replication origin recognition and deformation by a heterodimeric archaeal Orc1 complex. Science *317*, 1210-1213.

Erauso, G. et al. (1996). Sequence of plasmid pGT5 from the archaeon *Pyrococcus abyssi*: evidence for rolling-circle replication in a hyperthermophile. J. Bacteriol. *178*, 3232-3237.

Erauso, G. et al. (2006). Two novel conjugative plasmids from a single strain of *Sulfolobus*. Microbiology *152*, 1951-1968.

Gaudier, M. et al. (2007). Structural basis of DNA replication origin recognition by an ORC protein. Science *317*, 1213-1216.

Giraldo, R. (2003). Common domains in the initiators of DNA replication in Bacteria, Archaea and Eukarya: combined structural, functional and phylogenetic perspectives. FEMS Microbiol. Rev. *26*, 533-554.

Gomis, R.F. et al. (1998). The structure of plasmid-encoded transcriptional repressor CopG unliganded and bound to its operator. EMBO J. *17*, 7404-7415.

Grainge, I. et al. (2006). Biochemical analysis of a DNA replication origin in the archaeon *Aeropyrum pernix*. J. Mol. Biol. *363*, 355-369.

Greve, B. et al. (2004). Genomic comparison of archaeal conjugative plasmids from *Sulfolobus* . Archaea *1*, 231-239.

Greve, B. et al. (2005). Novel RepA-MCM proteins encoded in plasmids pTAU4, pORA1 and pTIK4 from *Sulfolobus neozealandicus*. Archaea *1*, 319-325.

Gropp, F. et al. (1992). The immunity-conferring plasmid p phi HL from the *Halobacterium salinarium* phage phi H: nucleotide sequence and transcription. Virology *190*, 45-54.

Hackett, N.R. et al. (1990). Nucleotide sequence of a high copy number plasmid from *Halobacterium* strain GRB. Nucleic Acids Res. *18*, 3408.

Hall, M.J. and Hackett, N.R. (1989). DNA sequence of a small plasmid from *Halobacterium* strain GN101. Nucleic Acids Res. *17*, 10501.

Holmes, M., Pfeifer, F., and Dyall-Smith, M. (1994). Improved shuttle vectors for *Haloferax volcanii* including a dual-resistance plasmid. Gene *146*, 117-121.

Ilyina, T.V. and Koonin, E.V. (1992). Conserved sequence motifs in the initiator proteins for rolling circle DNA replication encoded by diverse replicons from eubacteria, eucaryotes and archaebacteria. Nucleic. Acids. Res. *20*, 3279-3285.

Iyer, L.M. et al. (2005). Origin and evolution of the archaeo-eukaryotic primase superfamily and related palm-domain proteins: structural insights and new members. Nucleic Acids Res. *33*, 3875-3896.

Iyer, L.M. et al. (2004). Evolutionary history and higher order classification of AAA+ ATPases. J. Struct. Biol. *146*, 11-31.

Karner, M.B., DeLong, E.F., and Karl, D.M. (2001). Archaeal dominance in the mesopelagic zone of the Pacific Ocean. Nature *409*, 507-510.

Keeling, P.J. et al. (1996). Complete nucleotide sequence of the *Sulfolobus islandicus* multicopy plasmid pRN1. Plasmid *35*, 141-144.

Keeling, P.J. et al. (1998). *Sulfolobus islandicus* plasmids pRN1 and pRN2 share distant but common evolutionary ancestry. Extremophiles. *2*, 391-393.

Khan, S.A. (2000). Plasmid rolling-circle replication: recent developments. Mol. Microbiol. *37*, 477-484.

Klein, R. et al. (2002). *Natrialba magadii* virus phiCh1: first complete nucleotide sequence and functional organization of a virus infecting a haloalkaliphilic archaeon. Mol. Microbiol. *45*, 851-863.

Kletzin, A. et al. (1999). Molecular analysis of pDL10 from *Acidianus ambivalens* reveals a family of related plasmids from extremely thermophilic and acidophilic archaea. Genetics *152*, 1307-1314.

Lam, W.L. and Doolittle, W.F. (1989). Shuttle vectors for the archaebacterium *Halobacterium volcanii*. Proc. Natl. Acad. Sci. U. S. A *86*, 5478-5482.

Lipps, G. (2006). Plasmids and viruses of the thermoacidophilic crenarchaeote *Sulfolobus*. Extremophiles. *10*, 17-28.

Lipps, G. et al. (2001a). The protein ORF80 from the acidophilic and thermophilic archaeon *Sulfolobus islandicus* binds highly site-specifically to double-stranded DNA and represents a novel type of basic leucine zipper protein. Nucleic Acids Res. *29*, 4973-4982.

Lipps, G. et al. (2003). A novel type of replicative enzyme harbouring ATPase, primase and DNA polymerase activity. EMBO J. *22*, 2516-2525.

Lipps, G., Stegert, M., and Krauss, G. (2001b). Thermostable and site-specific DNA binding of the gene product ORF56 from the *Sulfolobus islandicus* plasmid pRN1, a putative archael plasmid copy control protein. Nucleic. Acids. Res. *29*, 904-913.

Lipps, G. et al. (2004). Structure of a bifunctional DNA primase-polymerase. Nat. Struct. Mol. Biol. *11*, 157-162.

Lucas, S. et al. (2002). Construction of a shuttle vector for, and spheroplast transformation of, the hyperthermophilic archaeon *Pyrococcus abyssi*. Appl. Environ. Microbiol. *68*, 5528-5536.

Lukashin, A.V. and Borodovsky, M. (1998). GeneMark.hmm: new solutions for gene finding. Nucleic Acids Res. *26*, 1107-1115.

Lysetska, M. et al. (2005). Site-specific binding of the 9.5 kilodalton DNA-binding protein ORF80 visualized by atomic force microscopy. Biomacromolecules. *6*, 1252-1257.

Maeder, D.L. et al. (2006). The *Methanosarcina barkeri* genome: comparative analysis with *Methanosarcina acetivorans* and *Methanosarcina mazei* reveals extensive rearrangement within *Methanosarcinal* genomes. J. Bacteriol. *188*, 7922-7931.

Marsin, S. and Forterre, P. (1998). A rolling circle replication initiator protein with a nucleotidyl-transferase activity encoded by the plasmid pGT5 from the hyperthermophilic archaeon *Pyrococcus abyssi*. Mol. Microbiol. *27*, 1183-1192.

Marsin, S. and Forterre, P. (1999). The active site of the rolling circle replication protein Rep75 is involved in site-specific nuclease, ligase and nucleotidyl transferase activities. Mol. Microbiol. *33*, 537-545.

Matsunaga, F. et al. (2007). Genomewide and biochemical analyses of DNA-binding activity of Cdc6/Orc1 and Mcm proteins in *Pyrococcus* sp. Nucleic Acids Res. *35*, 3214-3222.

Metcalf, W.W. et al. (1997). A genetic system for Archaea of the genus *Methanosarcina*: liposome-mediated transformation and construction of shuttle vectors. Proc. Natl. Acad. Sci. U. S. A *94*, 2626-2631.

Myllykallio, H. et al. (2000). Bacterial mode of replication with eukaryotic-like machinery in a hyperthermophilic archaeon. Science *288*, 2212-2215.

Ng, W.L. and Dassarma, S. (1993). Minimal replication origin of the 200-kilobase *Halobacterium* plasmid pNRC100. J. Bacteriol. *175*, 4584-4596.

Ng, W.V. et al. (2000). Genome sequence of *Halobacterium* species NRC-1. Proc. Natl. Acad. Sci. U. S. A *97*, 12176-12181.

Nolling, J. et al. (1992). Modular organization of related Archaeal plasmids encoding different restriction-modification systems in *Methanobacterium thermoformicicum*. Nucleic Acids Res. *20*, 6501-6507.

Norais, C. et al. (2007). Genetic and physical mapping of DNA replication origins in *Haloferax volcanii*. PLoS. Genet. *3*, e77.

Peng, X. et al. (2000). Evolution of the family of pRN plasmids and their integrase-mediated Insertion into the chromosome of the Crenarchaeon *Sulfolobus solfataricus*. J. Mol. Biol. *303*, 449-454.

Pfeifer, F. and Ghahraman, P. (1993). Plasmid pHH1 of *Halobacterium salinarium*: characterization of the replicon region, the gas vesicle gene cluster and insertion elements. Mol. Gen. Genet. *238*, 193-200.

Pfeifer, F., Weidinger, G., and Goebel, W. (1981). Characterization of plasmids in halobacteria. J. Bacteriol. *145*, 369-374.

Prangishvili, D. et al. (1998). Conjugation in archaea: frequent occurrence of conjugative plasmids in *Sulfolobus*. Plasmid *40*, 190-202.

Prato, S. et al. (2006). pIT3, a cryptic plasmid isolated from the hyperthermophilic crenarchaeon *Sulfolobus solfataricus* IT3. Plasmid *56*, 35-45.

Purschke, W.G. and Schaefer, G. (2001). Independent replication of the plasmids pRN1 and pRN2 in the archaeon *Sulfolobus islandicus*. FEMS Microbiol. Lett. *200*, 97-102.

Raumann, B.E. et al. (1994). DNA recognition by beta-sheets in the Arc repressor-operator crystal structure. Nature *367*, 754-757.

Robinson, N.P. and Bell, S.D. (2007). Extrachromosomal element capture and the evolution of multiple replication origins in archaeal chromosomes. Proc. Natl. Acad. Sci. U. S. A *104*, 5806-5811.

Robinson, N.P. et al. (2004). Identification of two origins of replication in the single chromosome of the archaeon *Sulfolobus solfataricus*. Cell *116*, 25-38.

Schleper, C. et al. (1995). A multicopy plasmid of the extremely thermophilic archaeon *Sulfolobus* effects its transfer to recipients by mating. J. Bacteriol. *177*, 4417-4426.

Schleper, C., Jurgens, G., and Jonuscheit, M. (2005). Genomic studies of uncultivated archaea. Nat. Rev. Microbiol. *3*, 479-488.

Schnabel, H. et al. (1982). *Halobacterium halobium* phage oH. EMBO J. *1*, 87-92.

She, Q. et al. (1998). Genetic profile of pNOB8 from *Sulfolobus*: the first conjugative plasmid from an archaeon. Extremophiles. 2, 417-425.

Soler, N. et al. (2007). The rolling-circle plasmid pTN1 from the hyperthermophilic archaeon *Thermococcus nautilus*. Mol. Microbiol. *66*, 357-370.

Stedman, K.M. et al. (2000). pING family of conjugative plasmids from the extremely thermophilic archaeon *Sulfolobus islandicus*: insights into recombination and conjugation in Crenarchaeota. J Bacteriol. *182*, 7014-7020.

Tumbula, D.L., Bowen, T.L., and Whitman, W.B. (1997). Characterization of pURB500 from the archaeon *Methanococcus maripaludis* and construction of a shuttle vector. J. Bacteriol. *179*, 2976-2986.

Ward, D.E. et al. (2002). Characterization of plasmid pRT1 from *Pyrococcus* sp. strain JT1. J. Bacteriol. *184*, 2561-2566.

Zillig, W. et al. (1994). Screening for Sulfolobales, their plasmids and their viruses in Icelandic solfataras. Syst. Appl. Microbiol. *16*, 609-628.

Zhou, M. et al. (2004). Complete sequence and molecular characterization of pNB101, a rolling-circle plasmid from the haloalhaliphilic archaeon *Natronobacterium* sp., strain AS7091. Extremophiles *8*, 91-98.

Chapter 3

Plasmid-based Expression Systems for Mammalian Cells

Armin Baiker, Rudolf Haase and Hans Joachim Lipps

ABSTRACT

In contrast to bacteria and yeast no natural plasmids are found in mammalian cells. Therefore many attempts to construct different expression vector systems for mammalian cells have been made in recent years. These vector systems can be categorized in terms of vector administration, mechanisms of vector replication and mechanisms to achieve nuclear persistence of the vectors. In this chapter we describe various transfection methods, construction principles for an optimal vector and the different strategies to achieve nuclear persistence. We then concentrate on the description of various plasmid-like expression systems for mammalian cells based either on viral or chromosomal functional elements. Finally, their application in biotechnology and gene therapy are discussed.

INTRODUCTION

Naturally occurring plasmids are not limited to prokaryotes. In 1976 Hollenberg et al. described the natural 2μm plasmid from *S. cerevisiae* (Hollenberg et al., 1976). This circular plasmid of around 6.3 kb in size does not give the yeast cells any apparent selective advantage, but is stably maintained at about 50-100 copies per haploid genome of the yeast cell. Further studies revealed that the 2μm plasmid is coated with nucleosomes and that its replication is initiated once per cell cycle by the cellular replication machinery. How the high copy number of 2μm plasmids (50-100 copies per yeast cell) is maintained, given that the origin of replication (ORI) initiates replication once per cell cycle, has been a long-standing mystery. Ten years after the 2μm plasmid discovery, this mystery could be finally solved (Futcher, 1986; Volkert et al., 1986): In parallel to initiation of DNA replication, the plasmid-encoded site-directed "flip" recombinase (FLP) is expressed. FLP in turn catalyzes the inversion between two inverted repeats (IRs) located on the 2μm plasmid backbone, which results in a switch from circular, replication fork collision amplification to rolling circle amplification, thus giving an explanation for the high copy numbers of 2μm plasmids. The rolling circle amplification is finally terminated when a second FLP-mediated inversion between the IRs occurs and the replication forks collide. Three 2μm plasmid-encoded repressor proteins (REP1, REP2 and D) tightly regulate this process by controlling the expression of FLP

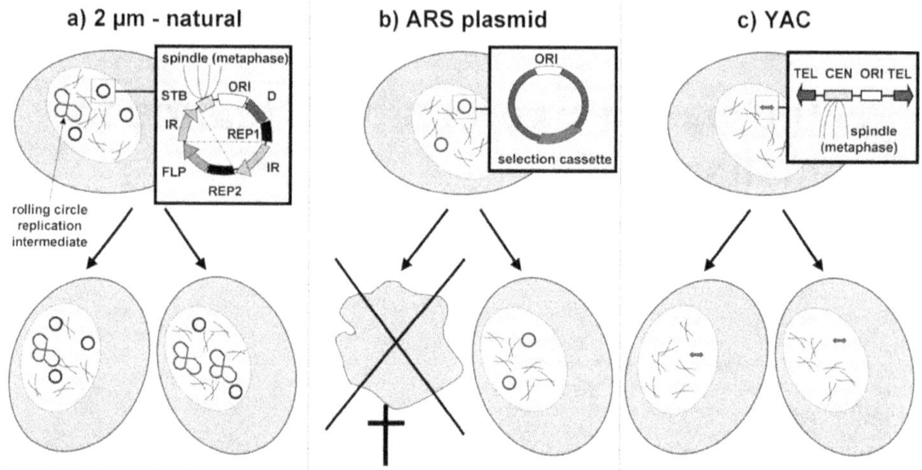

MCM 2-7 dependent	yes	yes	yes
nuclear retention	STB (CEN) / copy number	selection pressure	CEN / TEL
copy number	50-100	<10	<10
replication	2µm ORI	ORI	ORI

Fig. 1. Natural and artificial yeast vector systems. This figure depicts schematically the DNA replication and nuclear retention mechanisms of the 2μm plasmid, ARS plasmids and YACs in yeast. a: The 2μm plasmid ORI initiates replication once per cell cycle. In parallel, the plasmid encoded, site-directed "flip" recombinase (FLP) is expressed. FLP in turn catalyzes the inversion between two inverted repeats (IRs) located on the 2μm plasmid backbone, which results in a switch to rolling circle amplification. The rolling circle amplification is finally terminated when a second FLP mediated inversion between the IRs occurs and the replication forks collide. The STB locus is a functional centromere homologue and contributes to the nuclear retention of natural 2μm plasmids. b: ARS plasmids comprise an origin of replication (ORI) and a selection marker. The ORI is necessary for MCM 2-7 dependant replication initiation. Such vectors do not have a nuclear retention mechanism, but may be propagated under constant selection pressure. c: YACs are linearized ARS plasmids that contain centromeric regions (CEN) for segregation and telomeric regions (TEL) to overcome the end replication problem.

(Chen et al., 2003). It is also noteworthy that the natural 2μm plasmid contains a centromere-like region, which is known as the stability locus (STB) (Velmurugan et al., 2000). This STB locus is responsible for the correct segregation of 2μm plasmids during cell division (Fig. 1a).

In the years after the discovery of the naturally occurring 2μm plasmid, yeast geneticists focused on the identification of origin of replication (ORI) sequences necessary for the autonomous replication of plasmids in yeast (also called AR sequences or ARS) by subcloning chromosomal elements of *S. cerevisiae* into newly discovered *E.coli* pBR322 vectors that had been substituted with yeast selection markers. These efforts resulted in the construction of artificial (recombinant) *ARS plasmids*, such as YRp7, that allowed the propagation of a vector in yeast cells kept under steady selection pressure. In the following years, ARS plasmids were also used to identify chromosomal sequences that define centromeres (CEN), which are necessary for the correct mitotic segregation of chromosomes (Stinchcomb et al., 1979). Fig. 1b depicts the schematic structure of an ARS plasmid.

In the early 1980s, linear *yeast artificial chromosomes* (YACs) were developed by adding telomeric repeat (TEL) sequences to the ends of linearized ARS plasmids, along with centromeric region sequences (CEN) to enable correct segregation during mitosis. This strategy enabled the propagation of recombinant yeast vectors in a linear form and contributed to our current understanding of the end replication mechanism (Chan et al., 1983; Horowitz et al., 1985; Szostak et al., 1982). Fig. 1c depicts a schematic diagraph of a linear YAC.

In contrast to yeast, no naturally occurring plasmids have been found in mammalian cells. Therefore many different vector systems for mammalian cells have been developed in recent years. These systems can be categorized in terms of 1) methods of vector administration, 2) mechanisms of vector replication, and 3) strategies used to achieve nuclear persistence of the vector molecules.

Some vectors, for example, are administered as virus-like particles. These viral vectors are based on replication-deficient viruses and utilize the infection (transduction) and nuclear delivery strategies of the mammalian viruses from which they were derived. All other vectors have to be administered (transfected) as naked DNA.

Another criterion for the description of a vector system is the mechanism of its DNA replication. Some vectors have no way to amplify their genomes after administration into the nucleus of the target cell, which results in the loss of vector molecules during cell division and thus only short-term (transient) expression is achieved. Other vectors, however, are able to integrate their DNA into the host cell chromosomes, and use the cellular DNA replication machinery for their own amplification. Lastly, there are vector systems that replicate their DNA autonomously as extra-chromosomal elements (episomes). These vectors are either based on viral plasmid replicons or use chromosomal elements for their replication.

Different vector systems utilize different strategies to achieve nuclear persistence. Some vectors do not possess any mechanism to retain their genomes in the nucleus of the target cell, which results in transient transgene expression due to the loss of the vector molecules after cell division. Other vectors integrate into the host cell chromosomes and utilize the cellular mechanism of chromosomal segregation for their nuclear persistence. A third class of vectors exhibits different strategies to retain their genomes as extra-chromosomal elements in the nucleus of the target cell, among them the high-copy number strategy, the strategy of binding to the metaphase chromosomes (piggy-back strategy), or the centromere strategy.

All vectors currently used for the genetic modification of mammalian cells have their own advantages but also their own limitations. While replication-deficient viruses can deliver genetic material to the target cell with very high efficiency, the major risk of these virus-based vector systems are problems associated with insertional mutagenesis, immunogenicity and cytotoxicity. Therefore, their use must be judged on a careful risk/benefit analysis. On the other hand, delivery of naked DNA to the cell is very inefficient. Current protocols for transfection will be discussed within this article. Finally, all integrating vectors face the problem of unreliable gene expression, as integration of a transgene into the host genome often leads to its epigenetic silencing. In order to construct safe and efficient non-viral vectors, a detailed knowledge of the epigenetic regulation of transcription and replication is necessary. Vectors based on replication-deficient viruses, such as retroviral, lentiviral, adenoviral, and adeno-associated virus (AAV) vectors, will not be discussed within this article. In contrast, we try to present a comprehensive survey of the construction principles and modes of action of currently used plasmid vector systems for the *in vitro* and *in vivo* delivery of genes into mammalian cells. We also discuss the advantages and disadvantages of each system.

TRANSFECTION WITH (NAKED) DNA

The genetic manipulation of mammalian cells requires the efficient delivery of vector DNA into the nucleus of the desired target cells, followed by safe, reproducible and long-term gene expression. This paragraph discusses 1) currently used methods for the transfection of target cells with DNA, 2) construction principles of an optimal DNA vector to enable strong transgene expression and to prevent silencing phenomena, and 3) strategies to enable the persistence of vector DNA molecules in the host cell, as a prerequisite for a long-term gene expression.

TRANSFECTION METHODS OF (NAKED) DNA

In order to achieve gene expression from a DNA vector backbone, the respective construct has to be delivered efficiently to the nucleus of the target cell. To achieve this gene delivery, several strategies have been developed in the past few decades, involving physical and chemical transfection methods, and methods that utilize nuclear localization signals (NLS).

PHYSICAL TRANSFECTION METHODS

Needle injection of a DNA solution directly into a target tissue leads to low-level transgene expression , as demonstrated by experiments involving muscle (Wolff et al., 1990), liver (Budker et al., 1996; Hickman et al., 1994; Zhang et al., 1997) and skin tissue (Choate et al., 1997). Similar effects can be obtained by *airway instillation* of DNA solutions into the lung (Meyer et al., 1995). The detailed mechanism of this gene delivery remains unclear at this point. Specific or unspecific surface receptors on the target cells may bind and internalize DNA molecules in their proximity, facilitating their uptake and nuclear translocation. Gene delivery by local injection or instillation of DNA solutions is a fairly inefficient process, nevertheless it may be attractive for some applications due to its simplicity and low toxicity (Gao et al., 2007).

The *hydrodynamic gene delivery* of DNA solutions has been proven as a simple and highly efficient method to transfect target cells within highly perfused organs, such as the liver. In mice and rats for example, 30-40% of all hepatocytes were transfected by high-pressure injection of a large volume of a DNA solution (a volume that corresponds to a 7.5 liter injection into a human) into the tail vein (Liu et al., 1999). Mechanistically, the high pressure injection of any large volume of a physiological solution into a blood vein results in a back pressure at the interior vena cava leading to liver expansion, the disruption of liver fenestrae and the formation of transient membrane disruptions of hepatocytes, which in turn results in the unspecific uptake of dissolved molecules into the cytoplasm (Zhang et al., 2004a). Several efforts are currently underway to translate the hydrodynamic gene delivery protocols available for mice and rats into one that is applicable to humans (Alino et al., 2007; Yoshino et al., 2006).

Microinjection of a DNA solution directly into the nucleus provides a highly efficient gene delivery method for the manipulation of individual target cells. The usefulness of the technology, however, is limited by the small number of cells that can be microinjected within a reasonable amount of time (Shen et al., 1982). Nevertheless, DNA microinjection proved to be the first successful technology for the creation of transgenic animals (Gordon et al., 1981) and microinjection experiments involving *Xenopus* oocytes in the early 1970s contributed to our current understanding of basic biological processes, such as replication, transcription and translation (Amaya, 2005).

The *Gene Gun* transfer of gold-coated DNA molecules, a process invented in the early 1980s by John Sanford at Cornell University, was originally designed for plant cell transformation but was soon applied to mammalian organisms. It serves as a simple and efficient method for the transfection of exposed tissues, such as skin or mucosa, within a confined area. Heavy metal coated particles are accelerated by pressurized gas, for example helium, and expelled onto the respective target tissues. The particles are able to penetrate a few millimeters deep into the tissue and release the DNA into the cells (Yang et al., 1990). Gene Gun delivery may provide a promising strategy for DNA vaccination approaches (Gao et al., 2007).

Gene delivery of DNA solutions by brief, high-intensity, pulsed electrical fields, generally referred to as *Electroporation*, provides a powerful technology for the transfection of a variety of cell types *in vitro* and *in vivo*. The electrical pulse leads to the formation of transient permeable structures, so-called micropores, in the cellular membranes which enable DNA uptake into the cytoplasm and the nucleus of a target cell (Andre et al., 2004; Neumann et al., 1982). Electroporation protocols have been successfully used for the transfection of a variety of tissues *in vivo*, among them muscle (Dean, 2005), skin (Prausnitz, 1999), lung (Hasson et al., 2005), and even brain (Saito, 2006). Apparently the technology works in any tissue that can be exposed in a way that allows a pair of electrodes to be inserted. Safety issues, however, arise from the fact that high voltage treatment of tissue may result in damage to cells and to genomic and vector DNA.

CHEMICAL TRANSFECTION METHODS

The goal of a chemical transfection reagent is to introduce a foreign negatively charged molecule, for example DNA, into a eukaryotic cell that has a negatively charged membrane on its surface. Chemical delivery systems consist of a cationic component to compact the DNA molecule and to provide it with an overall net positive charge in order to facilitate cellular membrane cross-transportation. Furthermore, DNA compaction results in an increased resistance to cytosolic nucleases, thus extending the stability and half-life of any delivered vector DNA.

Calcium Phosphate Co-precipitation was invented in 1972 and has been shown to efficiently transfect a variety of different target cells *in vitro* (Graham et al., 1973). The protocol involves mixing DNA with calcium chloride and adding this mixture to a buffered phosphate solution to obtain a calcium phosphate / DNA co-precipitate. This precipitate can be dispersed onto cells and is subsequently taken up via endocytosis or phagocytosis. Calcium phosphate co-precipitation is a widely used technique, since the protocol is relatively simple and its components are available at a reasonable price.

Cationic Polymer based compounds, such as DEAE-dextran, poly-L-lysine, polyethyleneimine (PEI), protamine and polybrene (Gao et al., 2007), can condense DNA molecules into relatively small complexes, giving them a neutral or net-positive charge that allows them to cross cell membranes by endocytosis.

Cationic Lipid mediated gene delivery, using synthetic (non-bilayer forming) cationic lipids, was originally described by Felgner (Felgner et al., 1987). The cationic portion of the cationic lipid compound associates with the negatively charged nucleic acid to allow its compaction, while the lipid part of the molecule facilitates fusion of the DNA – polymer complex with the cell membrane. Many variations in the formulation of these compounds have been developed in recent years (Liu et al., 2003). Although some cationic lipids alone exhibit good transfection efficiencies, they are often formulated as liposomes.

Liposomes are generated by mixing cationic lipids with fusogenic noncharged phospholipids, such as dioleoylphosphatidylethanolamine (DOPE), and bilayer-forming cholesterol derivatives, such as dioleoylphosphatidylcholine. In combination they form spherical structures consisting of single or multiple concentric bilayers arising by spontaneous assembly of amphiphatic molecules in an aqueous environment. It is believed that the cholesterol compounds contribute to the bilayer formation of liposomes, whereas the fusogenic noncharged phospholipids facilitate their fusion at the outer cell membrane (Gao et al., 2007; Liu et al., 2003).

All chemical transfection methods have been proven to be effective tools for the *in vitro* delivery of genes. However, all of these systems have failed in *in vivo* experiments because of their low delivery efficiency and side effects, such as complement activation or acute inflammation. Apparently, polycations cause aggregates full of negatively charged molecules and further prevent the effective release of the negatively charged DNA molecules once inside the cell (Gao et al., 2007).

METHODS UTILIZING NUCLEAR LOCALIZATION SIGNALS (NLS)

The expression of any vector DNA requires as a prerequisite its efficient transportation to the nucleus. However, all the previously described physical and chemical delivery methods, possibly with the exception of direct microinjection into the nucleus and electroporation, enable uptake of the vector DNA only into the cytoplasm of the target cell. Here the DNA faces the nuclear envelope as a physical barrier. The nuclear envelope contains nuclear pores with a passive transportation limit of 70 kDa molecular mass or 10 nm of diameter, which is much smaller than the size of DNA, even if condensed with polycations. This barrier can be overcome during cell division, when the nuclear envelope is disassembled to enable efficient segregation of the genetic material. In this case residual DNA, such as vector DNA, may be transported more or less passively into the newly formed postmitotic nuclei of the daughter cells. From a practical standpoint, this means that only dividing cells can be transfected efficiently with naked DNA (Khalil et al., 2006; Lechardeur et al., 2006).

Nuclear proteins require a *Nuclear Localization Signal (NLS)* to mediate active transportation through the nuclear pore complex (Jans et al., 1996). During this active transportation process, the diameter of the nuclear pore complex is expanded to about 39 nm, which allows delivery of the protein to the nucleus (Dworetzky et al., 1988; Pante et al., 2002). Typically, a NLS consists of basic, positively charged oligopeptides containing several lysine (K) and arginine (R) residues. The Simian Virus 40 (SV40) T antigen NLS sequence for example is $P^{126}KKKRKV^{132}$. As a consequence of the positively charged nature of a NLS, which resembles the polycations used for DNA condensation (for example polylysine: $[K]_n$), synthetic NLS peptides are often able to directly bind to DNA by ionic interactions. Numerous studies have attempted to utilize NLS sequences for efficient gene delivery to the nucleus. These involve the covalent or non-covalent association of DNA (with or without condensing polycations) with NLS sequence-bearing synthetic peptides, or even whole proteins comprising NLS signals such as histones, transcription factors or HMG-box proteins (Cartier et al., 2002; Hebert, 2003; Sebestyen et al., 1998).

Some researchers have focused on the construction of *Pseudoviruses*, which utilize the nuclear translocation mechanism of small DNA viruses, such as polyomaviruses (for example SV40) or papillomaviruses. These viruses have evolved strategies to deliver their capsids actively into the nucleus of the infected cell, where the viral DNA is uncoated and released. Since the size of the SV40 capsids (45 nm) and papillomavirus capsids (55 nm) are bigger than the reported size of the maximally extended nuclear pore (39 nm) this process is not yet completely understood (Izaurralde et al., 1999; Pante et al., 2002). The construction of pseudoviruses relies on the ability of the viral capsid proteins to bind and package small DNA molecules efficiently and to self-assemble into virus-like particles that resemble wildtype viruses. Pseudoviruses based on the Human Papillomavirus (HPV), Bovine Papillomavirus (BPV) or SV40 have been generated by *in vitro* and *in vivo* packaging approaches, and have been successfully used to transduce target cells efficiently by introducing the packaged DNA into the nucleus of the target cells (Eisenberger et al., 2005; Liu et al., 2001; Roitman-Shemer et al., 2007).

CONSTRUCTION PRINCIPLES OF AN OPTIMAL DNA VECTOR

Any DNA vector used for gene delivery and expression consists of a minimum of three elements: 1) a promoter that enables the transcription of the encoded transgene, 2) a gene to be expressed and 3) a vector backbone consisting of residual elements necessary for bacterial propagation or selection.

GC CONTENT OF THE TRANSGENE

The composition of the four nucleotides A, G, C, and T is highly variable among the genomes of different organisms, and can be partially described in terms of *GC content*. The prokaryotes *Methanococcus jannashii* and *Deinococcus radiodurans,* for example, have genomes with GC contents of 31.3 % (GC-poor) and 66.9 % (GC-rich) respectively (Farias et al., 2003). The GC content of the human genome overall is about 41 %. Protein encoding regions in the human genome, however, have a significantly higher GC content of about 51 % (Lander et al., 2001; Zhang et al., 2004b). Due to the degenerate nature of the genetic code, every protein sequence can be encoded by a large number of different nucleotide sequences. The amino acid leucine, for example, can be encoded by the triplets TTA (GC content: 0 %), CTT (GC content: 33.3 %) or CTC (GC content: 66 %). Thus it is possible to construct the nucleotide sequence of a given transgene in GC-poor and GC-rich versions. It has been shown that gene expression is much more efficient when a GC-rich version of a transgene is used. The mechanism of how the GC content of a transgene influences its gene expression is not completely understood, but effects on the transcription efficiency, polyadenylation, nuclear export, stability (half-life) and translation rate of the transcribed mRNA's may be invovled (Kudla et al., 2006).

CODON USAGE IN THE TRANSGENE

The redundancy of the genetic code means that there are several isoaccepting tRNA's for each of the amino acids (with some exceptions). It has been shown that there is a direct correlation between the codon preference within an organism and the respective isoaccepting tRNA abundance (Ikemura, 1985). Individual organisms exhibit differences in their codon usage: in *E.coli*, for example, the amino acid proline is preferentially encoded by the triplet CCG (44 % of all cases), whereas the triplet CCC is used the least frequently (14 % of all cases) In humans the situation is turned around, as CCC and CCG are used for 35 % and 11 % of prolines, respectively (Zhou et al., 1999). This codon preference suggests that it may be possible to design the nucleotide sequence for a certain transgene by codon optimization, to improve its expression within a certain organism (Gustafsson et al., 2004). Leder et al. demonstrated that the expression of the HPV L1 capsid protein in human cell lines can be increased some 10.000 fold, by using a human codon optimized ("humanized") L1 transgene (Leder et al., 2001).

CpG ISLANDS IN THE VECTOR BACKBONE

A cytosine monophosphate (C) followed by a guanine monophosphate (G) in a nucleotide sequence is referred to as a CpG dinucleotide. In eukaryotic DNA, CpG dinucleotides appear at much lower frequencies than statistically expected—only one out of 60 dinucleotides is a CpG dinucleotide (the expected frequency is one out of 16). Furthermore, the cytosine residues of CpG dinucleotides are often methylated to 5-methyl-cytosine (^{m}CpG) in eukaryotes. In most bacterial genomes, however, CpG dinucleotides are represented according to their statistical expectation, and the cytosine residue is normally unmethylated (Bestor, 2000; Bird, 1986). The human (innate) immune system has evolved mechanisms to differentiate bacterial DNA from its own via Toll-like receptor 9 (TLR 9) signalling. The TLR9 protein is expressed in the endosomes of human B-cells and plasmacytoide dendritic cells and is able to interact with endocytosed DNA that comprises unmethylated CpG dinucleotides (Bauer et al., 2001). After recognition of the "foreign" DNA, TLR9 triggers downstream signalling via MyD88, IRAK and TRAF6 to increase NFκB and AP1 expression, which in turn results in the production of inflammatory cytokines (Akira et al., 2001).

CpG ISLANDS IN THE EUKARYOTIC PROMOTER REGION

As mentioned earlier, eukaryotic genomes are largely depleted of CpG dinucleotides and the residual CpG's are mainly methylated as ^{m}CpG. There are, however, genomic regions where the occurrence of CpG dinucleotides is significantly higher (close to the statistically expected value) and the methylation concentration is significantly lower than the overall genome. These so called CpG islands are typically GC-rich (>60 %), longer than 200 bp in size, and associated with promoters (Ioshikhes et al., 2000). In general, promoters can be divided into three classes: 1) high-CpG promoters, 2) low-CpG promoters and 3) non-CpG promoters. These classifications are based on CpG ratio, GC content and length (Weber et al., 2007). The majority of human promoters belong to the high-CpG class of promoters, for which it is well established that the methylation status of the surrounding CpG islands has a direct influence on the promoter activity, and on the transcription levels of the transgene. Unmethylated CpG islands in the promoter region, for example, correspond to active promoters, whereas heavily methylated ^{m}CpG islands correspond to inactive (silenced) promoters (Boyes et al., 1992). The promoter / enhancer element of the immediate early gene of the human cytomegalovirus (CMV-IEP) is most often used in commercial vectors, since it confers robust expression of a transgene in several cell types. This strong activity is due to the presence of several transcription factor binding sites within its sequence. The expression profile of CMV-IEP typically peaks at around 1-2 days after vector administration, followed by a steady decrease in its activity over a period of 1-2 weeks. Methylation of the CpG islands in the CMV-IEP region has been shown to be one explanation for this undesired silencing phenomenon (Brooks et al., 2004). The generation of CpG depleted CMV-IEP's or the *de novo* construction of synthetic promoters could help to overcome this problem (Yew et al., 2004). Taken together, these findings suggest that an optimal (state of the art) DNA vector for gene delivery into mammalian cells should be composed of 1) a promoter element that is depleted of CpG islands in order to minimize silencing phenomena, 2) a transgene that is codon-optimized to enable its efficient transcription and expression and 3) a CpG depleted vector backbone to prevent undesired stimulatory effects on the innate immune system.

STRATEGIES TO ENABLE THE PERSISTENCE OF TRANSFECTED VECTOR DNA

Long-term transgene expression after successful gene delivery depends on the persistence of the vector DNA in the nucleus. It has been shown that vector molecules are able to persist for up to several months in slowly dividing or non-dividing cells (such as skeletal muscle cells), enabling prolonged transgene expression *in vivo* (Dean, 2005; Trollet et al., 2006). For persistence in mitotically active, fast dividing target cells, however, the delivered vector molecule has to provide a mechanism to enable its replication and correct segregation to the daughter cells. One simple way to achieve this requirement is chromosomal integration of the vector molecule. Although any DNA can integrate randomly into the chromosomes of a cell, random integration is a very rare event that happens with a frequency of around 1 event per 10^4 -10^5 transfected cells (Athwal et al., 1985; Ledwith et al., 2000). Moreover, random integration may cause undesired genotoxic side effects such as insertional mutagenesis or cellular transformation by the disruption or misregulation of cellular genes. At present, several different integrating vector systems have been developed, which utilize homologous recombination, transposons, or phage integrases. The vector systems based on homologous recombination are described in detail in the chapter "Plasmids for gene therapy". An alternative approach to enable the stable nuclear persistence of vector molecules is the use of episomal vector systems that behave as additional (extra-) chromosomes. Such vectors replicate autonomously in the target cell and provide individual mechanisms to enable their correct segregation during cell division. Several different episomal vector systems have been constructed to date, including vectors based on viral plasmid replicons and non-viral episomal vectors. These vectors will be discussed in the following two paragraphs.

VECTORS BASED ON VIRAL PLASMID REPLICONS

Several animal viruses have evolved highly efficient strategies to infect cells and to deliver their respective genomes into the nucleus of mammalian cells, in order to enable their DNA replication and the generation of progeny. Vectors based on *replication deficient viruses*, such as retrovirus, lentivirus, adenovirus, or adenovirus assoviated virus (AAV), are assembled as virions (or virus-like particles) in separate producer systems. They exhibit all viral functions for the infection (transduction) of target cells and the nuclear translocation of their genomes, but lack the ability of the wildtype virus to replicate and produce progeny. *Viral plasmid replicon based vectors* are not assembled as virions and thus lack the ability of the original virus to infect (transduce) target cells. As a result, these vectors have to be administered as naked (plasmid) DNA. Once delivered into the nucleus, however, viral plasmid replicon-based vectors exhibit highly efficient strategies for their DNA replication and nuclear persistence. The replication strategies of mammalian viruses, such as Simian Virus 40 (SV40), Bovine Papillomavirus (BPV) and Epstein-Barr virus (EBV), have been investigated in great detail. Their respective plasmid replicons have been characterized, isolated and utilized for the construction of episomally replicating plasmids. These viral plasmid replicons consist of two elements: a well-defined origin of DNA replication and (at least) one *trans*-activating factor that binds to this replication origin. The latter protein enables the efficient initiation of DNA replication, and may even contribute to the nuclear retention of the viral genome.

THE SV40 (HIGH COPY NUMBER) PLASMID REPLICON

The SV40 origin of replication is a 64-bp DNA sequence and contains one binding site for the virally encoded large T-antigen. This protein is the only viral trans-activating factor required for replication, and functions as a replicative helicase, which allows replication of SV40 to be uncoupled from the genomic replication machinery, and the "once-per-cell-cycle" licensing of MCM2-7 mediated replication initiation (Friedrich et al., 2005). As a result, SV40 plasmid replicon-based vectors exhibit very high copy numbers of several thousand molecules per transfected cell (Cooper et al., 1997). This high copy number is responsible for the random, but highly efficient, segregation of SV40 plasmid replicon-based vector molecules during cell division and accounts for their episomal persistence. The large T-antigen, however, is a viral oncoprotein and is known to cause cell cycle stimulation and cell transformation (Valls et al., 2003). Furthermore, this protein has been shown to stimulate the humoral and cellular immune system (Strayer et al., 1999), limiting its use within a human gene delivery vector system (Fig. 2a).

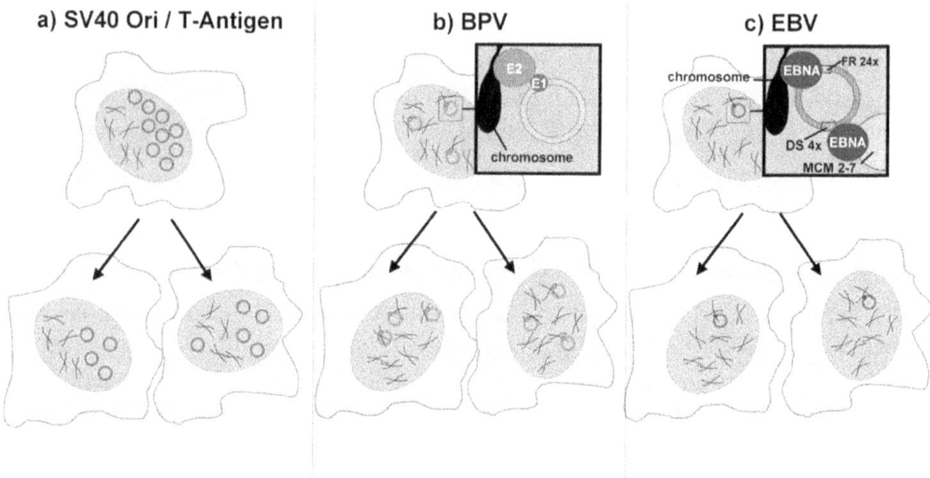

	a) SV40 Ori / T-Antigen	b) BPV	c) EBV
MCM 2-7 dependent	no	no	yes
nuclear retention	copy number	E2 - metaphase chromsome / copy number	EBV OriP (FR) + EBNA-1
copy number	100-1000	~ 50 - 150	~10
replication	SV40 ORI + T-Ag	BPV ORI + E1 + E2	EBV OriP (DS) + EBNA-1

Fig. 2. Viral plasmid replicons. This figure depicts schematically the DNA replication and nuclear retention mechanisms of several viral plasmid replicons. a: the SV40 plasmid replicon consists of the SV40 Ori and the viral T-antigen. DNA replication is initiated independently of the MCM 2-7 complex, which results in a high plasmid copy number. This high copy number of around 100-1000 molecules per cell is responsible for the nuclear retention of SV40 plasmid replicon based vectors. b: the BPV plasmid replicon consists of the BPV Ori and the viral proteins E1 and E2. DNA replication is initiated independently of the MCM 2-7 complex, which results in an intermediate plasmid copy number. This relatively high plasmid copy number of around 50-150 molecules per cell in combination with the E2 binding ability to metaphase chromosomes is responsible for the nuclear retention of BPV plasmid replicon-based vectors. c: Vectors carrying *oriP* and EBNA-1 replicate in a MCM 2-7 dependant way, once per cell cycle. The copy number of such vectors is around 10 molecules per cell. DNA replication is mediated via EBNA-1 binding to the DS-element (which contains 4 binding sites for EBNA-1) of the *oriP* and recruitment of the MCM 2-7 complex. Nuclear retention of such vectors is obtained by binding of EBNA-1 to both the metaphase chromosome and the FR-element (which contains 24 binding sites for EBNA-1) of the OriP.

THE BPV (INTERMEDIATE COPY NUMBER) PLASMID REPLICON

The BPV origin of replication contains binding sites for the virally encoded protein E1. E1 is a functional homologue of the SV40 large T-antigen and after binding to the ORI it assembles into hexameric helicases with the aid of the viral protein E2 (Wilson et al., 2002). As in SV40, the replication of BPV plasmid replicons is uncoupled from the MCM2-7 mediated replication initiation, and the "once per cell cycle" licensing (Ravnan et al., 1992). In contrast to SV40, however, BPV plasmid replicon-based vectors exhibit an intermediate copy number of around 50-150 copies per cell (Berg et al., 1986). Furthermore, the E2 protein of BPV is able to bind to the metaphase chromosomes, providing an explanation for the mitotic stability of the BPV plasmid replicon (Zheng et al., 2005) (Fig. 2b).

THE EBV (LOW COPY NUMBER) PLASMID REPLICON

Herpesviruses, such as Epstein-Barr virus (EBV), utilize different origins of DNA replication during lytic versus latent infection. During lytic DNA replication, the lytic origin (*ori-Lyt*) is bound by the viral origin binding protein (OBP), which in turn recruits the core replication machinery to initiate the rolling-circle amplification of the viral genome. During latent DNA replication, the EBV genome is replicated as a plasmid at around 5-20 copies per cell, utilizing *oriP* in combination with the *EBV nuclear antigen 1* (EBNA1) (Lin et al., 2003; Rose et al., 2002). Structurally, the *oriP* is a bipartite DNA sequence of 1700 bp, which consists of the *dyad symmetry* (DS) element and the *family of repeats* (FR) element. The DS element of the *oriP* consists of 4, and the FR element of 20, binding sites for the transactivating protein EBNA1. Replication from the *oriP* is dependent on host ORC and MCM2-7 complexes and is initiated only once per cell cycle, synchronously with the host chromosome (Wang et al., 2005). The binding of the transactivating protein EBNA1 to the bipartite *oriP* serves two functions. First of all, EBNA1 binding to the DS element recruits the cellular ORC in order to initiate DNA replication, and second, EBNA1 binding to the FR element is responsible for the correct EBV-plasmid segregation during cell division by interacting with metaphase chromosomes (Gerhardt et al., 2006; Ito et al., 2002; Avolio-Hunter et al., 2003). The use of EBV plasmid replicon-based vectors in gene therapy is limited by the fact that EBNA1 has been shown to enhance B cell immortalization up to several thousand-fold (Humme et al., 2003) and to stimulate both humoral and cellular immune responses (Taylor et al., 2004) (Fig. 2c).

Deletion of the DS element in the *oriP* region of the EBV plasmid replicon has led to the construction of "ori-fishing" vectors, which have been used successfully for the identification of mammalian origins of replication. Such DS element depleted vectors are segregated correctly during cell division by utilizing the ability of EBNA1 to bind both the FR element and the metaphase chromosome. However, because the DS element is no longer present, they do not have the ability to self-replicate and thus require chromosomal DNA sequences that support their episomal replication (Gerhardt et al., 2006; Stoll et al., 2002).

NON-VIRAL EPISOMAL VECTORS

Considerable efforts to construct vectors that behave as autonomous units in the cell and that arc composed exclusively from functional chromosomal elements have been undertaken within the past decades. Three different approaches have been described: 1) insertion of putative origins of replication into plasmids analogous to the ARS assay in yeast, 2) inclusion of chromosomal scaffold/matrix attachment region sequences (S/MAR) into plasmids encoding a transgene, 3) construction of mammalian or human artificial chromosomes (MAC, HAC) either by assembly of the functional elements of the chromosome, telomeres, origin of replication and centromere (bottom-up approach) or by fragmentation of natural chromosomes into minichromosomes (top-down approach).

VECTORS CONTAINING A MAMMALIAN ORIGIN OF REPLICATION

The replication of eukaryotic genomes is strictly regulated to ensure that an origin of replication is activated only once during the cell cycle. The first step in the establishment of an origin of replication and its activation during S-phase is the binding of the origin recognition complex (ORC) in G_1-phase of the cell cycle. The ORC complex consists of six subunits, ORC1-6, which are conserved in all eukaryotes. It recruits additional initiation factors, Cdc6p and Cdt1, and late in G_1-phase the ORC-Cdc6p-Cdt1-complex loads MCM2-7 onto chromatin. This pre-replication complex selects origins of replication and licenses them for subsequent activation. In S-phase, heterodimeric protein kinases such as cyclin/Cdk and Cdc7-Dbf4 are required for initiation of DNA replication. As S-phase proceeds, different origins of replication are activated according to a temporal and spatial programme that is determined by chromosome structure (Blow et al., 2005; Gilbert, 2002).

The first eukaryotic origins of replication were isolated in *Saccharomyces cerevisiae* (Stinchcomb et al., 1980), where short (about 125 bp) AT-rich sequences containing an 11 bp core sequence serve as origins. These sequences (ARS, autonomous replicating sequences) represent the binding site for the ORC and are able to support episomal replication in yeast cells when inserted into a plasmid (Newlon et al., 1993).

Most attempts to construct episomal vectors for mammalian cells by the insertion of mapped origins of replication have failed, resulting either in the loss of these constructs or integration into the genome (Jackson et al., 2006). Nonetheless, there are some reports in which the insertion of specific sequences into plasmids has led to episomal replication. After insertion of various short mammalian origins into a yeast artificial chromosome (YAC), some of these vectors were maintained as an episome in mammalian cells during cultivation in selective medium, although most of the constructs integrated (Nielsen et al., 2000) (Fig. 3a). Later, a 36 bp human consensus sequence was identified that supports episomal replication in mammalian cells (Price et al., 2003). A human homologue of the monkey replication origin ors8 was capable of conferring sustained episomal replication of a plasmid in HeLa cells. Interestingly, association of this origin with DNA replication initiator proteins, such as Orc2 and Cdc6, was confirmed by chromatin immunoprecipitation assays (Callejo et al., 2006). As outlined above, numerous chromosomal sequences were isolated that are able to support episomal replication when inserted into a vector encoding the viral EBNA-1 protein and the DS depleted OriP (Gerhardt et al.,

2006; Krysan et al., 1989; Stoll et al., 2002). This may suggest that the observed instability of replication origin containing plasmids is due to the lack of a retention sequence in these constructs.

In contrast to *S. cerevisiae*, mammalian origins of replication share no sequence homologies but rather a number of structural features such as AT-rich regions, CpG islands and bent DNA (Gilbert, 2004). It is therefore assumed that activation of a mammalian origin of replication depends on epigenetic factors, such as chromatin structure, ongoing transcription or global nuclear architecture, and thus these epigenetic factors have to be taken into account when constructing episomal replicating vectors.

S/MAR-BASED VECTORS

Based on the assumption that mammalian replication origins are selected and activated by epigenetic factors and the observation that replication origins bind to the nuclear scaffold or matrix prior to the onset of S-phase (Cook, 1999), a number of vectors containing scaffold-matrix attached regions (S/MAR) were constructed. S/MARs tethering DNA to a subnuclear structure are generally AT-rich sequences and are critical for the organization of chromatin loops that define the boundaries of chromatin domains (Bode et al., 2000; Jackson et al., 2006).

Incorporation of three putative mammalian origins of replication into a YAC vector containing a S/MAR produced two clones that were retained as an episome under selective conditions, and the binding of one clone to a nuclear matrix preparation was demonstrated (Cossons et al., 1997; Nielsen et al., 2000). The best characterized S/MAR-based vectors belong to the pEPI family (Piechaczek et al., 1999). In this vector system the strong human S/MAR from the γ-interferon gene cluster is linked to a transcription unit, and it has been shown that its functioning depends exclusively on the presence of a transcription unit upstream the S/MAR, and that transcription has to run into the S/MAR element (Jenke et al., 2004b; Stehle et al., 2003). This vector is retained at a copy number below 10 per cell in all mammalian cells tested, including primary cells, for several hundred generations in the absence of selection. It replicates once per cell cycle in early S-phase and chromatin immunoprecipitation analysis revealed that ORC assembly can occur at various regions on the vector (Schaarschmidt et al., 2004). Mitotic stability of this vector system seems to be achieved by a stable association with transcriptionally active sites in the chromosome via an interaction with the prominent nuclear matrix protein SAF-A (Baiker et al., 2000; Jenke et al., 2002; Stehle et al., 2007) (Fig. 3b). A series of variants of thc prototype vector have been generated through incorporation of various transgenes under the control of different promoters (Baiker et al., personal communication), incorporation of synthetic S/MAR sequences (Jenke et al., 2004b) or the generation of "minicircles" containing only a transcription unit linked to the S/MAR but lacking all bacterial sequences (Nehlsen et al., 2006). pEPI-vectors

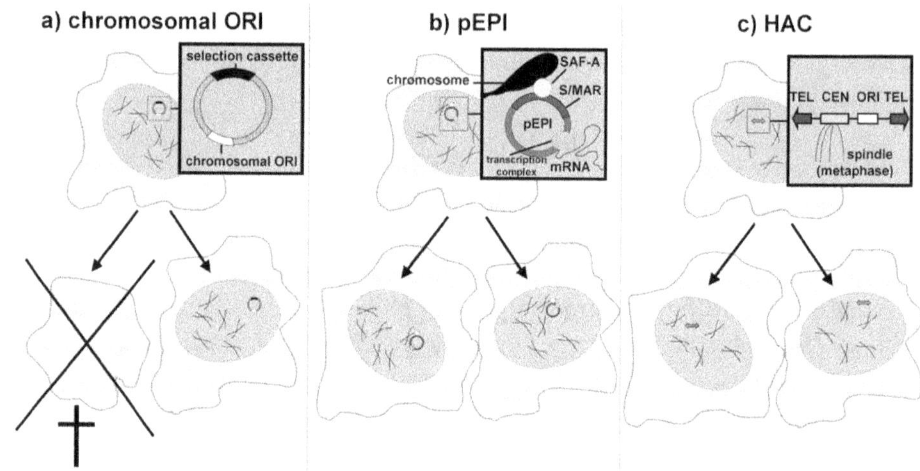

	a) chromosomal ORI	b) pEPI	c) HAC
MCM 2-7 dependent	yes	yes	yes
nuclear retention	selection pressure	S/MARs (via SAF-A)	CEN / TEL
copy number	<10	~10	<10
replication	ORI	transcription into S/MARs	ORI

Fig. 3. Non-viral plasmid replicons and artifical chromosomes. This figure depicts schematically the DNA replication and nuclear retention mechanisms of non-viral plasmid replicons and artificial chromosomes: a: Vectors based on plain chromosomal ORIs are able to replicate their DNA once per cell cycle in a MDM 2-7 dependant way. Such vectors do not have a nuclear retention mechanism, but may be propagated under constant selection pressure, when a selection marker is present. b: pEPI-based vectors exhibit copy numbers of around 10 molecules per cell. Their plasmid replicon consists of a transcription unit, which is directed into the S/MAR sequence. DNA-replication occurs once per cell cycle in a MCM 2-7 dependant way. The nuclear retention of pEPI-based vectors is due to the association of the S/MAR sequence with the metaphase chromosomes via SAF-A. c: artificial chromosomes for mammalian cells consist of an origin of replication (ORI), a centromeric region (CEN) and telomeres (TEL) at their linear ends.

have been used for long-term gene expression of various transgenes (including therapeutic genes), for long-term expression of shRNA and for the generation of genetically modified animals (Jenke et al., 2004a; Jenke et al., 2005; Manzini et al., 2006; Papapetrou et al., 2006). Interestingly, this vector system seems to have a virtually unlimited cloning capacity, allowing the insertion not only of a gene of interest, but a complete locus containing all regulatory control elements (Lufino et al., 2007).

A major limitation of all non-viral vector systems is their low transfection efficiency, and in the case of all autonomous replicons, including viral-based replicons, the low efficiency of establishment. Transfection efficiency issues likely can be overcome using novel transfection reagents or delivery as a "pseudovirus" (Eisenberger et al., 2005). Only a small percentage (between 1- and 10%) of transfected cells develop into stable clones. This is also true when plasmids are isolated from established cell clones and reintroduced into cells, demonstrating that the primary sequence of

DNA is not involved in establishment rate. It is now believed that establishment of clones is conferred by stochastic epigenetic events, for example acquiring the correct chromatin structure or reaching a specific nuclear compartment (Jackson et al., 2006). In fact, it has been shown that in all established cell clones containing pEPI, the vector molecules were exclusively found in nuclear compartments involved in active transcription (Stehle et al., 2007). In order to improve this establishment rate, *cis*- or *trans*- acting factors have to be identified that lead the vector molecule into a transcriptionally active nuclear compartment.

MAMMALIAN/HUMAN ARTIFICIAL CHROMOSOMES

The construction of the first mammalian artificial chromosome in 1997 (Harrington et al., 1997) was regarded as a breakthrough in gene therapy. It was believed that such mammalian or human artificial chromosomes (MAC, HAC) have major advantages over other vector systems: the use of a non-integrating DNA-molecule would ensure that it does not disturb the genome integrity of the host genome; the presence of a centromeric sequence should ensure mitotic stability of the construct; and artificial chromosomes should have an essentially unlimited cloning capacity, so that regulatory elements that are physically far apart from the gene can be inserted to ensure faithful expression of a therapeutic transgene.

There are two strategies used for the construction of human artificial chromosomes, the "bottom-up" approach and the "top-down" approach (Fig. 3c). In the bottom-up approach, the functional elements, telomeres, centromere and replication origin are isolated and assembled. In the top down approach, existing chromosomes are truncated in living cells to a minimal length by irradiation or telomere fragmentation (Farr et al., 1992), resulting in minichromosomes.

The first HAC was generated by the bottom-up approach. Long synthetic arrays of α-satellite DNA (a major component of natural human centromeres), PCR-generated telomeric DNA and about 100 kb of random genomic DNA (which should contain an origin of replication), were introduced in HT1080 cells. While many integration events were observed, a 6-10 Mbp linear HAC could be constructed (Harrington et al., 1997). Most of the HACs constructed by the bottom-up approach are structurally unstable and integration events are frequently observed (Basu et al., 2005; Basu et al., 2006). No structural instability and no integration events could be observed when using a circular HAC containing about 70kb α-satellite DNA (Ebersole et al., 2000).

The alternative top-down approach, involving human chromosome truncation to a minimal length by irradiation or telomere fragmentation, resulted in minichromosomes as small as 0.5-1 Mbp. These contain telomeric sequences at their termini and centromeric regions with long arrays of α-satellite DNA. While α-satellite DNA provides mitotic stability of both HACs and minichromosomes it is now clear that those sequences are not essential for the formation of fully functional human centromeres, since stable neocentromeric minichromosomes lacking α-satellite DNA have been described (Willard, 2001; Wong et al., 2002; Wong et al., 2005). Both types of minichromosomes seem to be structurally stable after propagation in cell lines and animals (Shen et al., 2000; Tomizuka et al., 1997; Voet et al., 2001). Human artificial chromosomes have been extensively used to characterize the structure and biological function of chromosomal elements, such as telomeres and centromeres, and still represent a model system for the study of chromosome biology (Basu et al., 2005; Masumoto et al., 2004; Monaco et al., 2006; Ren et al., 2005; Tsuduki et al., 2006). There are numerous reports demonstrating that HACs and minichromosomes can been used to achieve long-term and faithful transcription of transgenes. In most cases, a transgene is introduced into these large constructs by the use of targeted site-specific recombination between two minichromosome vectors using either the Cre-loxP (Kuroiwa et al., 2000) or the FLP-FRT systems (O'Gorman et al., 1991). The following examples illustrate the utility of these systems: HPRT (hypoxanthine-guanine phosphoribos yltranserase) deficiency was complemented by the expression of the HPRT locus encoded on HACs or minichromosomes, and stable expression of the transgene was observed over an extended period in the absence of selection (Grimes et al., 2001; Kotzamanis et al., 2005; Mejia et al., 2001; Moralli et al., 2006); expression of the human erythropoietin gene for at least 12 weeks was described in human primary fibroblasts (Kakeda et al., 2005); heat-regulated insulin production could be achieved when the insulin gene under the control of the heat shock protein 70 (Hsp70) promoter was introduced into a HAC (Suda et al., 2006); incorporation of the human cystic fibrosis transmembrane conductance regulator (CFTR) locus into a minichromosome resulted in a detectable chloride secretory response to cAMP in transfected cells (Auriche et al., 2002); and faithful expression of a reporter gene under the control of the osteopontin promoter was reported during differentiation into osteocytes (Ren et al., 2006).

HACs, minichromosomes and neocentromeric chromosomes have been introduced into mice either by microcell-mediated chromosome transfer or pronuclear injection (Basu et al., 2005; Jackson et al., 2006), and all were shown to be transmitted through the germline. More recently, a circular HAC containing the guanosine triphosphate cyclohydrolase I (GCHI) gene, and a linear HAC containing the human globin gene cluster, were introduced into mouse embryonic stem cells (ES) that were stably maintained for at least three months and used to create genetically modified chimeric mice (Suzuki et al., 2006).

In summary, non-viral episomally replicating vectors represent an interesting alternative to currently used replication-deficient virus-based vector systems. So far, none has been used routinely in biotechnology or gene therapy. Problems related to their use include the lack of efficient delivery systems, the low rate of their establishment in the transfected cell, and in the case of artificial chromosomes, the difficulty of their construction and delivery to the cell. However, with continued work on non-viral episomal vectors it can be expected that major improvements in these vector systems can be delivered within the next few years.

APPLICATION OF PLASMID-BASED VECTORS AND CONCLUSION

Episomal vectors, either based on viral plasmid replicons or on chromosomal elements, have been invaluable tools for basic and applied science. The biochemistry and cell cycle dependent regulation of mammalian DNA replication has been extensively studied using SV40-based vectors (DePamphilis, 2006) and more recently, EBV-based vectors have been used to isolate putative mammalian origins of replication (Gerhardt et al., 2006; Krysan et al., 1989). The non-viral vectors whose functioning depends on the insertion of a chromosomal S/MAR sequence represent a minimal system to study the epigenetic regulation of mammalian DNA replication and the relevance of global functional nuclear architecture (Schaarschmidt et al., 2004; Stehle et al., 2007). The construction of mammalian artificial chromosomes has had a considerable impact on our understanding of the functional elements of the eukaryotic chromosome, telomeres, replication origins and centromeres (Willard, 2001). Similarly, the analysis of the biochemistry and regulation of transcription was greatly facilitated by the use of episomal vectors (Amaya, 2005). These constructs will find even further applications for the analysis of basic biological phenomena, such as protein-protein interactions, signal transduction pathways, cell movements and many more.

One argument for developing extrachromosomal vectors was that transgene expression should be independent of the chromosomal context and therefore not subject to epigenetic silencing and *cis*-acting sequences. Although not always the case, long term expression of transgenes in the absence of selection has been reported for EBV-based vectors, S/MAR-based vectors and mammalian artificial chromosomes (Jackson et al., 2006), making these vectors attractive systems for the production of pharmaceutical relevant proteins in mammalian cells. Mammalian artificial chromosomes and S/MAR-based vectors have been successfully used for the genetic modification of mammalian organisms, such as mice and pigs, and have proven to be significantly more efficient than currently used integrating vectors (Basu et al., 2005; Manzini et al., 2006).

Finally, episomal vectors and especially vectors based on chromosomal elements may represent alternatives to the currently used viral vectors for genetic therapy. The major risks associated with virus-based vectors involve problems associated with insertional mutagenesis, imunogenicity and cytotoxicity (Thomas et al., 2003). These problems were highlighted by the first therapy related fatality, in which death was caused by an inflammatory reaction that was attributed to the use of adenoviral vectors (Hollon, 2000). Insertional mutagenesis events related to the use of integrating vectors have been reported for retroviral as well as AAV vectors (Donsante et al., 2007; Ott et al., 2006). Since episomal vectors based exclusively on chromosomal elements do not need any virus-encoded protein for their function and furthermore do not integrate into the host cell genome, they should avoid the major safety risks that have been associated with virus-based vectors. Their use has been demonstrated *in vitro* and some have been propagated in animal systems. Their present main limitation is their low transfection efficiency compared to systems based on replication deficient viruses. However, recent success in the construction of non-viral episomal vectors and steady improvement of DNA transfection techniques makes it likely that these limitations can be overcome in the near future, so that optimized episomal vectors will be available for the use in clinical trials.

REFERENCES

Akira, S., Takeda, K., and Kaisho, T. (2001). Toll-Like Receptors: Critical Proteins Linking Innate and Acquired Immunity. Nat.Immunol. *2*, 675-680.

Alino, S.F., Herrero, M.J., Noguera, I., Dasi, F., and Sanchez, M. (2007). Pig Liver Gene Therapy by Noninvasive Interventionist Catheterism. Gene Ther. *14*, 334-343.

Amaya, E. (2005). Xenomics. Genome Res. *15*, 1683-1691.

Andre, F. and Mir, L.M. (2004). DNA Electrotransfer: Its Principles and an Updated Review of Its Therapeutic Applications. Gene Ther. *11 Suppl 1*, S33-S42.

Athwal, R.S., Smarsh, M., Searle, B.M., and Deo, S.S. (1985). Integration of a Dominant Selectable Marker into Human Chromosomes and Transfer of Marked Chromosomes to Mouse Cells by Microcell Fusion. Somat.Cell Mol.Genet. *11*, 177-187.

Auriche, C., Carpani, D., Conese, M., Caci, E., Zegarra-Moran, O., Donini, P., and Ascenzioni, F. (2002). Functional Human CFTR Produced by a Stable Minichromosome. EMBO Rep. *3*, 862-868.

Avolio-Hunter, T.M. and Frappier, L. (2003). EBNA1 Efficiently Assembles on Chromatin Containing the Epstein-Barr Virus Latent Origin of Replication. Virology *315*, 398-408.

Baiker, A., Maercker, C., Piechaczek, C., Schmidt, S.B., Bode, J., Benham, C., and Lipps, H.J. (2000). Mitotic Stability of an Episomal Vector Containing a Human Scaffold/Matrix-Attached Region Is Provided by Association With Nuclear Matrix. Nat.Cell Biol. *2*, 182-184.

Basu, J. and Willard, H.F. (2005). Artificial and Engineered Chromosomes: Non-Integrating Vectors for Gene Therapy. Trends Mol.Med. *11*, 251-258.

Basu, J. and Willard, H.F. (2006). Human Artificial Chromosomes: Potential Applications and Clinical Considerations. Pediatr.Clin.North Am. *53*, 843-53, viii.

Bauer, S., Kirschning, C.J., Hacker, H., Redecke, V., Hausmann, S., Akira, S., Wagner, H., and Lipford, G.B. (2001). Human TLR9 Confers Responsiveness to Bacterial DNA Via Species-Specific CpG Motif Recognition. Proc.Natl.Acad.Sci.U.S.A *98*, 9237-9242.

Berg, L.J., Singh, K., and Botchan, M. (1986). Complementation of a Bovine Papilloma Virus Low-Copy-Number Mutant: Evidence for a Temporal Requirement of the Complementing Gene. Mol. Cell Biol. *6*, 859-869.

Bestor, T.H. (2000). The DNA Methyltransferases of Mammals. Hum.Mol.Genet. *9*, 2395-2402.

Bird, A.P. (1986). CpG-Rich Islands and the Function of DNA Methylation. Nature *321*, 209-213.

Blow, J.J. and Dutta, A. (2005). Preventing Re-Replication of Chromosomal DNA. Nat.Rev.Mol. Cell Biol. *6*, 476-486.

Bode, J., Benham, C., Knopp, A., and Mielke, C. (2000). Transcriptional Augmentation: Modulation of Gene Expression by Scaffold/Matrix-Attached Regions (S/MAR Elements). Crit Rev.Eukaryot. Gene Expr. *10*, 73-90.

Boyes, J. and Bird, A. (1992). Repression of Genes by DNA Methylation Depends on CpG Density and Promoter Strength: Evidence for Involvement of a Methyl-CpG Binding Protein. EMBO J. *11*, 327-333.

Brooks, A.R., Harkins, R.N., Wang, P., Qian, H.S., Liu, P., and Rubanyi, G.M. (2004). Transcriptional Silencing Is Associated With Extensive Methylation of the CMV Promoter Following Adenoviral Gene Delivery to Muscle. J.Gene Med. *6*, 395-404.

Budker, V., Zhang, G., Knechtle, S., and Wolff, J.A. (1996). Naked DNA Delivered Intraportally Expresses Efficiently in Hepatocytes. Gene Ther. *3*, 593-598.

Callejo, M., Sibani, S., Di, P.D., Price, G.G., and Zannis-Hadjopoulos, M. (2006). Identification and Functional Analysis of a Human Homologue of the Monkey Replication Origin Ors8. J.Cell Biochem. *99*, 1606-1615.

Cartier, R. and Reszka, R. (2002). Utilization of Synthetic Peptides Containing Nuclear Localization Signals for Nonviral Gene Transfer Systems. Gene Ther. *9*, 157-167.

Chan, C.S. and Tye, B.K. (1983). Organization of DNA Sequences and Replication Origins at Yeast Telomeres. Cell *33*, 563-573.

Chen, Y. and Rice, P.A. (2003). New Insight into Site-Specific Recombination From Flp Recombinase-DNA Structures. Annu.Rev.Biophys.Biomol.Struct. *32*, 135-159.

Choate, K.A. and Khavari, P.A. (1997). Direct Cutaneous Gene Delivery in a Human Genetic Skin Disease. Hum.Gene Ther. *8*, 1659-1665.

Cook, P.R. (1999). The Organization of Replication and Transcription. Science *284*, 1790-1795.

Cooper, M.J., Lippa, M., Payne, J.M., Hatzivassiliou, G., Reifenberg, E., Fayazi, B., Perales, J.C., Morrison, L.J., Templeton, D., Piekarz, R.L. et al. (1997). Safety-Modified Episomal Vectors for Human Gene Therapy. Proc.Natl.Acad.Sci.U.S.A *94*, 6450-6455.

Cossons, N., Nielsen, T.O., Dini, C., Tomilin, N., Young, D.B., Riabowol, K.T., Rattner, J.B., Johnston, R.N., Zannis-Hadjopoulos, M., and Price, G.B. (1997). Circular YAC Vectors Containing a Small Mammalian Origin Sequence Can Associate With the Nuclear Matrix. J.Cell Biochem. *67*, 439-450.

Dean, D.A. (2005). Nonviral Gene Transfer to Skeletal, Smooth, and Cardiac Muscle in Living Animals. Am.J.Physiol Cell Physiol *289*, C233-C245.

DePamphilis, M.L. (2006). DNA replication and human disease. (New York: Cold Spring Harbor Laboratory Press).

Donsante, A., Miller, D.G., Li, Y., Vogler, C., Brunt, E.M., Russell, D.W., and Sands, M.S. (2007). AAV Vector Integration Sites in Mouse Hepatocellular Carcinoma. Science *317*, 477-

Dworetzky, S.I., Lanford, R.E., and Feldherr, C.M. (1988). The Effects of Variations in the Number and Sequence of Targeting Signals on Nuclear Uptake. J.Cell Biol. *107*, 1279-1287.

Ebersole, T.A., Ross, A., Clark, E., McGill, N., Schindelhauer, D., Cooke, H., and Grimes, B. (2000). Mammalian Artificial Chromosome Formation From Circular Alphoid Input DNA Does Not Require Telomere Repeats. Hum.Mol.Genet. *9*, 1623-1631.

Eisenberger, T., Lipps, H.J., Mueller, M., and Baiker, A. (2005). A Human Papillomavirus (HPV)-Based Pseudoviral Gene Delivery System for the Non-Viral, Episomally Replicating Vector PEPI-1. Gene Ther.Mol.Biol. *9*, 371-376.

Farias, S.T. and Bonato, M.C. (2003). Preferred Amino Acids and Thermostability. Genet.Mol.Res. *2*, 383-393.

Farr, C.J., Stevanovic, M., Thomson, E.J., Goodfellow, P.N., and Cooke, H.J. (1992). Telomere-Associated Chromosome Fragmentation: Applications in Genome Manipulation and Analysis. Nat.Genet. *2*, 275-282.

Felgner, P.L., Gadek, T.R., Holm, M., Roman, R., Chan, H.W., Wenz, M., Northrop, J.P., Ringold, G.M., and Danielsen, M. (1987). Lipofection: a Highly Efficient, Lipid-Mediated DNA-Transfection Procedure. Proc.Natl.Acad.Sci.U.S.A *84*, 7413-7417.

Friedrich, T.D., Bedner, E., Darzynkiewicz, Z., and Lehman, J.M. (2005). Distinct Patterns of MCM Protein Binding in Nuclei of S Phase and Rereplicating SV40-Infected Monkey Kidney Cells. Cytometry A *68*, 10-18.

Futcher, A.B. (1986). Copy Number Amplification of the 2 Micron Circle Plasmid of Saccharomyces Cerevisiae. J Theor.Biol. *119*, 197-204.

Gao, X., Kim, K.S., and Liu, D. (2007). Nonviral Gene Delivery: What We Know and What Is Next. AAPS.J. *9*, E92-104.

Gerhardt, J., Jafar, S., Spindler, M.P., Ott, E., and Schepers, A. (2006). Identification of New Human Origins of DNA Replication by an Origin-Trapping Assay. Mol.Cell Biol. *26*, 7731-7746.

Gilbert, D.M. (2002). Replication Timing and Transcriptional Control: Beyond Cause and Effect. Curr.Opin.Cell Biol. *14*, 377-383.

Gilbert, D.M. (2004). In Search of the Holy Replicator. Nat.Rev.Mol.Cell Biol. *5*, 848-855.

Gordon, J.W. and Ruddle, F.H. (1981). Integration and Stable Germ Line Transmission of Genes Injected into Mouse Pronuclei. Science *214*, 1244-1246.

Graham, F.L. and van der Eb, A.J. (1973). A New Technique for the Assay of Infectivity of Human Adenovirus 5 DNA. Virology *52*, 456-467.

Grimes, B.R., Schindelhauer, D., McGill, N.I., Ross, A., Ebersole, T.A., and Cooke, H.J. (2001). Stable Gene Expression From a Mammalian Artificial Chromosome. EMBO Rep. *2*, 910-914.

Gustafsson, C., Govindarajan, S., and Minshull, J. (2004). Codon Bias and Heterologous Protein Expression. Trends Biotechnol. *22*, 346-353.

Harrington, J.J., Van, B.G., Mays, R.W., Gustashaw, K., and Willard, H.F. (1997). Formation of De Novo Centromeres and Construction of First-Generation Human Artificial Microchromosomes. Nat.Genet. *15*, 345-355.

Hasson, E., Slovatizky, Y., Shimoni, Y., Falk, H., Panet, A., and Mitrani, E. (2005). Solid Tissues Can Be Manipulated Ex Vivo and Used As Vehicles for Gene Therapy. J.Gene Med. *7*, 926-935.

Hebert, E. (2003). Improvement of Exogenous DNA Nuclear Importation by Nuclear Localization Signal-Bearing Vectors: a Promising Way for Non-Viral Gene Therapy? Biol.Cell *95*, 59-68.

Hickman, M.A., Malone, R.W., Lehmann-Bruinsma, K., Sih, T.R., Knoell, D., Szoka, F.C., Walzem, R., Carlson, D.M., and Powell, J.S. (1994). Gene Expression Following Direct Injection of DNA into Liver. Hum.Gene Ther. *5*, 1477-1483.

Hollenberg, C.P., Degelmann, A., Kustermann-Kuhn, B., and Royer, H.D. (1976). Characterization of 2-Mum DNA of Saccharomyces Cerevisiae by Restriction Fragment Analysis and Integration in an Escherichia Coli Plasmid. Proc.Natl.Acad.Sci.U.S.A *73*, 2072-2076.

Hollon, T. (2000). Researchers and Regulators Reflect on First Gene Therapy Death. Nat.Med *6*, 6-

Horowitz, H. and Haber, J.E. (1985). Identification of Autonomously Replicating Circular Subtelomeric Y' Elements in Saccharomyces Cerevisiae. Mol.Cell Biol. *5*, 2369-2380.

Humme, S., Reisbach, G., Feederle, R., Delecluse, H.J., Bousset, K., Hammerschmidt, W., and Schepers, A. (2003). The EBV Nuclear Antigen 1 (EBNA1) Enhances B Cell Immortalization Several Thousandfold. Proc.Natl.Acad.Sci.U.S.A *100*, 10989-10994.

Ikemura, T. (1985). Codon Usage and TRNA Content in Unicellular and Multicellular Organisms. Mol.Biol.Evol. *2*, 13-34.

Ioshikhes, I.P. and Zhang, M.Q. (2000). Large-Scale Human Promoter Mapping Using CpG Islands. Nat.Genet. *26*, 61-63.

Ito, S., Gotoh, E., Ozawa, S., and Yanagi, K. (2002). Epstein-Barr Virus Nuclear Antigen-1 Is Highly Colocalized With Interphase Chromatin and Its Newly Replicated Regions in Particular. J.Gen. Virol. *83*, 2377-2383.

Izaurralde, E., Kann, M., Pante, N., Sodeik, B., and Hohn, T. (1999). Viruses, Microorganisms and Scientists Meet the Nuclear Pore. Leysin, VD, Switzerland, February 26-March 1, 1998. EMBO J. *18*, 289-296.

Jackson, D.A., Juranek, S., and Lipps, H.J. (2006). Designing Nonviral Vectors for Efficient Gene Transfer and Long-Term Gene Expression. Mol.Ther. *14*, 613-626.

Jans, D.A. and Hubner, S. (1996). Regulation of Protein Transport to the Nucleus: Central Role of Phosphorylation. Physiol Rev. *76*, 651-685.

Jenke, A.C., Eisenberger, T., Baiker, A., Stehle, I.M., Wirth, S., and Lipps, H.J. (2005). The Nonviral Episomal Replicating Vector PEPI-1 Allows Long-Term Inhibition of Bcr-Abl Expression by ShRNA. Hum.Gene Ther. *16*, 533-539.

Jenke, A.C., Scinteie, M.F., Stehle, I.M., and Lipps, H.J. (2004a). Expression of a Transgene Encoded on a Non-Viral Episomal Vector Is Not Subject to Epigenetic Silencing by Cytosine Methylation. Mol.Biol.Rep. *31*, 85-90.

Jenke, A.C., Stehle, I.M., Herrmann, F., Eisenberger, T., Baiker, A., Bode, J., Fackelmayer, F.O., and Lipps, H.J. (2004b). Nuclear Scaffold/Matrix Attached Region Modules Linked to a Transcription Unit Are Sufficient for Replication and Maintenance of a Mammalian Episome. Proc.Natl.Acad. Sci.U.S.A *101*, 11322-11327.

Jenke, B.H., Fetzer, C.P., Stehle, I.M., Jonsson, F., Fackelmayer, F.O., Conradt, H., Bode, J., and Lipps, H.J. (2002). An Episomally Replicating Vector Binds to the Nuclear Matrix Protein SAF-A in Vivo. EMBO Rep. *3*, 349-354.

Kakeda, M., Hiratsuka, M., Nagata, K., Kuroiwa, Y., Kakitani, M., Katoh, M., Oshimura, M., and Tomizuka, K. (2005). Human Artificial Chromosome (HAC) Vector Provides Long-Term Therapeutic Transgene Expression in Normal Human Primary Fibroblasts. Gene Ther. *12*, 852-856.

Khalil, I.A., Kogure, K., Akita, H., and Harashima, H. (2006). Uptake Pathways and Subsequent Intracellular Trafficking in Nonviral Gene Delivery. Pharmacol.Rev. *58*, 32-45.

Kotzamanis, G., Cheung, W., Abdulrazzak, H., Perez-Luz, S., Howe, S., Cooke, H., and Huxley, C. (2005). Construction of Human Artificial Chromosome Vectors by Recombineering. Gene *351*, 29-38.

Krysan, P.J., Haase, S.B., and Calos, M.P. (1989). Isolation of Human Sequences That Replicate Autonomously in Human Cells. Mol.Cell Biol. *9*, 1026-1033.

Kudla, G., Lipinski, L., Caffin, F., Helwak, A., and Zylicz, M. (2006). High Guanine and Cytosine Content Increases MRNA Levels in Mammalian Cells. PLoS.Biol. *4*, e180-

Kuroiwa, Y., Tomizuka, K., Shinohara, T., Kazuki, Y., Yoshida, H., Ohguma, A., Yamamoto, T., Tanaka, S., Oshimura, M., and Ishida, I. (2000). Manipulation of Human Minichromosomes to Carry Greater Than Megabase-Sized Chromosome Inserts. Nat.Biotechnol. *18*, 1086-1090.

Lander, E.S., Linton, L.M., Birren, B., Nusbaum, C., Zody, M.C., Baldwin, J., Devon, K., Dewar, K., Doyle, M., FitzHugh, W. et al. (2001). Initial Sequencing and Analysis of the Human Genome. Nature *409*, 860-921.

Lechardeur, D. and Lukacs, G.L. (2006). Nucleocytoplasmic Transport of Plasmid DNA: a Perilous Journey From the Cytoplasm to the Nucleus. Hum.Gene Ther. *17*, 882-889.

Leder, C., Kleinschmidt, J.A., Wiethe, C., and Muller, M. (2001). Enhancement of Capsid Gene Expression: Preparing the Human Papillomavirus Type 16 Major Structural Gene L1 for DNA Vaccination Purposes. J.Virol. *75*, 9201-9209.

Ledwith, B.J., Manam, S., Troilo, P.J., Barnum, A.B., Pauley, C.J., Griffiths, T.G., Harper, L.B., Beare, C.M., Bagdon, W.J., and Nichols, W.W. (2000). Plasmid DNA Vaccines: Investigation of Integration into Host Cellular DNA Following Intramuscular Injection in Mice. Intervirology *43*, 258-272.

Lin, C.L., Li, H., Wang, Y., Zhu, F.X., Kudchodkar, S., and Yuan, Y. (2003). Kaposi's Sarcoma-Associated Herpesvirus Lytic Origin (Ori-Lyt)-Dependent DNA Replication: Identification of the Ori-Lyt and Association of K8 BZip Protein With the Origin. J.Virol. *77*, 5578-5588.

Liu, D., Ren, T., and Gao, X. (2003). Cationic Transfection Lipids. Curr.Med.Chem. *10*, 1307-1315.

Liu, F., Song, Y., and Liu, D. (1999). Hydrodynamics-Based Transfection in Animals by Systemic Administration of Plasmid DNA. Gene Ther. *6*, 1258-1266.

Liu, Y., Frazer, I.H., Liu, W.J., Liu, X.S., McMillan, N., and Zhao, K.N. (2001). Efficiency of Delivery of DNA to Cells by Bovine Papillomavirus Type-1 L1/L2 Pseudovirions. Appl.Microbiol. Biotechnol. *56*, 150-156.

Lufino, M.M., Manservigi, R., and Wade-Martins, R. (2007). An S/MAR-Based Infectious Episomal Genomic DNA Expression Vector Provides Long-Term Regulated Functional Complementation of LDLR Deficiency. Nucleic Acids Res.

Manzini, S., Vargiolu, A., Stehle, I.M., Bacci, M.L., Cerrito, M.G., Giovannoni, R., Zannoni, A., Bianco, M.R., Forni, M., Donini, P. et al. (2006). Genetically Modified Pigs Produced With a Nonviral Episomal Vector. Proc.Natl.Acad.Sci.U.S.A *103*, 17672-17677.

Masumoto, H., Nakano, M., and Ohzeki, J. (2004). The Role of CENP-B and Alpha-Satellite DNA: De Novo Assembly and Epigenetic Maintenance of Human Centromeres. Chromosome.Res. *12*, 543-556.

Mejia, J.E., Willmott, A., Levy, E., Earnshaw, W.C., and Larin, Z. (2001). Functional Complementation of a Genetic Deficiency With Human Artificial Chromosomes. Am.J.Hum.Genet. *69*, 315-326.

Meyer, K.B., Thompson, M.M., Levy, M.Y., Barron, L.G., and Szoka, F.C., Jr. (1995). Intratracheal Gene Delivery to the Mouse Airway: Characterization of Plasmid DNA Expression and Pharmacokinetics. Gene Ther. *2*, 450-460.

Monaco, Z.L. and Moralli, D. (2006). Progress in Artificial Chromosome Technology. Biochem.Soc. Trans. *34*, 324-327.

Moralli, D., Simpson, K.M., Wade-Martins, R., and Monaco, Z.L. (2006). A Novel Human Artificial Chromosome Gene Expression System Using Herpes Simplex Virus Type 1 Vectors. EMBO Rep. *7*, 911-918.

Nehlsen, K., Broll, S., and Bode, J. (2006). Replicating Minicircles: Generation of Nonviral Episomes Fort the Efficient Modification of Dividing Cells. Gene Ther.Mol.Biol. *10*, 233-244.

Neumann, E., Schaefer-Ridder, M., Wang, Y., and Hofschneider, P.H. (1982). Gene Transfer into Mouse Lyoma Cells by Electroporation in High Electric Fields. EMBO J. *1*, 841-845.

Newlon, C.S. and Theis, J.F. (1993). The Structure and Function of Yeast ARS Elements. Curr.Opin. Genet.Dev. *3*, 752-758.

Nielsen, T.O., Cossons, N.H., Zannis-Hadjopoulos, M., and Price, G.B. (2000). Circular YAC Vectors Containing Short Mammalian Origin Sequences Are Maintained Under Selection As HeLa Episomes. J.Cell Biochem. *76*, 674-685.

O'Gorman, S., Fox, D.T., and Wahl, G.M. (1991). Recombinase-Mediated Gene Activation and Site-Specific Integration in Mammalian Cells. Science *251*, 1351-1355.

Ott, M.G., Schmidt, M., Schwarzwaelder, K., Stein, S., Siler, U., Koehl, U., Glimm, H., Kuhlcke, K., Schilz, A., Kunkel, H. et al. (2006). Correction of X-Linked Chronic Granulomatous Disease by Gene Therapy, Augmented by Insertional Activation of MDS1-EVI1, PRDM16 or SETBP1. Nat.Med *12*, 401-409.

Pante, N. and Kann, M. (2002). Nuclear Pore Complex Is Able to Transport Macromolecules With Diameters of About 39 Nm. Mol.Biol.Cell *13*, 425-434.

Papapetrou, E.P., Ziros, P.G., Micheva, I.D., Zoumbos, N.C., and Athanassiadou, A. (2006). Gene Transfer into Human Hematopoietic Progenitor Cells With an Episomal Vector Carrying an S/MAR Element. Gene Ther. *13*, 40-51.

Piechaczek, C., Fetzer, C., Baiker, A., Bode, J., and Lipps, H.J. (1999). A Vector Based on the SV40 Origin of Replication and Chromosomal S/MARs Replicates Episomally in CHO Cells. Nucleic Acids Res. *27*, 426-428.

Prausnitz, M.R. (1999). A Practical Assessment of Transdermal Drug Delivery by Skin Electroporation. Adv.Drug Deliv.Rev. *35*, 61-76.

Price, G.B., Allarakhia, M., Cossons, N., Nielsen, T., az-Perez, M., Friedlander, P., Tao, L., and Zannis-Hadjopoulos, M. (2003). Identification of a Cis-Element That Determines Autonomous DNA Replication in Eukaryotic Cells. J.Biol.Chem. *278*, 19649-19659.

Ravnan, J.B., Gilbert, D.M., Ten Hagen, K.G., and Cohen, S.N. (1992). Random-Choice Replication of Extrachromosomal Bovine Papillomavirus (BPV) Molecules in Heterogeneous, Clonally Derived BPV-Infected Cell Lines. J Virol. *66*, 6946-6952.

Ren, X., Katoh, M., Hoshiya, H., Kurimasa, A., Inoue, T., Ayabe, F., Shibata, K., Toguchida, J., and Oshimura, M. (2005). A Novel Human Artificial Chromosome Vector Provides Effective Cell Lineage-Specific Transgene Expression in Human Mesenchymal Stem Cells. Stem Cells *23*, 1608-1616.

Ren, X., Tahimic, C.G., Katoh, M., Kurimasa, A., Inoue, T., and Oshimura, M. (2006). Human Artificial Chromosome Vectors Meet Stem Cells: New Prospects for Gene Delivery. Stem Cell Rev. *2*, 43-50.

Roitman-Shemer, V., Stokrova, J., Forstova, J., and Oppenheim, A. (2007). Assemblages of Simian Virus 40 Capsid Proteins and Viral DNA Visualized by Electron Microscopy. Biochem.Biophys. Res.Commun. *353*, 424-430.

Rose, C., Green, M., Webber, S., Kingsley, L., Day, R., Watkins, S., Reyes, J., and Rowe, D. (2002). Detection of Epstein-Barr Virus Genomes in Peripheral Blood B Cells From Solid-Organ Transplant Recipients by Fluorescence in Situ Hybridization. J.Clin.Microbiol. *40*, 2533-2544.

Saito, T. (2006). In Vivo Electroporation in the Embryonic Mouse Central Nervous System. Nat. Protoc. *1*, 1552-1558.

Schaarschmidt, D., Baltin, J., Stehle, I.M., Lipps, H.J., and Knippers, R. (2004). An Episomal Mammalian Replicon: Sequence-Independent Binding of the Origin Recognition Complex. EMBO J. *23*, 191-201.

Sebestyen, M.G., Ludtke, J.J., Bassik, M.C., Zhang, G., Budker, V., Lukhtanov, E.A., Hagstrom, J.E., and Wolff, J.A. (1998). DNA Vector Chemistry: the Covalent Attachment of Signal Peptides to Plasmid DNA. Nat.Biotechnol. *16*, 80-85.

Shen, M.H., Mee, P.J., Nichols, J., Yang, J., Brook, F., Gardner, R.L., Smith, A.G., and Brown, W.R. (2000). A Structurally Defined Mini-Chromosome Vector for the Mouse Germ Line. Curr.Biol. *10*, 31-34.

Shen, Y.M., Hirschhorn, R.R., Mercer, W.E., Surmacz, E., Tsutsui, Y., Soprano, K.J., and Baserga, R. (1982). Gene Transfer: DNA Microinjection Compared With DNA Transfection With a Very High Efficiency. Mol.Cell Biol. *2*, 1145-1154.

Stehle, I.M., Postberg, J., Rupprecht, S., Cremer, T., Jackson, D.A., and Lipps, H.J. (2007). Establishment and Mitotic Stability of an Extra-Chromosomal Mammalian Replicon. BMC.Cell Biol. *8*, 33-

Stehle, I.M., Scinteie, M.F., Baiker, A., Jenke, A.C., and Lipps, H.J. (2003). Exploiting a Minimal System to Study the Epigenetic Control of DNA Replication: the Interplay Between Transcription and Replication. Chromosome.Res. *11*, 413-421.

Stinchcomb, D.T., Struhl, K., and Davis, R.W. (1979). Isolation and Characterisation of a Yeast Chromosomal Replicator. Nature *282*, 39-43.

Stinchcomb, D.T., Thomas, M., Kelly, J., Selker, E., and Davis, R.W. (1980). Eukaryotic DNA Segments Capable of Autonomous Replication in Yeast. Proc.Natl.Acad.Sci.U.S.A *77*, 4559-4563.

Stoll, S.M. and Calos, M.P. (2002). Extrachromosomal Plasmid Vectors for Gene Therapy. Curr. Opin.Mol.Ther. *4*, 299-305.

Strayer, D.S. and Zern, M.A. (1999). Gene Delivery to the Liver Using Simian Virus 40-Derived Vectors. Semin.Liver Dis. *19*, 71-81.

Suda, T., Katoh, M., Hiratsuka, M., Takiguchi, M., Kazuki, Y., Inoue, T., and Oshimura, M. (2006). Heat-Regulated Production and Secretion of Insulin From a Human Artificial Chromosome Vector. Biochem.Biophys.Res.Commun. *340*, 1053-1061.

Suzuki, N., Nishii, K., Okazaki, T., and Ikeno, M. (2006). Human Artificial Chromosomes Constructed Using the Bottom-Up Strategy Are Stably Maintained in Mitosis and Efficiently Transmissible to Progeny Mice. J.Biol.Chem. *281*, 26615-26623.

Szostak, J.W. and Blackburn, E.H. (1982). Cloning Yeast Telomeres on Linear Plasmid Vectors. Cell *29*, 245-255.

Taylor, G.S., Haigh, T.A., Gudgeon, N.H., Phelps, R.J., Lee, S.P., Steven, N.M., and Rickinson, A.B. (2004). Dual Stimulation of Epstein-Barr Virus (EBV)-Specific CD4+- and CD8+-T-Cell Responses by a Chimeric Antigen Construct: Potential Therapeutic Vaccine for EBV-Positive Nasopharyngeal Carcinoma. J.Virol. *78*, 768-778.

Thomas, C.E., Ehrhardt, A., and Kay, M.A. (2003). Progress and Problems With the Use of Viral Vectors for Gene Therapy. Nat.Rev.Genet. *4*, 346-358.

Tomizuka, K., Yoshida, H., Uejima, H., Kugoh, H., Sato, K., Ohguma, A., Hayasaka, M., Hanaoka, K., Oshimura, M., and Ishida, I. (1997). Functional Expression and Germline Transmission of a Human Chromosome Fragment in Chimaeric Mice. Nat.Genet. *16*, 133-143.

Trollet, C., Bloquel, C., Scherman, D., and Bigey, P. (2006). Electrotransfer into Skeletal Muscle for Protein Expression. Curr.Gene Ther. *6*, 561-578.

Tsuduki, T., Nakano, M., Yasuoka, N., Yamazaki, S., Okada, T., Okamoto, Y., and Masumoto, H.

(2006). An Artificially Constructed De Novo Human Chromosome Behaves Almost Identically to Its Natural Counterpart During Metaphase and Anaphase in Living Cells. Mol.Cell Biol. *26*, 7682-7695.

Valls, E., de, l.C., X, and Martinez-Balbas, M.A. (2003). The SV40 T Antigen Modulates CBP Histone Acetyltransferase Activity. Nucleic Acids Res. *31*, 3114-3122.

Velmurugan, S., Yang, X.M., Chan, C.S., Dobson, M., and Jayaram, M. (2000). Partitioning of the 2-Microm Circle Plasmid of Saccharomyces Cerevisiae. Functional Coordination With Chromosome Segregation and Plasmid-Encoded Rep Protein Distribution. J Cell Biol. *149*, 553-566.

Voet, T., Vermeesch, J., Carens, A., Durr, J., Labaere, C., Duhamel, H., David, G., and Marynen, P. (2001). Efficient Male and Female Germline Transmission of a Human Chromosomal Vector in Mice. Genome Res. *11*, 124-136.

Volkert, F.C. and Broach, J.R. (1986). Site-Specific Recombination Promotes Plasmid Amplification in Yeast. Cell *46*, 541-550.

Wang, J. and Sugden, B. (2005). Origins of Bidirectional Replication of Epstein-Barr Virus: Models for Understanding Mammalian Origins of DNA Synthesis. J.Cell Biochem. *94*, 247-256.

Weber, M., Hellmann, I., Stadler, M.B., Ramos, L., Paabo, S., Rebhan, M., and Schubeler, D. (2007). Distribution, Silencing Potential and Evolutionary Impact of Promoter DNA Methylation in the Human Genome. Nat.Genet. *39*, 457-466.

Willard, H.F. (2001). Neocentromeres and Human Artificial Chromosomes: an Unnatural Act. Proc. Natl.Acad.Sci.U.S.A *98*, 5374-5376.

Wilson, V.G., West, M., Woytek, K., and Rangasamy, D. (2002). Papillomavirus E1 Proteins: Form, Function, and Features. Virus Genes *24*, 275-290.

Wolff, J.A., Malone, R.W., Williams, P., Chong, W., Acsadi, G., Jani, A., and Felgner, P.L. (1990). Direct Gene Transfer into Mouse Muscle in Vivo. Science *247*, 1465-1468.

Wong, L.H., Saffery, R., Anderson, M.A., Earle, E., Quach, J.M., Stafford, A.J., Fowler, K.J., and Choo, K.H. (2005). Analysis of Mitotic and Expression Properties of Human Neocentromere-Based Transchromosomes in Mice. J.Biol.Chem. *280*, 3954-3962.

Wong, L.H., Saffery, R., and Choo, K.H. (2002). Construction of Neocentromere-Based Human Minichromosomes for Gene Delivery and Centromere Studies. Gene Ther. *9*, 724-726.

Yang, N.S., Burkholder, J., Roberts, B., Martinell, B., and McCabe, D. (1990). In Vivo and in Vitro Gene Transfer to Mammalian Somatic Cells by Particle Bombardment. Proc.Natl.Acad.Sci.U.S.A *87*, 9568-9572.

Yew, N.S. and Cheng, S.H. (2004). Reducing the Immunostimulatory Activity of CpG-Containing Plasmid DNA Vectors for Non-Viral Gene Therapy. Expert.Opin.Drug Deliv. *1*, 115-125.

Yoshino, H., Hashizume, K., and Kobayashi, E. (2006). Naked Plasmid DNA Transfer to the Porcine Liver Using Rapid Injection With Large Volume. Gene Ther. *13*, 1696-1702.

Zhang, G., Gao, X., Song, Y.K., Vollmer, R., Stolz, D.B., Gasiorowski, J.Z., Dean, D.A., and Liu, D. (2004a). Hydroporation As the Mechanism of Hydrodynamic Delivery. Gene Ther. *11*, 675-682.

Zhang, G., Vargo, D., Budker, V., Armstrong, N., Knechtle, S., and Wolff, J.A. (1997). Expression of Naked Plasmid DNA Injected into the Afferent and Efferent Vessels of Rodent and Dog Livers. Hum.Gene Ther. *8*, 1763-1772.

Zhang, L., Kasif, S., Cantor, C.R., and Broude, N.E. (2004b). GC/AT-Content Spikes As Genomic Punctuation Marks. Proc.Natl.Acad.Sci.U.S.A *101*, 16855-16860.

Zheng, P.S., Brokaw, J., and McBride, A.A. (2005). Conditional Mutations in the Mitotic Chromosome Binding Function of the Bovine Papillomavirus Type 1 E2 Protein. J Virol. *79*, 1500-1509.

Zhou, J., Liu, W.J., Peng, S.W., Sun, X.Y., and Frazer, I. (1999). Papillomavirus Capsid Protein Expression Level Depends on the Match Between Codon Usage and TRNA Availability. J.Virol. *73*, 4972-4982.

Chapter 4

Bioinformatics Tools and Methods for Plasmid Sequence Analysis and Annotation

Jason R. Grant and Paul Stothard

ABSTRACT

The past decade has seen tremendous growth in the development and application of bioinformatics databases and software. This chapter provides an overview of some of the current bioinformatics tools and methods available for the analysis of plasmid sequences, including resources for obtaining sequences, performing routine analysis operations, and annotating novel sequences. Specific topics covered include sequence databases, restriction analysis, plasmid construction, gene prediction, similarity searches, sequence alignment, annotation pipelines, and visualization tools. In addition, the most popular sequence formats are described, with an emphasis on open formats. Many of the resources presented are applicable to naturally occurring plasmids and to plasmid vectors, from prokaryotic or eukaryotic systems.

INTRODUCTION

This chapter describes bioinformatics programs and databases that can be used to analyze plasmid sequences. Although many commercial software packages are available for this purpose, they are usually very expensive, costing hundreds or even thousands of dollars per license. Therefore, we have decided to focus on free software, with an emphasis on open source and open standards. Open source refers to software that is available in source-code form with a license that allows for unrestricted redistribution and the creation of royalty-free derived works. Open source projects can make a lasting contribution to the scientific community, since the code they produce can be reused to form more complete or advanced programs. Another advantage of open source software is that it favors the use of open standards for data storage. Commercial programs often use proprietary file formats, which can be problematic for archiving data for long periods. Will that sequence analysis software be available in the future to open those files? As biology becomes more data intensive, it is important to ensure that information will remain accessible in the future. How many people have old 3½ inch floppies with files that cannot be opened by today's computers or software?

Bioinformatics tools are commonly available as web servers or stand-alone programs. Both types have their advantages and disadvantages. Web-based tools are generally free, and they can be accessed from any computer with a web browser and an internet connection. Software does not have to be downloaded and configured, which makes web-based tools convenient for large labs or classrooms, where the majority of users do not have administrative privileges. One drawback of web-based software is that users wishing to perform multiple operations often have to submit sequences to several different specialized servers. Another disadvantage is that web servers may be modified or become unavailable without warning. Users may suddenly find that their favorite server is no longer available, or that the output is no longer in the format they prefer. In comparison, stand-alone programs are often platform-dependent (*i.e.* they require a particular operating system), and they need to be downloaded and installed on each computer before use. Installation can be quite tedious, as some bioinformatics tools are dependent on several other programs, each of which must also be installed. If the software is a commercial product, a costly license may have to be purchased, although some commercial programs, such as Invitrogen's Vector NTI, are free for academic or government use. An advantage of some stand-alone tools, especially the commercial ones, is that many analysis methods are integrated into a single program. For example, a user may be able to perform a BLAST search to identify a sequence element of interest, followed by a virtual ligation between a plasmid backbone and the sequence element, without having to copy and paste any sequence data. Also, many people prefer the graphical user interfaces (GUIs) provided by stand-alone programs, as opposed to the more limited capabilities for user interaction supported by web-based programs. These advantages of stand-alone software will likely become less important as web browsers and web-based programs become more advanced. Indeed, feature-rich web-based spreadsheet and word processor programs now exist, which use techniques like AJAX (Asynchronous JavaScript and XML) to increase functionality and usability. The problem of having to visit several servers to perform a set of analysis operations will also be mitigated by systems such as BioMoby (Wilkinson and Links, 2002) and Taverna (Oinn et al., 2004), which increase server interoperability by specifying standardized biological data formats and protocols for program communication. Fortunately, it is becoming increasingly common for providers of bioinformatics software to offer both web-based and stand-alone versions of the same program. This solution offers the best of both worlds, as individuals can choose the version that best meets their requirements. Several of the programs and databases described below can be used online or downloaded.

PLASMID SEQUENCE SOURCES

There are many databases that contain plasmid and vector DNA sequences (Table 1). One of the best sequence resources is GenBank (Benson et al., 2007), a comprehensive nucleotide database maintained by the National Center for

Biotechnology Information (NCBI), a division of the US National Library of Medicine (NLM). GenBank is a member of the International Nucleotide Sequence Database Collaboration, which also includes the DNA Data Bank of Japan (DDBJ) (Okubo et al., 2006) and the European Molecular Biology Laboratory (EMBL) nucleotide sequence database (Kulikova et al., 2007). Data is exchanged between these three organizations on a daily basis, and thus they are all excellent sources of sequence information. NCBI also provides a database of popular cloning vectors called UniVec, which can be searched for vectors of interest. As the general repository for all nucleotide sequences, GenBank/EMBL/DDBJ is also the best place to search for naturally occurring plasmid sequences.

Although GenBank/EMBL/DDBJ contain many plasmid sequences, these sequences are scattered among the more than 60 million records contained in these databases. Not surprisingly, it can be difficult to find the specific sequence you are looking for, even when using carefully constructed text queries. Instead, it is generally much easier to find sequences in databases dedicated to plasmid vectors. In addition, these specialized databases often provide additional information about how plasmids were derived and their original functions. Dedicated vector databases can be particularly useful when searching for a plasmid with a particular characteristic or function, for example, a mammalian expression vector with a green fluorescent protein (GFP) tag. One such database is VectorDB, which contains information on many common vectors. VectorDB is hosted by the *Saccharomyces* Genome Database (Cherry et al., 1998) and conveniently has organism subsets of vectors for *Drosophila*, *C. elegans*, and yeast. Although VectorDB includes a large collection of sequences, it has not been updated in several years.

Plasmid repositories are dedicated to archiving and sharing plasmids with the scientific community. In addition to serving as a convenient source of vectors for use in the lab, they also provide plasmid sequences and maps. Two popular repositories are PlasmID (Zuo et al., 2007) and Addgene (Fan et al., 2005). PlasmID, formerly the Dana-Farber/Harvard Cancer Center DNA Resource Core plasmid repository, can be searched using common gene names, keywords, vector features, author names and PubMed IDs. Addgene is a non-profit organization that facilitates the sharing of plasmid constructs described in published articles. It has most, if not all, the vectors found in VectorDB, and it provides search capabilities similar to those in PlasmID. Addgene also includes tools for analyzing plasmid sequences.

There are also specialized databases dedicated to vectors designed for specific functions. These are usually hosted by labs or research groups interested in a particular research area or model system. For example, the Forsburg lab studies fission yeast, which require vectors with specialized replication origins, promoters, and splicing mechanisms. For the benefit of the fission yeast research community,

Table 1. Sources of plasmid sequences. Additional columns indicate whether maps or plasmid DNA are also available.

Collections	Maps	DNA*
General		
GenBank *http://www.ncbi.nlm.nih.gov/Genbank* Annotated collection of all publicly available DNA sequences (data is shared among GenBank, EMBL, and DDBJ) (Benson et al., 2007).		
EMBL *http://www.ebi.ac.uk/embl* Annotated collection of all publicly available DNA sequences (data is shared among GenBank, EMBL, and DDBJ) (Kulikova et al., 2007).		
DDBJ *http://www.ddbj.nig.ac.jp* Annotated collection of all publicly available DNA sequences (data is shared among GenBank, EMBL, and DDBJ) (Okubo et al., 2006).		
GenBank Natural plasmids *http://www.ncbi.nlm.nih.gov/genomes/static/o.html* Subset of GenBank specific for naturally occurring plasmids. Can be further subdivided into archaeal, bacterial, eukaryotic, and viral specific plasmids.		
NCBI UniVec *http://www.ncbi.nlm.nih.gov/VecScreen/replist.html* Subset of GenBank specific for vectors. Contains sequences of plasmids that NCBI uses to verify that submitted sequences are free from vector contamination.		
VectorDB *http://seq.yeastgenome.org/vectordb* Contains information about many common vectors. Has not been updated since 1997.		
Plasmid Genome Database *http://www.genomics.ceh.ac.uk/plasmiddb/* Database built from plasmid information found in GenBank.		
Repositories		
Addgene *http://www.addgene.org* Non-Profit organization that is dedicated to archiving and distributing plasmids from published papers (Fan et al., 2005).	X	X
PlasmID *http://plasmid.hms.harvard.edu* Centralized repository for plasmids and gene collections. Primarily eukaryotic plasmids (Zuo et al., 2007).	X	X

Specialized		
BacMap *http://wishart.biology.ualberta.ca/BacMap* An interactive atlas for exploring bacterial genomes that also contains naturally occurring bacterial plasmid sequences (Stothard et al., 2005).	X	
Cloning Vector Collection *http://www.shigen.nig.ac.jp/cvector/cvector.html* *Escherichia coli* cloning vectors.	X	X
Fission Yeast Vectors *http://www-rcf.usc.edu/~forsburg/vectors.html* A list of fission yeast vectors.	X	
I.M.A.G.E. Consortium Vectors *http://image.llnl.gov/image/html/vectors.shtml* Contains information about plasmids commonly used in library construction.	X	
FlyBase *http://flybase.org/static_pages/lists/vectors.html* Information on plasmids used in *Drosophila* research (Crosby et al., 2007).	X	
Sanger Plasmid Sequencing Project *http://www.sanger.ac.uk/Projects/Plasmids/* Sequence data from the Sanger Institute Pathogen Sequencing Unit which is sequencing a number of plasmids from various sources		
ACLAME (A CLAssification of genetic Mobile Elements) *http://aclame.ulb.ac.be/* Database dedicated to the collection and classification of mobile genetic elements from various sources including plasmids (Leplae et al., 2004).		
IS Finder *http://www-is.biotoul.fr/* List of insertion sequences isolated from bacteria.		
Islander: Database of Genomic Islands *http://kementari.bioinformatics.vt.edu/cgi-bin/islander.cgi* Database of genomic islands in bacterial genomes.		
Commercial		
ATCC *http://atcc.org/comon/products/vectors.cfm*	X	X
Clontech *http://clontech.com/support/vectors.asp*	X	X
Invitrogen *http://invitrogen.com*	X	X
Promega *http://www.promega.com/vectors/allvectors.htm*	X	X
Stratagene *http://stratagene.com/lit/vector.aspx*	X	X
*Plasmids can be ordered from these suppliers.		

they have compiled a list of suitable vectors. FlyBase (Crosby et al., 2007) maintains a list of vectors useful for research related to *Drosophila*. The I.M.A.G.E. consortium (Lennon et al., 1996), which represents the world's largest public cDNA collection, has information about vectors commonly used to construct cDNA libraries. The National Institute of Genetics in Japan, hosts the Cloning Vector Collection, a collection of *Escherichia coli* cloning vectors.

Lastly, there are commercial sources of plasmids. Commercial vendors typically provide sequence and map data for their plasmids online. If the sequence is not available, it can often be obtained by emailing the company, although proof of purchase may sometimes be required.

SEQUENCE FORMATS

An understanding of standard sequence formats is important, as many programs require or generate data in a specific format, and formats differ in terms of what types of information or annotation can be included with the sequence. Generally, sequence files should be text files containing only ASCII characters. ASCII, which stands for American Standard Code for Information Interchange, is a character encoding for the English alphabet (letter, numbers, punctuation marks, space and carriage return). ASCII format essentially means plain text with no formatting. Files in ASCII format have the advantage of being human readable and easy to move between different programs and different operating systems. When manually editing sequence files, it is best to use a simple text editor (*e.g.* NotePad on Windows systems or TextEdit on Macintosh systems). Binary formats such as the Microsoft Word .doc format are not directly readable by sequence analysis programs, and should be avoided. The sequence itself should use the standard code for nucleic acids, as established by the Nomenclature Committee of the International Union of Biochemistry (1985). The way the sequence letters and supporting information are organized in the text file differs among the standard formats. Here we will discuss some of the most widely used formats.

RAW (PLAIN TEXT)

In its most basic form, a sequence is usually stored as a series of letters representing nucleic acids. This format is called "raw" or "plain". There is no annotation or heading describing the sequence source or characteristics. Depending on the analysis program used, numbers or blank spaces appearing in a raw file may be ignored or signal an error. A file using the raw format can only contain one sequence, as there is no way to distinguish one sequence from the next.

An example of the raw sequence format:

```
GATAAGGGTTACGATAAGGATTATGATAAGGCTTAAGATGATAAGGATTAAGATAAGGTTATGGTGTGAC
CTATAGGGAACACATGGTGCTTATAATATTTAGTAGGGCTCATAATAATGAAACAACCCCGGGTGTACCC
ACATACACCCGGGGAAAAATCCAAAAGAAAGGCGGTGTAACTACCCGATGTCAAAAGAGATTATAACACA
AAATAGGGTCGAGCAAGGTTCACTTGATAGTAAATGTCATAAGTCGCCGCTGGCGTCTACAAGGCGAAAA
AAGAGATTCTACCAGAGGGCTGTAAGTGGTGTAACTGCTGGTAAGCATAGGAACGAATTTATAGCCTTCC
TGACGCTTACCTCTTCATTGGAATCTCCTGCCGATATCACACATAGCTGGGAAAAGCTTAAAAAACGAAT
CCGGAGAAGATATGGTAATTTTGAATACATTTGGGTACGGGAACGTACTCAAAGTGGTTTGGTGCATATG
CATATATTATTTAGAGGACCTTATATTCCCCAAGATTGGATAAGTAAGAATTGGGAAGAGATTCATAAGG
CAAAAATCGTTTATGTAGAAGCAGTATGGGACACAGGTAAAGCAATACGATATATGATGAAGTACTTATC
AAAGGAGATGGAAGGAAGATTCGGGTATAGTTGGAAGTGGATTTTCAAAGGGGCTGCGCAAGTGTGGAAG
TGGTTATGTAGAGCATTAAGGTACGAGATGAAAGAGATTATCAAAATATGGGAGAAGTTGATTATTGAGA
TACCGCCAGAGGGTATAAGGTGGGGAAGAATTTGGGAACTAGTGGGATATGGTTAAGAATGTGGTTAGTG
TGATTA
```

FASTA

The FASTA format supports the ability to name and describe sequences. A description line located before the sequence text provides this basic level of annotation. The line begins with a greater-than sign (">") followed by an alphanumeric string. The first word after the ">" is usually the accession number of the sequence or some other sequence identifier, while the rest of the line is often a description of the sequence. In the example below, the sequence identifier includes everything up to the first space, *i.e.* "gi|56687495|ref|NC_006528.1| ". A sequence file in FASTA format can contain several sequences, with each new record starting with a ">".

An example of the FASTA sequence format:

```
>gi|56687495|ref|NC_006528.1| Thermotoga petrophila plasmid pRKU1, complete sequence
GATAAGGGTTACGATAAGGATTATGATAAGGCTTAAGATGATAAGGATTAAGATAAGGTTATGGTGTGAC
CTATAGGGAACACATGGTGCTTATAATATTTAGTAGGGCTCATAATAATGAAACAACCCCGGGTGTACCC
ACATACACCCGGGGAAAAATCCAAAAGAAAGGCGGTGTAACTACCCGATGTCAAAAGAGATTATAACACA
AAATAGGGTCGAGCAAGGTTCACTTGATAGTAAATGTCATAAGTCGCCGCTGGCGTCTACAAGGCGAAAA
AAGAGATTCTACCAGAGGGCTGTAAGTGGTGTAACTGCTGGTAAGCATAGGAACGAATTTATAGCCTTCC
TGACGCTTACCTCTTCATTGGAATCTCCTGCCGATATCACACATAGCTGGGAAAAGCTTAAAAAACGAAT
CCGGAGAAGATATGGTAATTTTGAATACATTTGGGTACGGGAACGTACTCAAAGTGGTTTGGTGCATATG
CATATATTATTTAGAGGACCTTATATTCCCCAAGATTGGATAAGTAAGAATTGGGAAGAGATTCATAAGG
CAAAAATCGTTTATGTAGAAGCAGTATGGGACACAGGTAAAGCAATACGATATATGATGAAGTACTTATC
AAAGGAGATGGAAGGAAGATTCGGGTATAGTTGGAAGTGGATTTTCAAAGGGGCTGCGCAAGTGTGGAAG
TGGTTATGTAGAGCATTAAGGTACGAGATGAAAGAGATTATCAAAATATGGGAGAAGTTGATTATTGAGA
TACCGCCAGAGGGTATAAGGTGGGGAAGAATTTGGGAACTAGTGGGATATGGTTAAGAATGTGGTTAGTG
TGATTA
```

GENBANK

Files in raw or FASTA format, while containing the sequence itself, provide little if any other information about the sequence, such as which segments encode proteins. In order to store features related to a sequence, a more advance format is required.

Probably the most commonly encountered format that includes feature information is the GenBank format. A GenBank flat file is the basic unit of sequence information in the GenBank database (Benson et al., 2007). Each record contains three sections: the header, which contains information about the record; the feature table, which contains sequence annotations; and sequence itself. Given the widespread use of GenBank files it is worth understanding the various ways in which information is stored in this format. For a complete description of the GenBank database and file format, see the GenBank release notes (ftp://ftp.ncbi.nih.gov/genbank/gbrel.txt). The first line of the header section in a GenBank file is the locus line. This line contains the locus name (often the same as the accession number) the length of the sequence, the molecule type (*e.g.* DNA), the division code (*e.g.* "BCT" for bacterial), and the date the record was last modified. The next line is the definition line, which provides a brief description of the sequence. This is followed by the accession number. The accession number is the primary key associated with a sequence record and does not change, even if the sequence is updated. On the forth line is the sequence version, which does change with each modification to the nucleotide sequence. The version is simply the accession number followed by a period and a version number. The accession portion remains constant over time, but each time the sequence is modified the version increases by one (*e.g.* AJ810485.1 becomes AJ810485.2). The GI number, which is also given in the version section, serves as a second unique identifier for the sequence (each version is given a distinct GI). Keywords describe the sequence, but are not based on a controlled vocabulary and are generally only found in older records. The source line may contain the common or abbreviated form of the organism name, while the organism line contains the formal scientific name as well as its lineage (based on the NCBI taxonomy database). All DNA sequence records must contain source organism information, even plasmid vectors, whose source may simply be "synthetic construct", or the name of the plasmid. Following the organism section are several lines providing information about who submitted the sequence and what publications are associated with the sequence submission. The final part of the header is an optional comment section, which sometimes contains information about the history of a sequence when it has been modified.

The second major section of a GenBank record is the feature table, which contains information about sequence features (also called annotations). The only required feature is the "source" feature, which identifies the biological source for the specified span of the sequence. A record can have more than one source. For example, a derived plasmid sequence record may contain multiple source keys, which together span the entire sequence. One of the most commonly used features is "CDS" (coding sequence), which is used to describe nucleotide sequences that are translated into proteins (including the start and stop codon). Other features often found in plasmid sequences include "promoter", "mRNA", "gene", "polyA_signal", "rep_origin" (origin of replication), "operon", and "oriT" (origin of transfer). Every feature in the feature table includes a location range, to indicate which portion of the sequence contains the feature. For example, the range "186..826" indicates that the feature spans bases 186 to 826, inclusive. Usually the location is a single base, a continuous span of bases, or a joining of sequence spans. Locations referring to the reverse strand include the term "complement." In addition to the location text, feature entries can include qualifiers that provide additional information about the feature (the qualifiers begin with a "/" character). Some qualifiers are mandatory for certain feature types (*e.g.* "/organism" and "/mol_type" are required for the "source" feature). One common optional qualifier is the database cross-reference, "/db_xref". This qualifier points to related information in another database.

The last section of a GenBank record is the sequence itself. The sequence begins on the line immediately below ORIGIN and is reported in the 5' to 3' direction. The end of the sequence as well as the GenBank record is marked by two slashes ("//"). A GenBank file can contain one or more sequence records, each beginning with the LOCUS line and ending with "//".

An example of the GenBank sequence format:

```
LOCUS AJ810485 846 bp DNA circular BCT 01-MAR-2006
DEFINITION Thermotoga petrophila plasmid pRKU1.
ACCESSION AJ810485
VERSION AJ810485.1 GI:56561746
KEYWORDS complete genome; rep gene; Rep protein.
SOURCE Thermotoga petrophila RKU-1
ORGANISM Thermotoga petrophila RKU-1
Bacteria; Thermotogae; Thermotogales; Thermotogaceae; Thermotoga.
REFERENCE 1
AUTHORS Nesbo,C.L., Dlutek,M. and Doolittle,W.F.
TITLE Recombination in Thermotoga: implications for species concepts
and
biogeography
JOURNAL Genetics 172 (2), 759-769 (2006)
PUBMED 16322518
REFERENCE 2 (bases 1 to 846)
AUTHORS Nesbo,C.L.
TITLE Direct Submission
JOURNAL Submitted (12-AUG-2004) Nesbo C.L., Biochemistry and Molecular
Biology, Dalhousie University, 5850 College Street, Nova Scotia,
B3H 1X5, CANADA
FEATURES Location/Qualifiers
source 1..846
/organism="Thermotoga petrophila RKU-1"
/mol_type="genomic DNA"
/strain="RKU1"
/db_xref="taxon:390874"
/plasmid="pRKU1"
gene 188..826
```

```
/gene="rep"
CDS 188..826
/gene="rep"
/codon_start=1
/transl_table=11
/product="putative Rep protein"
/protein_id="CAH17919.1"
/db_xref="GI:56561747"
/db_xref="GOA:Q5NE25"
/db_xref="InterPro:IPR004191"
/db_xref="UniProtKB/TrEMBL:Q5NE25"
/translation="MSKEIITQNRVEQGSLDSKCHKSPLASTRRKKRFYQRAVSGVTA
GKHRNEFIAFLTLTSSLESPADITHSWEKLKKRIRRRYGNFEYIWVRERTQSGLVHMH
ILFRGPYIPQDWISKNWEEIHKAKIVYVEAVWDTGKAIRYMMKYLSKEMEGRFGYSWK
WIFKGAAQVWKWLCRALRYEMKEIIKIWEKLIIEIPPEGIRWGRIWELVGYG"
ORIGIN
  1 gataagggtt acgataagga ttatgataag gcttaagatg ataaggatta agataaggtt
 61 atggtgtgac ctatagggaa cacatggtgc ttataatatt tagtagggct cataataatg
121 aaacaacccc gggtgtaccc acatacaccc ggggaaaaat ccaaaagaaa ggcggtgtaa
181 ctacccgatg tcaaaagaga ttataacaca aaatagggtc gagcaaggtt cacttgatag
241 taaatgtcat aagtcgccgc tggcgtctac aaggcgaaaa aagagattct accagagggc
301 tgtaagtggt gtaactgctg gtaagcatag gaacgaattt atagccttcc tgacgcttac
361 ctcttcattg gaatctcctg ccgatatcac acatagctgg gaaaagctta aaaaacgaat
421 ccggagaaga tatggtaatt ttgaatacat ttgggtacgg gaacgtactc aaagtggttt
481 ggtgcatatg catatattat ttagaggacc ttatattccc caagattgga taagtaagaa
541 ttgggaagag attcataagg caaaaatcgt ttatgtagaa gcagtatggg acacaggtaa
601 agcaatacga tatatgatga agtacttatc aaaggagatg gaaggaagat tcgggtatag
661 ttggaagtgg attttcaaag gggctgcgca agtgtggaag tggttatgta gagcattaag
721 gtacgagatg aaagagatta tcaaaatatg ggagaagttg attattgaga taccgccaga
781 gggtataagg tggggaagaa tttgggaact agtgggatat ggttaagaat gtggttagtg
841 tgatta
//
```

EMBL

Data in the DDBJ, EMBL and GenBank databases is interchangeable, and a record of interest can be accessed in each database using the same accession number. DDBJ records are nearly identical to the GenBank format. EMBL records on the other hand, while having the same information, have a distinctive format (Kulikova et al., 2007). Each line begins with a two-character line code, which indicates the type of information present in the line (*e.g.* "FT" for feature table).

One difference between GenBank and EMBL records that should be pointed out is in how version numbers are presented. EMBL records potentially have two version numbers. The first is the sequence version (SV), located on the ID line. This number corresponds to the version number in GenBank records. The SV number only changes if the nucleotide sequence is altered. On the other hand, if a record is updated for any reason, through the addition of a new reference for example, there will be a DT (date) line giving the last date of modification along with an additional version number. This DT version number does not exist in GenBank records and should not be confused with the sequence version number.

XML

Extensible Markup Language (XML) is a method to describe structured information in a relatively human-readable form. Structured information contains both content and some indication of how the content should be interpreted. The example below shows a date in XML. The content, which in the example includes the year "2004", is contained within a pair of tags describing the content, in this case as "Date-std year" or a standard date year. Although somewhat more difficult to read (for humans), XML files can be robustly parsed by computer software. NCBI, EMBL (Wang et al., 2002) and DDBJ (Miyazaki et al., 2003) have made efforts to offer their sequence data in XML format, and recently the International Nucleotide Sequence Database Collaboration (www.insdc.org) has defined INSDseq XML as the official XML format for the components of a GenBank/EMBL/DDBJ sequence record. XML files can be very large due to the verbosity of XML syntax. In the example below, it takes five lines to describe a date.

Example of a portion of a sequence record describing a date in XML format:

```
<Date-std>
<Date-std_year>2004</Date-std_year>
<Date-std_month>8</Date-std_month>
<Date-std_day>12</Date-std_day>
</Date-std>
```

GFF

GFF (General Feature Format) is a simple format developed for describing DNA sequence features. Although it is possible to include the underlying DNA sequence in a GFF file, it is not a common practice. Instead, a sequence identifier is often included in the file, so that the sequence can be retrieved from elsewhere (GenBank for example). Alternatively, the sequence is included in an accompanying FASTA file. GFF files are commonly accepted and generated by bioinformatics programs, and they can be viewed and manipulated using spreadsheet programs like Excel. Each line in a GFF file corresponds to a sequence feature. The characteristics of each feature are described using a series of tab-delimited fields. Additional information pertaining to the entire sequence can be included as comments (comment lines are preceded by "#"). The full specification of the GFF format is available online (http://www.sanger.ac.uk/Software/formats/GFF/GFF_Spec.shtml).

CONVERTING SEQUENCE FORMATS WITH READSEQ

Some analysis programs may require that sequences be supplied in a particular format. Fortunately, there are several programs available for converting sequence files from one standard format into another. One of the most comprehensive is Readseq (see http://iubio.bio.indiana.edu/cgi-bin/readseq.cgi). Readseq currently supports over 20 sequence formats, including the ones discussed above, albeit with limited support for XML. Readseq can be accessed online, or it can be downloaded and run locally. It should be noted that converting sequence files sometimes results in the loss of information. For example, a FASTA record generated from a GenBank record will lack the feature information described in the original file.

BASIC SEQUENCE ANALYSIS AND MANIPULATION

This section describes useful tools for performing some of the more routine plasmid analysis and manipulation operations (Table 2), usually in preparation for standard molecular biology procedures, such as restriction digestion and PCR amplification. Some of the tools are designed specifically for plasmids while others are for DNA sequences in general. Both stand-alone and web-based tools are discussed. It should be noted that many web-based tools are actually online interfaces to command-line programs, which can be downloaded and installed locally. The decision of whether to use the command line or online version depends on the relative importance of the issues discussed at the beginning of this chapter.

RESTRICTION ANALYSIS

The primary database of restriction enzyme information is REBASE (Roberts et al., 2007). This comprehensive resource contains detailed information about recognition and cleavage sites, commercial availability, methylation sensitivity, and related publications. Software tools that perform restriction analysis usually obtain their restriction enzyme data from REBASE.

Two useful online tools for restriction analysis of plasmids are NEBcutter from New England BioLabs (Vincze et al., 2003) and REviewer from Fermentas. Both programs can display a sequence as text or as a linear or circular diagram showing ORFs and restriction sites. Useful restriction sites can be selected, based on how many times they cut a particular sequence. Both programs can also perform virtual digests, as well as display a hypothetical gel of what the digest products would look like after electrophoresis. NEBcutter can identify sites for excising a particular ORF. In addition, a silent mutagenesis function can be used to determine which restriction sites can be introduced into an ORF without changing the resulting protein sequence. REviewer includes tools for designing and analyzing PCR primers, and for performing common biochemical calculations. Both programs provide mechanisms for exporting or saving data for later analysis.

Restriction analysis functionality is built in to nearly all multi-purpose stand-alone sequence analysis software packages. An advantage of using a stand-alone package is that restriction fragments of interest can be visualized or analyzed using other functions without the need for copying and pasting. Fragments can also be ligated to other sequences *in silico*, using the plasmid-building tools that are becoming more commonplace in these packages.

PRIMER DESIGN

Molecular biologists design oligonucleotide primers on a regular basis for DNA sequencing and PCR amplification. Although it is possible to design primers by hand, computer software greatly simplifies the process. Programs are able to take into account many factors when choosing the most suitable primers, including product size and melting temperature, the potential for primer dimer formation, the difference in melting temperature between primers, the 3' stability of annealed primers, and primer secondary structure. Some programs can also interpret sequence feature information, so that primers can be designed to span features of interest.

Primer3 is a popular primer design program that has been in use for more then a decade (Rozen and Skaletsky, 2000). It has recently been updated, and now includes an improved formula for calculating melting temperature (Koressaar and Remm, 2007). Primer3 is available as a command-line program, or as a web server. Because of the many options available, the web interface can be quite daunting. A more intuitive interface to Primer3 is provided by Primer3Plus (Untergasser et al., 2007). Primer3Plus can also be used to automate design tasks, such as selecting primers for step-wise sequencing. As with restriction analysis, most stand-alone analysis packages support primer design. However, in many cases the included primer design tools lack the flexibility of Primer3.

PLASMID GENERATION

The process of plasmid construction typically involves ligating a digested plasmid with a DNA fragment excised or PCR amplified from another source. There are several tools that can aid in designing construction strategies, including Serial Cloner, PlasmaDNA (Angers-Loustau et al., 2007), and pDRAW32. Each allows virtual DNA fragments to be generated and joined. The user first supplies one or more plasmid or insert sequences. From these sequences, fragments are created by selecting restriction enzymes, or by providing PCR primers that will amplify a segment of interest. The predicted fragments can then be joined together based on complementary DNA overhangs or through blunt end ligations. The programs save time and effort by automatically selecting compatible enzymes, and by preventing the errors that sometimes occur when sequence segments are manually combined.

More advanced solutions for generating cloning strategies include BioInfoMan and Vector NTI. BioInfoMan is a web-based software suite containing many molecular cloning and sequence analysis tools. It includes a series of cloning guides that cover several methods including sub-cloning with restriction enzymes, PCR cloning, Gateway cloning, and gene synthesis. Each guide consists of a series of steps specific for that procedure. For example, when cloning by PCR, the first step is to select the appropriate vector and insert DNA. Next, restriction enzyme sites on the vector are chosen, while for the insert, a particular region of interest is selected.

A

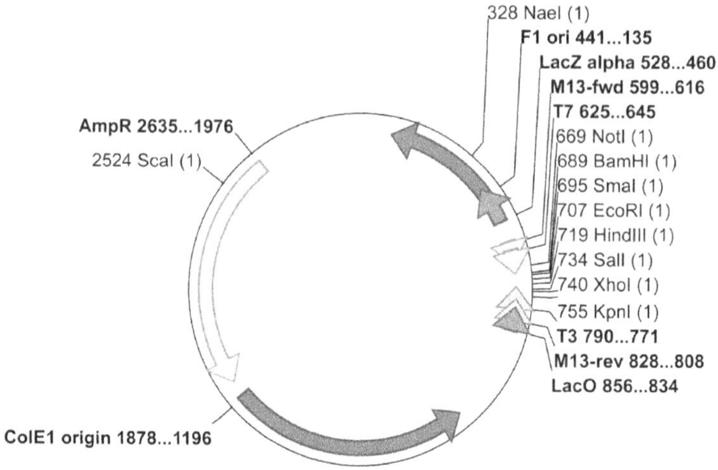
328 NaeI (1)
F1 ori 441...135
LacZ alpha 528...460
M13-fwd 599...616
T7 625...645
669 NotI (1)
689 BamHI (1)
695 SmaI (1)
707 EcoRI (1)
719 HindIII (1)
734 SalI (1)
740 XhoI (1)
755 KpnI (1)
T3 790...771
M13-rev 828...808
LacO 856...834
AmpR 2635...1976
2524 ScaI (1)
ColE1 origin 1878...1196

B

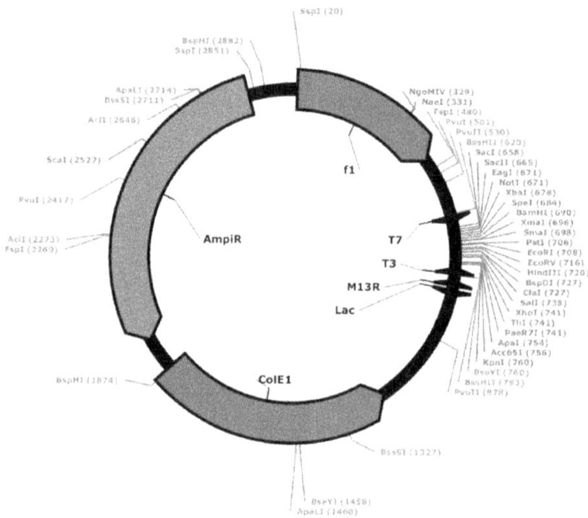
f1
AmpiR
T7
T3
M13R
Lac
ColE1

C

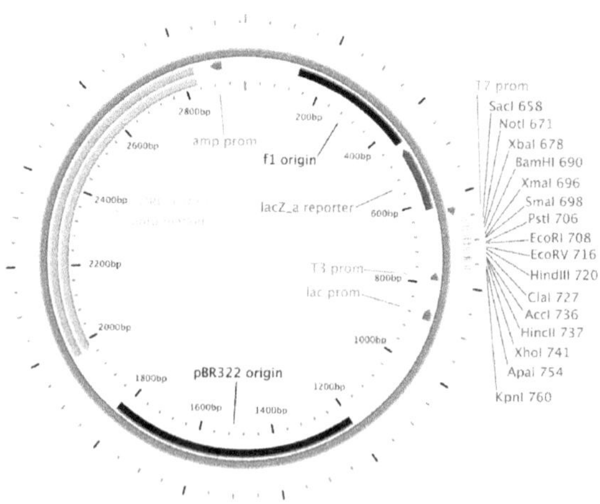
T7 prom
SacI 658
NotI 671
XbaI 678
BamHI 690
XmaI 696
SmaI 698
PstI 706
EcoRI 708
EcoRV 716
HindIII 720
ClaI 727
AccI 736
HincII 737
XhoI 741
ApaI 754
KpnI 760
amp prom
f1 origin
lacZ_a reporter
T3 prom
lac prom
pBR322 origin

Any ORFs or coding region annotations on the insert are automatically detected and presented in a list as possible options. The server then automatically generates PCR primers that will amplify the selected region on the insert DNA. The primers are also automatically appended with the appropriate sequence containing the restriction sites required for cloning. The resulting construct, primers, and the cloning strategy itself can be saved to the BioInfoMan database for future use. The construct-building tool included with Vector NTI provides two modes for creating new molecules: construction and design. In construction mode, the user makes decisions regarding which methods and reagents are used. In design mode, the user specifies the donor and recipient fragments and Vector NTI determines the best method to create the construct. In addition to building a virtual construct, the program returns a detailed protocol describing how to create the molecule in the lab. Many aspects of both modes can be tweaked by altering the many available options.

PLASMID DRAWING

Plasmid maps are routinely used for visualizing plasmid features, and for planning cloning strategies, and they often appear in lab notebooks, publications, and presentations (Fig. 1). Tools for generating plasmid maps generally support several output formats. Common formats include JPG, PNG and GIF. While acceptable for viewing on the computer, images of these types are not always suitable for publication purposes, because they are raster images (a rectangular grid of pixels) of a fixed resolution, which can appear to lose quality when resized. In contrast, vector graphics formats, in which elements are described by geometrical primitives such as points, lines, curves, and polygons, allow images to be scaled without any degradation. An increasingly popular vector graphics format is SVG (Scalable Vector Graphics). An added benefit of using the SVG format for plasmid maps is that vector graphics-capable software can be used to easily modify or reposition individual features and labels on the map.

Both NEBcutter and REviewer (discussed earlier) support plasmid map creation. NEBcutter allows maps to be saved in PDF, GIF, or EPS format, while REviewer can generate SVG maps. Another more specialized resource for generating plasmid maps is PlasMapper (Dong et al., 2004). PlasMapper is designed to automatically create high quality annotated plasmid maps using only a DNA sequence as input. Features are identified through comparisons with a database of common plasmid components (replication origins, promoters, selectable markers, reporter genes, etc.). PlasMapper generates fully labeled maps in PNG, JPG, or SVG format. Most stand-alone plasmid programs are also able to generate plasmid maps. Often these

Fig. 1. Examples of Plasmid Maps and Annotations. Plasmid maps and feature annotations as generated by (A) ApE, (B) PlasmaDNA, and (C) PlasMapper. The unannotated DNA sequence of pBlueScript II KS+ (Accession X52327.1) was supplied to each program.

programs have an intuitive graphical interface that allows selection and editing of map features directly on the map itself.

MULTI-PURPOSE TOOLS

Multi-purpose plasmid analysis tools are programs or web-based packages that perform most, if not all, of the functions described above. There are several free tools available, including pDRAW32, ApE, PlasmaDNA, and Serial Cloner. CLC Free Workbench, which is a free and stripped down version of a more powerful commercial package, includes several useful functions. A very popular full-featured program is Vector NTI from Invitrogen. Invitrogen has made this program available at no cost to academic and government researchers. While having all the tools necessary for plasmid manipulation and analysis, it can take time for first-time users to become familiar with the unique organizational structure of Vector NTI (Lu and Moriyama, 2004). In addition to supporting most analysis methods, Vector NTI allows large collections of molecules to be managed in a local database. This capability makes Vector NTI an attractive solution for larger research groups.

An alternative to locally run software packages is the free web-based BioInfoMan, a suite of molecular cloning and sequence analysis tools integrated with a data management system. Many common constructs and oligonucleotides are already present in the online database, and users can easily create their own database of sequences or newly created constructs. It is also possible to create a user group, so that data can be shared among lab personnel or colleagues. There are useful cloning guides for creating new constructs and there is an automated annotation tool that will predict common features in a new sequence. The main drawback of this system and others like it is that data is stored on a remote server, which may not always be accessible. Another issue is that web-based sites like BioInfoMan are somewhat slower than equivalent locally run programs.

Although not specifically designed for plasmid analysis, two web-accessible tool collections also worth mentioning are EMBOSS and the Sequence Manipulation Suite. EMBOSS (European Molecular Biology Open Software Suite) is a package of open source sequence analysis tools (Rice et al., 2000). It consists of around 150 programs organized into 46 groups based on their function. Some of the areas covered are sequence alignment, nucleotide pattern searches, codon usage, and gene identification. Each EMBOSS program includes detailed documentation describing which input and output formats are accepted, the available options, and what underlying calculations or algorithms are used. The Sequence Manipulation Suite is suitable for quickly performing simple tasks, such as DNA translation and restriction analysis (Stothard, 2000). It contains over 50 programs organized into four main groups: format conversion, sequence analysis, sequence figures, and random sequences. Several of the programs include a setting to specify whether the input sequence is circular or linear, making them applicable to plasmid sequences.

SOFTWARE FOR CHARACTERIZING NOVEL SEQUENCES

The previous section discussed bioinformatics procedures and tools that are routinely applied to existing plasmid sequences. Here we describe approaches and software for the characterization of newly sequenced plasmids, such as those that occur naturally in bacteria. Although the programs discussed in the previous section are applicable to novel sequences, they are usually used after the gene prediction and database searching steps outlined below. Due to continuous improvements in reference databases and annotation tools, this section is also relevant to sequences that were annotated in the past--re-annotation may reveal information that was missed previously. Generally, the procedure for the analysis of plasmid sequences includes gene prediction, database similarity searches and alignment, gene annotation, manual editing, and data visualization. Programs specifically designed for each of these steps are available (Table 3), as are pipelines that handle the entire analysis process (Stothard and Wishart, 2006). Free annotation services are also offered that allow users to submit a plasmid or genome sequence and obtain in-depth annotations.

GENE PREDICTION

Understanding the function of a novel plasmid sequence often begins with gene prediction programs, which scan DNA sequences for regions that likely encode proteins or functional RNA products (depending on the program). Although the position information itself is not very useful, it allows the predicted gene products to be generated and characterized through sequence similarity searches. The accuracy of gene prediction is generally better for prokaryotic sequences than for eukaryotic sequences. This is largely due to the simple gene structure, high density of coding regions, and well-known signal sequences employed by bacteria, whereas in eukaryotes, the promoters and other regulatory signals are not as well-understood, and the protein coding regions can be split over several exons. Two popular prokaryotic gene prediction programs are Glimmer (Delcher et al., 1999; Salzberg et al., 1998) and GeneMark (Lukashin and Borodovsky, 1998). Both use complex probabilistic models to distinguish coding regions from non-coding DNA. Glimmer returns the coordinates of predicted genes, while GeneMark can generate the coordinates and provide the predicted translations and coding sequences. Both programs can be accessed online through a web interface, or downloaded and run locally. Specialized versions of Glimmer and GeneMark are also available for eukaryotic sequences. Another more recently developed prokaryotic gene prediction program is MED (Zhu et al., 2007). It uses a novel algorithm that performs well on GC-rich genomes and archaeal genomes. The MED program is currently available as a command-line tool for Windows or in source code form. Transfer RNA (tRNA) genes lack the protein-coding signature required for detection by the standard gene prediction programs. Instead, dedicated tRNA gene identification programs are available. The most widely used is tRNAscan-SE, which is capable of detecting

Table 2. Sequence analysis tools.

Analysis Tool Name	Access					Features								
	Web-Based	Windows GUI*	Mac OS GUI*	Linux GUI*	Command Line	Restriction Analysis	Virtual Gels	Plasmid Maps	Feature Editing	Primer Design	Construct Design	Pairwise Alignment	Multiple Alignment	Auto-Annotate
REBASE *http://rebase.neb.com* Restriction enzyme database. Contains a comprehensive collection of information about restriction enzymes and related proteins.	X					X								
NEBcutter *http://tools.neb.com/NEBcutter* Finds ORFs and restriction sites in DNA sequences and draws plasmids maps. Maps can be saved in GIF, EPS, or PDF format. Can locate sites to excise ORFs or introduce sites through silent mutagenesis.	X					X	X	X						
REviewer *http://www.fermentas.com/reviewer* Finds ORFs and restriction sites in DNA sequences and draws plasmids maps. Maps can be saved in SVG format. Additional tools are available for designing primers and performing common biochemical calculations.	X					X	X	X	X	X				
PlasMapper *http://wishart.biology.ualberta.ca/PlasMapper* Automatically generates and annotates plasmid maps using only the plasmid DNA sequence as input. Maps can be saved in PNG, JPG, or SVG format.	X					X		X	X					X
EnzymeX *http://mekentosj.com/enzymex/* Mac OS program for restriction analysis. Includes buffer information for selecting compatible enzymes.			X			X	X							
Primer3 *http://primer3.sourceforge.net* Widely used program for the design of PCR and sequencing primers.	X				X					X				

ORFprimer *http://www.proteinstrukturfabrik.de/ORFprimer* Java application for automatic primer design.	X	X	X	X	X					X				
FastPCR *http://www.biocenter.helsinki.fi/bi/Programs/fastpcr.htm* Automatic and manual PCR primer or probe design.		X								X				
EMBOSS *http://emboss.sourceforge.net* The European Molecular Biology Open Software Suite. Contains over 150 sequence analysis tools, many of which are applicable to plasmid sequences.	X				X	X				X		X	X	
Sequence Manipulation Suite *http://bioinformatics.org/sms2* Collection of web-based programs for generating, formatting, and analyzing DNA and protein sequences.	X					X						X		
pDRAW32 *http://www.acaclone.com* Windows program for graphical display and editing of plasmid sequences. Files are saved in a custom text format.		X				X	X	X	X	X	X			
ApE (A Plasmid Editor) http://www.biology.utah.edu/jorgensen/wayned/ape Program for graphical display and editing of plasmid sequences. Also performs BLAST searches. Files can be saved in GenBank format.		X	X	X		X	X	X	X	X	X			X
CLC Free Workbench 4 *http://www.clcbio.com/* Free stripped down version of the CLC DNA Workbench.		X	X	X		X		X				X	X	
Vector NTI *http://www.invitrogen.com/VNTIGateway* Feature-rich sequence analysis application free for academic and government use.		X	X			X	X	X	X	X	X	X	X	
Serial Cloner *http://serialbasics.free.fr* Construct-building, sequence analysis, and visualization tool.		X	X			X		X			X	X		
PlasmaDNA *http://research.med.helsinki.fi/plasmadna* Primarily a plasmid-building tool.		X	X			X	X	X	X		X			X
BioInfoMan *http://www.bioinfoman.com/bm/* Web-based software suite with many cloning and sequence analysis tools integrated with data management.	X					X		X	X	X	X	X	X	X
*GUI-Graphic User Interface														

Table 3. Software for characterizing novel sequences.

Analysis Tool Name	Access					References
	Web-Based	Windows GUI*	Mac OS GUI*	Linux GUI*	Command Line	
Gene Prediction						
Glimmer *http://cbcb.umd.edu/software/glimmer* Microbial gene prediction.	X				X	Salzberg et al., 1998 Delcher et al., 2007
GeneMark *http://exon.gatech.edu/GeneMark* Family of programs for general gene prediction.	X				X	Lukashin and Borodovsky, 1998
MED *http://ctb.pku.edu.cn/main/SheGroup/Software/MED2.htm* Good for gene prediction in GC-rich or archaeal genomes.					X	Zhu et al., 2007
tRNAscan-SE *http://lowelab.ucsc.edu/tRNAscan-SE* Detects tRNA genes in genomic sequences.	X				X	Lowe and Eddy, 1997 Schattner et al., 2005
GeneHacker *http://www-btls.jst.go.jp/cgi-bin/Tools/GeneHacker/index.cgi* Gene prediction in microbial genomes.	X					Yada and Hirosawa, 1996
Sequence Similarity Searches						
BLAST *http://www.ncbi.nlm.nih.gov/blast* Software for sequence database searching.	X				X	Altschul et al., 1990 Altschul et al., 1997
FASTA *http://www.ebi.ac.uk/fasta* Software for sequence database searching.	X				X	Pearson and Lipman, 1988 Pearson, 1990
SAM *http://www.soe.ucsc.edu/research/compbio/sam.html* Software for sensitive sequence database searching.	X				X	Hughey and Krogh, 1996
HMMER *http://hmmer.janelia.org* Software for sensitive sequence database searching.	X				X	Eddy, 1998
Sequence Alignment						
Clustal *http://www.ebi.ac.uk/clustalw* *http://bips.u-strasbg.fr/fr/Documentation/ClustalX/* Popular multiple alignment programs.	X	X	X	X	X	Chenna et al., 2003
T-Coffee *http://www.ebi.ac.uk/t-coffee* Multiple alignment tool suitable for distantly related proteins.	X				X	Notredame et al., 2000
MUSCLE *http://www.drive5.com/muscle* Multiple alignment software.	X				X	Edgar, 2004
Jalview *http://www.jalview.org* Java-based multiple alignment editor.	X	X	X	X		Clamp et al., 2004

CINEMA *http://www.bioinf.man.ac.uk/dbbrowser/CINEMA2.1* Java-based multiple alignment editor.	X	X	X	X		Parry-Smith et al., 1998
GeneDoc *http://www.nrbsc.org/gfx/genedoc* Multiple alignment editor for Windows.		X				Nicholas et al., 1997
Annotation Pipelines						
BASys *http://wishart.biology.ualberta.ca/basys* Performs automated annotation of bacterial sequences.	X					Van Domselaar et al., 2005
TIGR Annotation Engine *http://www.tigr.org/AnnotationEngine* Annotation service for bacterial sequences.	X					-
PUMA2 *http://compbio.mcs.anl.gov/puma2* Annotation system that provides metabolic pathway information.	X					Maltsev et al., 2006
MAGPIE *http://magpie.ucalgary.ca/* Annotation system that can be run locally.					X	Gaasterland and Sensen, 1996
GenDB *http://www.cebitec.uni-bielefeld.de/groups/brf/software/gendb_info/* Annotation system that can be run locally.				X	X	Meyer et al., 2003
SABIA *http://www.sabia.lncc.br/* Annotation system that can be run locally.					X	Almeida et al., 2004
Sequence Visualization						
Artemis *http://www.sanger.ac.uk/Software/Artemis* DNA sequence visualization and annotation program.		X	X	X		Berriman and Rutherford, 2003
ACT (Artemis Comparison Tool) *http://www.sanger.ac.uk/Software/ACT* DNA sequence comparison viewer.		X	X	X		Carver et al., 2005
CGView *http://wishart.biology.ualberta.ca/cgview* DNA sequence visualization program.					X	Stothard and Wishart, 2005
Integrated Microbial Genomes *http://img.jgi.doe.gov/cgi-bin/pub/main.cgi* Tools for viewing and analyzing genomes.	X					Markowitz et al., 2006
Other Tools and Sites						
Gene Ontology *http://www.geneontology.org* Controlled vocabulary to describe gene and gene products.	X					Ashburner et al., 2000
Sequence Ontology http://www.sequenceontology.org/ Controlled vocabulary to describe biological sequences.	X					Eilbeck et al., 2005
Codon Usage Database *http://www.kazusa.or.jp/codon/* Codon usage database.	X					Nakamura et al., 2000
ORF Finder *http://www.ncbi.nlm.nih.gov/gorf/gorf.html* Tool for finding open reading frames.	X					Wheeler et al., 2003
*GUI-Graphic User Interface						

99-100% of the tRNA genes in a DNA sequence (Lowe and Eddy, 1997). The tRNAscan-SE program is available with a command-line interface or it can be accessed as a web server (Schattner et al., 2005).

In some cases, particularly when short sequences are involved, gene prediction programs may not be needed. Protein-coding segments can be identified as open reading frames, using one of the many tools listed in Tables 2 and 3. Alternatively, DNA sequences can be compared to a protein sequence database using a translated BLAST search. The results will indicate which regions, if any, are similar to known proteins when translated. This strategy is particularly effective when a closely related plasmid has previously been characterized. Genes that encode functional RNAs like tRNA can easily be detected through sequence similarity searches at the nucleotide level (blastn for example).

SEQUENCE SIMILARITY SEARCHES

Once a predicted gene sequence is available, the next step is to learn more about its potential function through sequence comparisons. This process involves searching a database for similar sequences of known function. Searches can be performed using either the gene sequence or the corresponding protein sequence. DNA searches typically yield fewer significant hits, and thus for protein-coding genes, protein (or translated DNA) searches are preferred (Pearson, 1996).

Two widely used programs for sequence similarity searches are FASTA (Pearson, 1990; Pearson and Lipman, 1988) and BLAST (Altschul et al., 1990; Altschul et al., 1997). Both increase search speed by using heuristic algorithms. These algorithms are not guaranteed to find the optimal match, but can be used to quickly search very large databases. Both BLAST and FASTA have their advantages, but here we focus on BLAST because of its speed, popularity, and flexibility.

There are several types of BLAST, which differ in terms of which sequence types are accommodated. The blastp program searches a protein database using a protein query, while blastn searches a nucleotide database using a nucleotide query. Nucleotide sequences can be compared to a protein database using blastx, while protein sequences can be compared to a DNA database using tblastn. In both cases, the DNA sequences are automatically translated to proteins (in all six reading frames) prior to the search. The most computationally intensive BLAST is tblastx, which is used to compare a translated DNA query to a translated DNA database. There are two specialized blastp algorithms available for finding more distant sequence relatives: psi-blast and phi-blast (Altschul et al., 1997). Both use a standard blastp search to identify related protein sequences. These sequence hits are then used to build a position-specific scoring matrix, or profile, for use as the query in subsequent searches. As further searches are performed, any hits over a certain quality threshold are used to update the profile. The procedure continues until no new sequences are added to the profile. The phi-blast algorithm uses the same series of steps as psi-blast, but restricts the search to sequences containing a user-specified pattern. The pattern can be used, for example, to restrict the search to a formally defined family.

BLAST can be accessed online (http://www.ncbi.nlm.nih.gov/blast/), or command-line versions can be downloaded for several platforms. The web interface is the most convenient option for most people, since there is no need to download and format a large database. Furthermore, the web version allows searches to be restricted to subsets of a database of interest. This database filtering uses NCBI's Entrez text-based search system, which allows sequences to be identified by numerous criteria, such as source organism, sequence length, gene name, publication date, and sequence type. Once a particular BLAST search and parameter set is chosen, it is possible to save the settings as a strategy. Multiple strategies can be saved for later use, simplifying the characterization of several sequences.

Pairwise search methods such as FASTA and BLAST perform well for most purposes, but there are more advanced approaches for finding distantly related proteins. As mentioned previously, psi-blast and phi-blast use a position-specific scoring matrix that describes the degree of conservation at various positions in an alignment of several sequences (Park et al., 1998). A related but more sensitive strategy is to use profile hidden Markov models (HMMs), which provide a richer description of a protein family. Two freely available HHM programs are SAM (Hughey and Krogh, 1996) and HMMER (Eddy, 1998). These programs can create a profile HMM representing a domain or sequence family of interest and then use the profile to search a sequence database to identify distant homologs.

SEQUENCE ALIGNMENT

The comparison of sequences by alignment can provide a wealth of information regarding sequence function and evolutionary history. While similarity searches can identify related sequences, an alignment may reveal more detail regarding which residues or motifs are conserved. Once constructed, alignments can also be used to build phylogenetic trees, which can be used to better understand the relationships between sequences or species (Page and Holmes, 1998). A sequence of interest can be compared to one other sequence (pairwise alignment) or several sequences (multiple alignment). Both types of alignments are discussed below.

There are two types of pairwise sequence alignment: global and local. Global alignments are used to compare sequences over their entire length, while local alignments are used to identify conserved sequence blocks or motifs. The standard algorithms used to create global and local alignments are very similar. They both use a technique called dynamic programming to determine the optimal alignment for a given scoring matrix and gap penalty (Needleman and Wunsch, 1970; Smith and Waterman, 1981). These algorithms and others are implemented in the EMBOSS package. Multiple sequence alignment tools such as ClustalW can also be used to generate global pairwise alignments.

Multiple sequence alignments show the relationship among three or more sequences simultaneously. The alignment of several related proteins can be used to define protein family signatures, establish sequence similarity, and determine ortholog / paralog relationships. Unlike the situation for pairwise alignments, there is no practical algorithm available for determining the optimal multiple alignment. Hence, heuristic approaches are used instead. The most common of these is the progressive method (Feng and Doolittle, 1987; Taylor, 1988), which involves performing a series of pairwise alignments to align larger and larger groups of sequences. The Clustal series of programs are probably the most popular progressive alignment tools (Chenna et al., 2003; Higgins and Sharp, 1988). They are freely available, have an easy-to-use interface, and run on most major operating systems. ClustalW (Higgins et al., 1996; Thompson et al., 1994) is a command-line program, but several web versions can be found (for example http://www.ebi.ac.uk/clustalw/). ClustalX provides the same features as ClustalW, but with a graphical user interface (Thompson et al., 1997). T-Coffee is another, more recently developed progressive alignment program. Although slower than ClustalW, T-Coffee generally produces better alignments for more distantly related proteins (Notredame et al., 2000). One problem with progressive alignment techniques is that decisions made early in the alignment process get preserved and propagated to the final alignment, regardless of whether subsequently added sequences suggest a better alternative alignment. Iterative methods address this problem by repeatedly realigning the initial sequences as new sequences are added to the multiple alignment (Gotoh, 1996). An example of a program using this technique is MUSCLE (Edgar, 2004). Like ClustalW, T-Coffee and MUSCLE are available as command-line tools, or they can be accessed online.

Once a sequence alignment has been generated, alignment-editing software can be used to rearrange sequences, or to add and remove gaps. Gap placement may need to be adjusted to properly visualize conserved motifs or domains, particularly in the case of multiple alignments. Some tools allow color to be added, to highlight alignment columns containing a certain proportion of identical or physiochemically similar residues. Full-featured editors such as Jalview (Clamp et al., 2004), CINEMA (Parry-Smith et al., 1998) and GeneDoc (Nicholas et al., 1997) can generate high quality images of alignments for use in publications and presentations. Jalview and CINEMA are available as online Java applets or stand alone programs that run on all major operating systems, while GeneDoc is available for Windows systems.

ANNOTATION PIPELINES

Using separate programs to predict genes, perform similarity searches, and prepare alignments can be very time consuming for anything but the smallest plasmids. Furthermore, depending on the programs used, the formats produced by one program may not be easily supplied to the next. For example, some gene prediction programs provide gene coordinates but not the gene sequences or protein translations. Similarity search tools like BLAST require a sequence as input, so users would have to prepare these manually, or find or write a program to do the conversion. Fortunately, there are software pipelines available that serve to connect analysis tools like Glimmer and BLAST, so that a single genome or plasmid sequence can be submitted at one end, and a set of fully annotated genes retrieved at the other. These pipelines are constantly evolving, and can perform many detailed analyses not covered in this chapter, such as metabolic pathway reconstruction and operon identification.

Naturally occurring bacterial and archaeal plasmids can be annotated using many of the same methods applied to prokaryotic genomes. Several prokaryotic genome annotation pipelines are available (Stothard and Wishart, 2006). The most convenient are those that can be accessed online or as a service. TIGR (The Institute for Genomic Research) provides a free annotation service (the Annotation Engine Service) for prokaryotic sequences that returns an extensive collection of results including coordinates of protein and functional RNA genes, GO (Gene Ontology) terms for proteins (Ashburner et al., 2000), paralogous families, gene names and symbols, EC (Enzyme Commission) numbers (IUBMB, 1992), BLAST search results, HMMER search results, transmembrane domain predictions from TMHMM (Krogh et al., 2001), and signal peptide predictions from SignalP (Bendtsen et al., 2004). These automatically identified results can be manually inspected and edited using Manatee, a web-based annotation editor (http://manatee.sourceforge.net/). Another powerful online annotation system is PUMA2 (Maltsev et al., 2006). Available for both prokaryotic and eukaryotic sequences, this pipeline can be used to assign gene functions and reconstruct metabolic pathways. As with the TIGR system, users can review and edit the annotations using a web-based tool. BASys is another web-based pipeline that can annotate a raw prokaryotic plasmid or genome sequence (Van Domselaar et al., 2005). BASys can perform gene prediction using Glimmer, or users can supply their own gene coordinates in a variety of formats. The gene sequences are then submitted to a variety of analysis programs, and the results returned to the user. A major drawback of BASys is its lack of integration with a manual curation tool.

Some disadvantages of using web-based annotation services are that it may take several weeks to receive the results and there is little control over what programs and databases are used in the annotation process. One solution to these issues is to install an annotation system locally. Examples of stand-alone annotation pipelines include

A
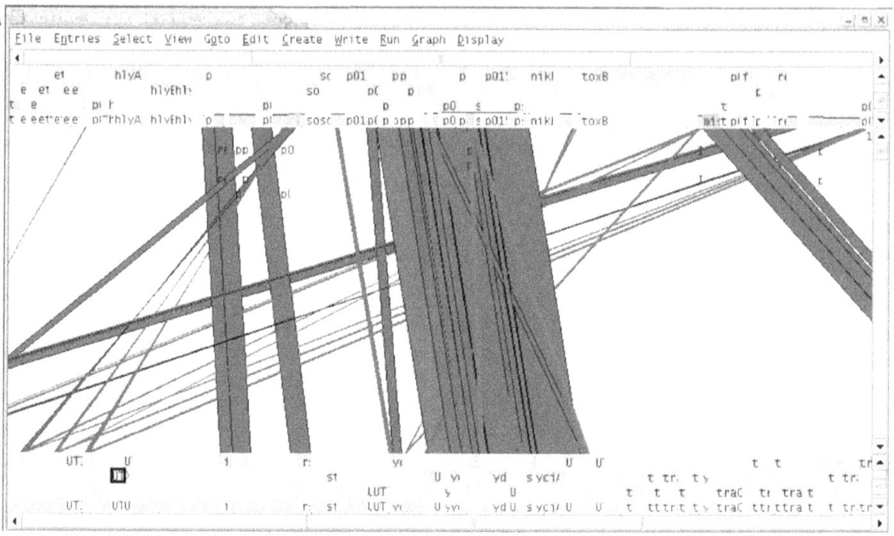

Fig. 2. Plasmid Sequence Comparisons. (A) Sequence comparison using ACT. The top sequence is plasmid p0157 from Escherichia coli 0157:H7 (accession NC_002128), while the bottom sequence is plasmid pUTI89 from E. coli UTI89 (accession NC_007941). The sequences were compared using blastn (E value = 0.00001; minimum score = 100). The lines show regions of similarity between the sequences, as identified by the blastn search. Genes are shown on the sequences as rectangles.

MAGPIE (Gaasterland and Sensen, 1996), GenDB (Meyer et al., 2003), and SABIA (Almeida et al., 2004). Although these systems offer more control and flexibility, the installation process can be complicated, as several bioinformatics programs and sequence databases usually have to be obtained and installed separately.

Some additional considerations when choosing a tool for annotation, independent of whether the tool is web-based or stand-alone, include which databases are used during the similarity searches, and whether the results are going to be submitted to GenBank. In regards to the first issue, some sequence databases contain fewer sequences, but are generally of higher quality, because they are manually curated. Examples of curated databases include Swiss-Prot (Boeckmann et al., 2003) and HAMAP (Gattiker et al., 2003). In contrast, NCBI's nr database contains coding sequence translations from GenBank records, many of which have been assigned predicted functions based on similarity searches. Although the larger nr database will likely provide more annotation information for a new genome sequence than Swiss-Prot, there is a greater chance that some of the annotations will be incorrect. Which reference database is preferable depends on the goals of the investigator. The issue of whether the sequence is going to be submitted to GenBank is important, as NCBI specifies requirements regarding what types of features need to be identified, how proteins should be named, and how the data should be formatted (see http://

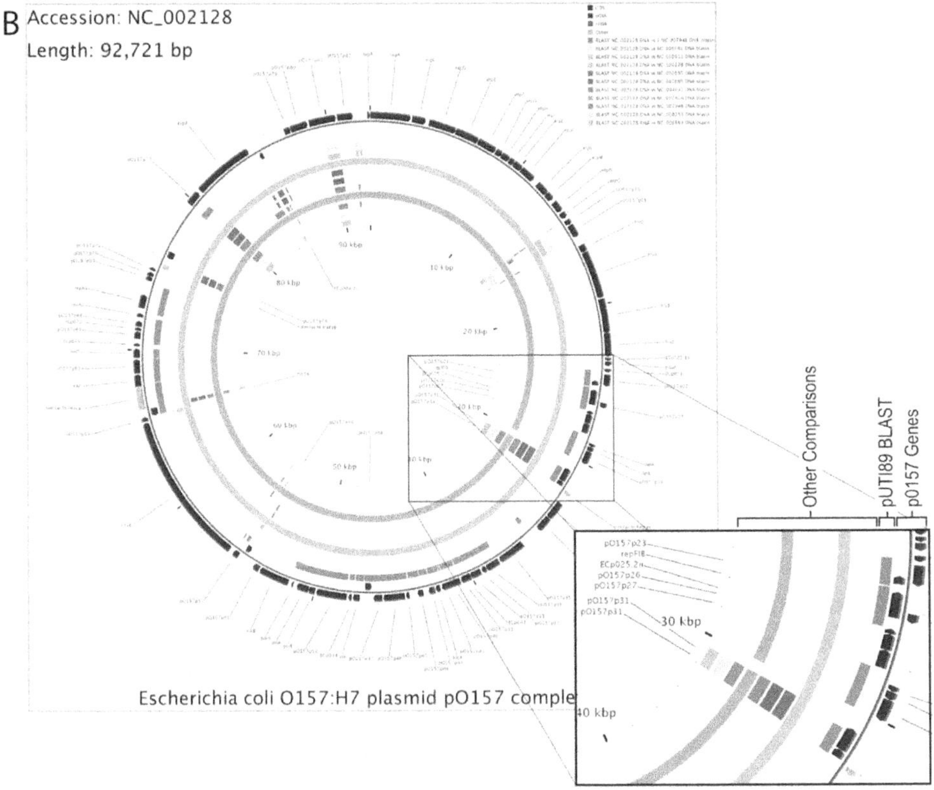

Fig. 2 (B) Sequence comparison using CGView. The outer two rings represent genes from p0157, in the forward and reverse orientations. The next ring (from outside to inside) represents a BLAST comparison against pUTI89. The next nine rings represent BLAST comparisons against various E. coli genomes and plasmids. The sequences were compared using blastn (E value = 0.00001; minimum score = 100). Regions of similarity are displayed as solid arcs on each ring.

www.ncbi.nlm.nih.gov/Genbank/). The output from some pipelines may need to be filtered or reformatted to conform to the GenBank requirements.

The annotation of cloning plasmids is fairly straightforward, by virtue of their small size and tendency to contain well-characterized sequence elements. Plasmid annotation is built into several plasmid analysis programs, including PlasMapper, BioInfoMan, ApE, and PlasmaDNA. These programs use sequence comparisons and open reading frame analysis to identify common plasmid domains, including resistance genes, origins of replication, promoters, and reporter genes. ApE and PlasmaDNA allow users to expand the reference databases used in the similarity searches, so that more specialized sequences can be properly annotated.

SEQUENCE VISUALIZATION

Sequence visualization tools can be used to quickly find sequence regions or annotations of interest. They also allow evidence from several different analyses to be viewed and evaluated simultaneously. Plasmid sequences and features can be viewed using many of the tools developed for larger genomes. A powerful stand-alone program suitable for prokaryotic or eukaryotic sequences of any length is Artemis. Using Artemis it is possible to view the results of numerous sequence analyses, along with the underlying DNA sequence (Berriman and Rutherford, 2003; Rutherford et al., 2000). Analysis results and sequence feature information can be read in from files, in GenBank, EMBL, or GFF format. In addition, Artemis can generate and display several types of plots, including GC content, GC skew, and codon usage. In cases where there is disagreement between analysis results and annotations, or if annotations suggest the presence of a sequencing error, revisions can be made within the program. Another useful tool is ACT (Artemis Comparison Tool), a DNA sequence comparison viewer based on Artemis (Carver et al., 2005). ACT can display two different DNA sequences at the same time, along with any available feature information. When supplied with a BLAST results file, ACT connects regions of similarity between the sequences using coloured lines. (Fig. 2A). These lines can reveal conservation in terms of sequence and genome organization. Artemis and ACT are written in Java and therefore can run on any Java-enabled platform, including Windows, Mac OS, and Linux. A program specifically designed for visualizing and comparing circular sequences is CGView (Stothard and Wishart, 2005). CGView can be used to display sequence features, base content plots, and BLAST results (Fig. 2B). Several sequences can be compared to a reference sequence on a single plot, and plots can be generated at very high resolutions, making them suitable for posters or publications. CGView is a available as a Java command-line program.

CONCLUSIONS AND FUTURE TRENDS

This chapter reviewed current tools and methods for the analysis and annotation of plasmid sequences. With the growing number of free programs, it is unlikely that the average lab will have to purchase commercial sequence analysis software. Many of the latest programs are web accessible, allowing for convenient access from any platform. These programs usually support open data formats, meaning that users can easily switch to new software if the need arises. For plasmid vector sequences, several options are available for performing routine analysis operations, such as creating restriction maps. For larger plasmids, a combination of programs for gene prediction, similarity searches, and alignment may be required, or complete pipelines can be accessed online or downloaded.

It can be very difficult to keep up with the vast number of new bioinformatics tools and databases. Excellent resources for recent advances in this area are two special issues of Nucleic Acids Research. The first issue of each year is dedicated to biological databases, and the July issue is devoted to web-based software. The Bioinformatics Links Directory (http://bioinformatics.ca/links_directory) maintains a list of servers published in issues of Nucleic Acids Research together with other useful bioinformatics tools and databases (Fox et al., 2007).

With many labs performing high-throughput biology, and accumulating numerous plasmid sequences, there is a need for plasmid management systems. Plasmid management capabilities can be built into current laboratory information management systems (LIMS), but there is a lack of free or open source LIMS. For naturally occurring plasmids, future work will likely focus on improving analysis pipelines, through the addition of more advanced features such as metabolic network predictions. Sequence databases will also continually improve through the incorporation of more sequences and new experimentally determined functional information. Continued development of tools like ACT will be important for comparing plasmid sequences, to better understand their relationships and evolutionary history.

WEB RESOURCES

ACLAME	S	http://aclame.ulb.ac.be/
ACT	V	http://www.sanger.ac.uk/Software/ACT
Addgene	S	http://www.addgene.org
ApE	M	http://www.biology.utah.edu/jorgensen/wayned/ape
Artemis	V	http://www.sanger.ac.uk/Software/Artemis
BacMap	S	http://wishart.biology.ualberta.ca/BacMap
BASys	N	http://wishart.biology.ualberta.ca/basys
BioInfoMan	M	http://bioinfo4u.org
Bioinformatics Links Directory	O	http://bioinformatics.ca/links_directory
BioMoby	O	http://biomoby.org
BLAST	D	http://www.ncbi.nlm.nih.gov/blast
CGView	V	http://wishart.biology.ualberta.ca/cgview
CINEMA	A	http://www.bioinf.man.ac.uk/dbbrowser/CINEMA2.1
CLC Free Workbench	M	http://www.clcbio.com/
Cloning Vector Collection	S	http://www.shigen.nig.ac.jp/cvector/cvector.html
ClustalW	A	http://www.ebi.ac.uk/clustalw
ClustalX	A	http://bips.u-strasbg.fr/fr/Documentation/ClustalX/
Codon Usage Database	O	http://www.kazusa.or.jp/codon/
DDBJ	S	http://www.ddbj.nig.ac.jp
EMBL	S	http://www.ebi.ac.uk/embl
EMBOSS	M	http://emboss.sourceforge.net
EnzymeX	R	http://mekentosj.com/enzymex
FASTA	D	http://www.ebi.ac.uk/fasta
FastPCR	P	http://www.biocenter.helsinki.fi/bi/Programs/fastpcr.htm

Fission Yeast Vectors	S	http://www-rcf.usc.edu/~forsburg/vectors.html
FlyBase Vectors	S	http://flybase.org/static_pages/lists/vectors.html
GenBank	S	http://www.ncbi.nlm.nih.gov/Genbank
GenBank Natural plasmids	S	http://www.ncbi.nlm.nih.gov/genomes/static/o.html
GenDB	N	http://www.cebitec.uni-bielefeld.de/groups/brf/software/gendb_info/
Gene Ontology	O	http://www.geneontology.org
GeneDoc	A	http://www.nrbsc.org/gfx/genedoc
GeneHacker	G	http://www-btls.jst.go.jp/cgi-bin/Tools/GeneHacker/index.cgi
GeneMark	G	http://exon.gatech.edu/GeneMark/
GFF	F	http://www.sanger.ac.uk/Software/formats/GFF/index.shtml
Glimmer	G	http://cbcb.umd.edu/software/glimmer
HAMAP	N	http://www.expasy.org/sprot/hamap
HMMER	D	http://hmmer.janelia.org
IMAGE Consortium Vectors	S	http://image.llnl.gov/image/html/vectors.shtml
Integrated Microbial Genomes	V	http://img.jgi.doe.gov/cgi-bin/pub/main.cgi
International Nucleotide Sequence Database Collaboration	O	http://insdc.org
IS Finder	S	http://www-is.biotoul.fr/
Islander	S	http://kementari.bioinformatics.vt.edu/cgi-bin/islander.cgi
Jalview	A	http://www.jalview.org
MAGPIE	N	http://magpie.ucalgary.ca/
Manatee	N	http://manatee.sourceforge.net
MED	G	http://ctb.pku.edu.cn/main/SheGroup/Software/MED2.htm
MUSCLE	A	http://www.drive5.com/muscle
NEBcutter	R	http://tools.neb.com/NEBcutter
ORF Finder	O	http://www.ncbi.nlm.nih.gov/gorf/gorf.html
ORFprimer	P	http://www.proteinstrukturfabrik.de/ORFprimer
pDRAW32	M	http://www.acaclone.com
PlasmaDNA	M	http://research.med.helsinki.fi/plasmadna
PlasMapper	M	http://wishart.biology.ualberta.ca/PlasMapper
PlasmID	S	http://plasmid.hms.harvard.edu
Plasmid Genome Database	S	http://www.genomics.ceh.ac.uk/plasmiddb/
Primer3	P	http://primer3.sourceforge.net
PUMA2	N	http://compbio.mcs.anl.gov/puma2
Readseq	F	http://iubio.bio.indiana.edu/cgi-bin/readseq.cgi
REBASE	R	http://rebase.neb.com
REviewer	R	http://www.fermentas.com/reviewer
SABIA	N	http://www.sabia.lncc.br/
SAM	D	http://www.soc.ucsc.edu/research/compbio/sam.html
Sanger Plasmid Sequencing Project	S	http://www.sanger.ac.uk/Projects/Plasmids/
Sequence Manipulation Suite	M	http://bioinformatics.org/sms2

Sequence Ontology	O	http://www.sequenceontology.org/
Serial Cloner	M	http://serialbasics.free.fr
SignalP	T	http://www.cbs.dtu.dk/services/SignalP
Swiss-Prot	T	http://www.expasy.org/sprot
Taverna	O	http://taverna.sourceforge.net
T-Coffee	A	http://www.ebi.ac.uk/t-coffee
TIGR Annotation Engine	N	http://www.tigr.org/AnnotationEngine
TMHMM	T	http://www.cbs.dtu.dk/services/TMHMM
tRNAscan-SE	G	http://lowelab.ucsc.edu/tRNAscan-SE
UniVec	S	http://www.ncbi.nlm.nih.gov/VecScreen/replist.html
Vector NTI	M	http://www.invitrogen.com/VNTIGateway
VectorDB	S	http://seq.yeastgenome.org/vectordb
XML	F	http://www.w3.org/XML

Legend

A – Alignment of sequences	O – Other
D – Database similarity search	P – Primer design
F – Format conversion tools	R – Restriction analysis
G – Gene prediction	S – Source of plasmid sequences
M – Multipurpose	T – Protein analysis
N – Annotation	V – Visualization of sequences

REFERENCES

Almeida, L.G., Paixão, R., Souza, R.C., Costa, G.C., Barrientos, F.J., Santos, M.T., Almeida, D.F., and Vasconcelos, A.T. (2004). A System for Automated Bacterial (genome) Integrated Annotation--SABIA. Bioinformatics *20*, 2832-2833.

Altschul, S.F., Gish, W., Miller, W., Myers, E.W., and Lipman, D.J. (1990). Basic local alignment search tool. J Mol Biol *215*, 403-410.

Altschul, S.F., Madden, T.L., Schäffer, A.A., Zhang, J., Zhang, Z., Miller, W., and Lipman, D.J. (1997). Gapped BLAST and PSI-BLAST: a new generation of protein database search programs. Nucleic Acids Res *25*, 3389-3402.

Angers-Loustau, A., Rainy, J., and Wartiovaara, K. (2007). PlasmaDNA: a free, cross-platform plasmid manipulation program for molecular biology laboratories. BMC Mol Biol *8*, 77.

Ashburner, M., Ball, C.A., Blake, J.A., Botstein, D., Butler, H., Cherry, J.M., Davis, A.P., Dolinski, K., Dwight, S.S., Eppig, J.T., et al. (2000). Gene ontology: tool for the unification of biology. The Gene Ontology Consortium. Nat Genet *25*, 25-29.

Bendtsen, J.D., Nielsen, H., von Heijne, G., and Brunak, S. (2004). Improved prediction of signal peptides: SignalP 3.0. J Mol Biol *340*, 783-795.

Benson, D.A., Karsch-Mizrachi, I., Lipman, D.J., Ostell, J., and Wheeler, D.L. (2007). GenBank. Nucleic Acids Res *35*, D21-25.

Berriman, M., and Rutherford, K. (2003). Viewing and annotating sequence data with Artemis. Brief Bioinformatics *4*, 124-132.

Boeckmann, B., Bairoch, A., Apweiler, R., Blatter, M.C., Estreicher, A., Gasteiger, E., Martin, M.J., Michoud, K., O'Donovan, C., Phan, I., et al. (2003). The SWISS-PROT protein knowledgebase and its supplement TrEMBL in 2003. Nucleic Acids Res *31*, 365-370.

Carver, T.J., Rutherford, K.M., Berriman, M., Rajandream, M.A., Barrell, B.G., and Parkhill, J. (2005). ACT: the Artemis Comparison Tool. Bioinformatics *21*, 3422-3423.

Chenna, R., Sugawara, H., Koike, T., Lopez, R., Gibson, T.J., Higgins, D.G., and Thompson, J.D. (2003). Multiple sequence alignment with the Clustal series of programs. Nucleic Acids Res *31*, 3497-3500.

Cherry, J.M., Adler, C., Ball, C., Chervitz, S.A., Dwight, S.S., Hester, E.T., Jia, Y., Juvik, G., Roe, T., Schroeder, M., et al. (1998). SGD: Saccharomyces Genome Database. Nucleic Acids Res *26*, 73-79.

Clamp, M., Cuff, J., Searle, S.M., and Barton, G.J. (2004). The Jalview Java alignment editor. Bioinformatics *20*, 426-427.

Crosby, M.A., Goodman, J.L., Strelets, V.B., Zhang, P., Gelbart, W.M., and FlyBase Consortium (2007). FlyBase: genomes by the dozen. Nucleic Acids Res *35*, D486-491.

Delcher, A.L., Bratke, K.A., Powers, E.C., and Salzberg, S.L. (2007). Identifying bacterial genes and endosymbiont DNA with Glimmer. Bioinformatics *23*, 673-679.

Delcher, A.L., Harmon, D., Kasif, S., White, O., and Salzberg, S.L. (1999). Improved microbial gene identification with GLIMMER. Nucleic Acids Res *27*, 4636-4641.

Dong, X., Stothard, P., Forsythe, I.J., and Wishart, D.S. (2004). PlasMapper: a web server for drawing and auto-annotating plasmid maps. Nucleic Acids Res *32*, W660-664.

Eddy, S.R. (1998). Profile hidden Markov models. Bioinformatics *14*, 755-763.

Edgar, R.C. (2004). MUSCLE: multiple sequence alignment with high accuracy and high throughput. Nucleic Acids Res *32*, 1792-1797.

Eilbeck, K., Lewis, S.E., Mungall, C.J., Yandell, M., Stein, L., Durbin, R., and Ashburner, M. (2005). The Sequence Ontology: a tool for the unification of genome annotations. Genome biology *6*, R44.

Fan, M., Tsai, J., Chen, B., Fan, K., and LaBaer, J. (2005). A central repository for published plasmids. Science *307*, 1877.

Feng, D.F., and Doolittle, R.F. (1987). Progressive sequence alignment as a prerequisite to correct phylogenetic trees. J Mol Evol *25*, 351-360.

Fox, J.A., McMillan, S., and Ouellette, B.F. (2007). Conducting research on the web: 2007 update for the bioinformatics links directory. Nucleic Acids Res *35*, W3-5.

Gaasterland, T., and Sensen, C.W. (1996). MAGPIE: automated genome interpretation. Trends Genet *12*, 76-78.

Gattiker, A., Michoud, K., Rivoire, C., Auchincloss, A.H., Coudert, E., Lima, T., Kersey, P., Pagni, M., Sigrist, C.J., Lachaize, C., et al. (2003). Automated annotation of microbial proteomes in SWISS-PROT. Computational biology and chemistry *27*, 49-58.

Gotoh, O. (1996). Significant improvement in accuracy of multiple protein sequence alignments by iterative refinement as assessed by reference to structural alignments. J Mol Biol *264*, 823-838.

Higgins, D.G., and Sharp, P.M. (1988). CLUSTAL: a package for performing multiple sequence alignment on a microcomputer. Gene *73*, 237-244.

Higgins, D.G., Thompson, J.D., and Gibson, T.J. (1996). Using CLUSTAL for multiple sequence alignments. Meth Enzymol *266*, 383-402.

Hughey, R., and Krogh, A. (1996). Hidden Markov models for sequence analysis: extension and analysis of the basic method. Comput Appl Biosci *12*, 95-107.

International Union of Biochemistry and Molecular Biology. Nomenclature Committee., and Webb, E.C. (1992). Enzyme nomenclature 1992 : recommendations of the Nomenclature Committee of the International Union of Biochemistry and Molecular Biology on the nomenclature and classification of enzymes (San Diego ; Toronto, Published for the International Union of Biochemistry and Molecular Biology by Academic Press).

Koressaar, T., and Remm, M. (2007). Enhancements and modifications of primer design program Primer3. Bioinformatics *23*, 1289-1291.

Krogh, A., Larsson, B., von Heijne, G., and Sonnhammer, E.L. (2001). Predicting transmembrane protein topology with a hidden Markov model: application to complete genomes. J Mol Biol *305*, 567-580.

Kulikova, T., Akhtar, R., Aldebert, P., Althorpe, N., Andersson, M., Baldwin, A., Bates, K., Bhattacharyya, S., Bower, L., Browne, P., et al. (2007). EMBL Nucleotide Sequence Database in 2006. Nucleic Acids Res *35*, D16-20.

Lennon, G., Auffray, C., Polymeropoulos, M., and Soares, M.B. (1996). The I.M.A.G.E. Consortium: an integrated molecular analysis of genomes and their expression. Genomics *33*, 151-152.

Leplae, R., Hebrant, A., Wodak, S.J., and Toussaint, A. (2004). ACLAME: a CLAssification of Mobile genetic Elements. Nucleic Acids Res *32*, D45-49.

Lowe, T.M., and Eddy, S.R. (1997). tRNAscan-SE: a program for improved detection of transfer RNA genes in genomic sequence. Nucleic Acids Res *25*, 955-964.

Lu, G., and Moriyama, E.N. (2004). Vector NTI, a balanced all-in-one sequence analysis suite. Brief Bioinformatics *5*, 378-388.

Lukashin, A.V., and Borodovsky, M. (1998). GeneMark.hmm: new solutions for gene finding. Nucleic Acids Res *26*, 1107-1115.

Maltsev, N., Glass, E., Sulakhe, D., Rodriguez, A., Syed, M.H., Bompada, T., Zhang, Y., and D'Souza, M. (2006). PUMA2--grid-based high-throughput analysis of genomes and metabolic pathways. Nucleic Acids Res *34*, D369-372.

Markowitz, V.M., Korzeniewski, F., Palaniappan, K., Szeto, E., Werner, G., Padki, A., Zhao, X., Dubchak, I., Hugenholtz, P., Anderson, I., et al. (2006). The integrated microbial genomes (IMG) system. Nucleic Acids Res *34*, D344-348.

Meyer, F., Goesmann, A., McHardy, A.C., Bartels, D., Bekel, T., Clausen, J., Kalinowski, J., Linke, B., Rupp, O., Giegerich, R., et al. (2003). GenDB--an open source genome annotation system for prokaryote genomes. Nucleic Acids Res *31*, 2187-2195.

Miyazaki, S., Sugawara, H., Gojobori, T., and Tateno, Y. (2003). DNA Data Bank of Japan (DDBJ) in XML. Nucleic Acids Res *31*, 13-16.

Nakamura, Y., Gojobori, T., and Ikemura, T. (2000). Codon usage tabulated from international DNA sequence databases: status for the year 2000. Nucleic Acids Res *28*, 292.

Needleman, S.B., and Wunsch, C.D. (1970). A general method applicable to the search for similarities in the amino acid sequence of two proteins. J Mol Biol *48*, 443-453.

Nicholas, K.B., H.B.Jr., N., and Deerfield, D.W.I. (1997). GeneDoc: Analysis and Visualization of Genetic Variation. EMBNEWNEWS *4*, 14.

Nomenclature Committee of the International Union of Biochemistry (NC-IUB). (1985). Nomenclature for incompletely specified bases in nucleic acid sequences. Recommendations 1984. Biochem J *229*, 281-286.

Notredame, C., Higgins, D.G., and Heringa, J. (2000). T-Coffee: A novel method for fast and accurate multiple sequence alignment. J Mol Biol *302*, 205-217.

Oinn, T., Addis, M., Ferris, J., Marvin, D., Senger, M., Greenwood, M., Carver, T., Glover, K., Pocock, M.R., Wipat, A., et al. (2004). Taverna: a tool for the composition and enactment of bioinformatics workflows. Bioinformatics *20*, 3045-3054.

Okubo, K., Sugawara, H., Gojobori, T., and Tateno, Y. (2006). DDBJ in preparation for overview of research activities behind data submissions. Nucleic Acids Res *34*, D6-9.

Page, R.D.M., and Holmes, E.C. (1998). Molecular evolution : a phylogenetic approach (Oxford ; Malden, MA, Blackwell Science).

Park, J., Karplus, K., Barrett, C., Hughey, R., Haussler, D., Hubbard, T., and Chothia, C. (1998). Sequence comparisons using multiple sequences detect three times as many remote homologues as pairwise methods. J Mol Biol *284*, 1201-1210.

Parry-Smith, D., Payne, A.W., Michie, A.D., and Attwood, T.K. (1998). CINEMA--a novel colour INteractive editor for multiple alignments. Gene *221*, GC57-63.

Pearson, W.R. (1990). Rapid and sensitive sequence comparison with FASTP and FASTA. Meth Enzymol *183*, 63-98.

Pearson, W.R. (1996). Effective protein sequence comparison. Meth Enzymol *266*, 227-258.

Pearson, W.R., and Lipman, D.J. (1988). Improved tools for biological sequence comparison. Proc Natl Acad Sci USA *85*, 2444-2448.

Rice, P., Longden, I., and Bleasby, A. (2000). EMBOSS: the European Molecular Biology Open Software Suite. Trends Genet *16*, 276-277.

Roberts, R.J., Vincze, T., Posfai, J., and Macelis, D. (2007). REBASE--enzymes and genes for DNA restriction and modification. Nucleic Acids Res *35*, D269-270.

Rozen, S., and Skaletsky, H. (2000). Primer3 on the WWW for general users and for biologist programmers. Methods Mol Biol *132*, 365-386.

Rutherford, K., Parkhill, J., Crook, J., Horsnell, T., Rice, P., Rajandream, M.A., and Barrell, B. (2000). Artemis: sequence visualization and annotation. Bioinformatics *16*, 944-945.

Salzberg, S.L., Delcher, A.L., Kasif, S., and White, O. (1998). Microbial gene identification using interpolated Markov models. Nucleic Acids Res *26*, 544-548.

Schattner, P., Brooks, A.N., and Lowe, T.M. (2005). The tRNAscan-SE, snoscan and snoGPS web servers for the detection of tRNAs and snoRNAs. Nucleic Acids Res *33*, W686-689.

Smith, T.F., and Waterman, M.S. (1981). Identification of common molecular subsequences. J Mol Biol *147*, 195-197.

Stothard, P. (2000). The sequence manipulation suite: JavaScript programs for analyzing and formatting protein and DNA sequences. BioTechniques *28*, 1102, 1104.

Stothard, P., Van Domselaar, G., Shrivastava, S., Guo, A., O'Neill, B., Cruz, J., Ellison, M., and Wishart, D.S. (2005). BacMap: an interactive picture atlas of annotated bacterial genomes. Nucleic Acids Res *33*, D317-320.

Stothard, P., and Wishart, D.S. (2005). Circular genome visualization and exploration using CGView. Bioinformatics *21*, 537-539.

Stothard, P., and Wishart, D.S. (2006). Automated bacterial genome analysis and annotation. Curr Opin Microbiol *9*, 505-510.

Taylor, W.R. (1988). A flexible method to align large numbers of biological sequences. J Mol Evol *28*, 161-169.

Thompson, J.D., Gibson, T.J., Plewniak, F., Jeanmougin, F., and Higgins, D.G. (1997). The CLUSTAL_X windows interface: flexible strategies for multiple sequence alignment aided by quality analysis tools. Nucleic Acids Res *25*, 4876-4882.

Thompson, J.D., Higgins, D.G., and Gibson, T.J. (1994). CLUSTAL W: improving the sensitivity of progressive multiple sequence alignment through sequence weighting, position-specific gap penalties and weight matrix choice. Nucleic Acids Res *22*, 4673-4680.

Untergasser, A., Nijveen, H., Rao, X., Bisseling, T., Geurts, R., and Leunissen, J.A. (2007). Primer3Plus, an enhanced web interface to Primer3. Nucleic Acids Res *35*, W71-74.

Van Domselaar, G.H., Stothard, P., Shrivastava, S., Cruz, J.A., Guo, A., Dong, X., Lu, P., Szafron, D., Greiner, R., and Wishart, D.S. (2005). BASys: a web server for automated bacterial genome annotation. Nucleic Acids Res *33*, W455-459.

Vincze, T., Posfai, J., and Roberts, R.J. (2003). NEBcutter: A program to cleave DNA with restriction enzymes. Nucleic Acids Res *31*, 3688-3691.

Wang, L., Riethoven, J.J., and Robinson, A. (2002). XEMBL: distributing EMBL data in XML format. Bioinformatics *18*, 1147-1148.

Wheeler, D.L., Church, D.M., Federhen, S., Lash, A.E., Madden, T.L., Pontius, J.U., Schuler, G.D., Schriml, L.M., Sequeira, E., Tatusova, T.A., et al. (2003). Database resources of the National Center for Biotechnology. Nucleic Acids Res *31*, 28-33.

Wilkinson, M.D., and Links, M. (2002). BioMOBY: an open source biological web services proposal. Brief Bioinformatics *3*, 331-341.

Yada, T., and Hirosawa, M. (1996). Detection of short protein coding regions within the cyanobacterium genome: application of the hidden Markov model. DNA Res *3*, 355-361.

Zhu, H., Hu, G.Q., Yang, Y.F., Wang, J., and She, Z.S. (2007). MED: a new non-supervised gene prediction algorithm for bacterial and archaeal genomes. BMC Bioinformatics *8*, 97.

Zuo, D., Mohr, S.E., Hu, Y., Taycher, E., Rolfs, A., Kramer, J., Williamson, J., and LaBaer, J. (2007). PlasmID: a centralized repository for plasmid clone information and distribution. Nucleic Acids Res *35*, D680-684.

Chapter 5

Horizontal Gene Transfer Mediated by Plasmids

Masahiro Sota and Eva M. Top

ABSTRACT

The ever-increasing information from bacterial genome sequences has clearly revealed frequent acquisition of genetic material from phylogenetically distant bacteria and even other organisms such as eukaryotes. The event, generally recognized as horizontal gene transfer (HGT), is now considered as a strong driving force for the evolution of bacterial genome organization and for rapid adaptation to the surrounding environments. Among the mobile elements and mechanisms of HGT, plasmids are undoubtedly critical players because of their ability to transfer by conjugation among both closely and very distantly related bacterial hosts. This feature allows them to broadly distribute genes or gene clusters that code for various host-beneficial phenotypes. This chapter introduces the features of recently characterized plasmids, with an emphasis on horizontal transfer, and the domain Bacteria (mostly Gram-positive and Gram-negative bacteria). We also describe some examples of transposable elements on plasmids, which have greatly contributed to the dissemination of phenotypic traits by HGT. The chapter ends with current research and suggestions for future studies in the areas of molecular biology, ecology and evolution of plasmids.

INTRODUCTION

There is no doubt that horizontal gene transfer (HGT) is a driving force in the evolution and rapid adaptation of Bacteria and Archaea, including those that infect humans (Gogarten and Townsend, 2005; Jain et al., 2003; Ochman et al., 2000). Recently published analyses of prokaryotic gene and whole genome sequences have revealed that HGT between very distantly related organisms plays a far more important role in the evolution of these organisms than had been previously recognized (Daubin et al., 2003; Frigaard et al., 2006; Gogarten et al., 2002; Gogarten and Townsend, 2005; Lawrence and Hendrickson, 2003; Lawrence and Ochman, 2002; Lerat et al., 2005). Disparities in % G+C composition, codon usage bias, incongruent phylogenies of various genes within the same species, and other criteria suggest that a substantial fraction of genome sequences have been acquired by HGT (Daubin et al., 2003; Frigaard et al., 2006; Gogarten and Townsend, 2005;

Kurland et al., 2003; Lawrence and Ochman, 2002; Nelson et al., 1999; Ochman et al., 2000). For example, more than a quarter of the genome of a vancomycin-resistant clinical isolate of the species *Enterococcus feacalis* was shown to consist of potentially mobile or foreign DNA (Paulsen et al., 2003). Moreover, several other reports have demonstrated that the genome sequence similarity among bacteria that belong to the same species is sometimes less than 50% (Abby and Daubin, 2007; Welch et al., 2002). A typical example of the enormous consequences of within-species genome sequence divergence is the difference between the pathogenic and non-pathogenic *Escherichia coli* strains, O157:H7 and K-12 (Hayashi et al., 2001; Perna et al., 2001). In addition to this retrospective evidence for HGT between chromosomes, it is now clear that most prokaryotes carry mobile genetic elements (MGEs), and that those elements actively transfer host-beneficial genes between closely and distantly related organisms, not only in the laboratory but also in microcosms and in the field (Dröge et al., 1999; Frost et al., 2005; Hill and Top, 1998; Sørensen et al., 2005; van Elsas and Bailey, 2002) (see Plasmid ecology section). This gene transfer results in the wholesale acquisition of new traits that are sometimes essential (e.g., antibiotic or heavy metal resistance) or otherwise confer a selective advantage (e.g., the ability to metabolize a novel carbon source).

How do bacteria acquire this foreign genetic material, and where do plasmids fit in this pool of horizontally transferred DNA? Three main mechanisms for horizontal gene transfer between prokaryotes have been identified: (a) Transformation, a process whereby cell-free DNA is taken up by a cell and usually integrates in the chromosome; (b) transduction, the transfer of DNA by bacteriophages; and (c) conjugation, a process whereby plasmids or conjugative transposons mediate cell-cell contact and transfer DNA from donor to recipient cells (Jain et al., 2002; Thomas and Nielsen, 2005). The genetic information exchanged is referred to as 'the horizontal gene pool' (Thomas, 2000b): Genetic information that is shared by a broad spectrum of prokaryotic species representing distinct, phylogenetic lineages. It can be considered as a 'virtual genome': A collection of genes that are potentially available to many genomes but do not form a separate entity. Although the relative contribution of these three gene transfer mechanisms to overall HGT in prokaryotes in natural and clinical environments is not really known, conjugative gene transfer mediated by plasmids is generally believed to be one of the most common and widespread mechanism for the gene flow among prokaryotic species (Jain et al., 2002; Thomas, 2000b; Thomas and Nielsen, 2005). Specifically gene exchange across wide phylogenetic ranges may be largely mediated by transferable plasmids. Several lines of evidence support this premise. First, the majority of bacteria isolated from a wide range of habitats carry plasmids that are known to transfer between closely or distantly related bacteria. Second, some conjugative plasmids are known to transfer rather long stretches of chromosomal DNA, and recent studies have provided evidence of past integrations of plasmids and typical plasmid genes in bacterial chromosomes (Chiu and Thomas, 2004; Lavigne et al., 2005). The scope of this chapter is to provide an overview of bacterial plasmids and their ability to transfer genetic material among members of the Bacteria. We summarize recent advances in our knowledge on the biology, ecology and evolution of these mobile elements, the various phenotypic traits that are dispersed by them, and the factors that determine the extent of HGT. Considering the aim of this book (current work and future trends), we focus mostly on findings published since 2000 and discuss the future research directions that we find most exciting and challenging.

PLASMID CLASSIFICATION

There have been various classification systems for plasmids. Grouping of plasmids of Gram-negative (G^-) bacteria has been mostly based on incompatibility, which is defined as the inability of two specific plasmids to be propagated stably in the same cell line (Couturier et al., 1988). While it is possible for a bacterial cell to harbor two cognate plasmids temporally, under non-selective conditions the cell will usually lose one of them over a few generations by the mechanisms of plasmid replication and partitioning. Practically, incompatibility testing involves introduction of a plasmid into a strain carrying another plasmid, testing for presence of the resident non-selected plasmid after selection for the incoming plasmid. Plasmids that cannot co-exist belong to the same incompatibility (Inc) group. More recently, however, a plasmid is often classified in an Inc group based on DNA sequence information of (at least) the replication region. In G^- bacteria, the names of Inc groups for plasmids from *Enterobacteriaceae* follow the characters of the alphabet (IncA to IncZ) (Jacob et al., 1977), while for plasmids form *Pseudomonas* strains all Inc group names consist of P followed by a number (IncP-1 to IncP-14) (Jacoby, 1986; Thomas and Haines, 2004). Several groups have subgroups, for instance, IncHI and IncHII, IncP-1α to IncP-1ε (Couturier et al., 1988; Schlüter et al., 2007; Taylor et al., 2004). Overlaps within/between the two Inc classification systems have also been confirmed: IncA = IncC (IncA/C) = IncP-3, IncB = IncO (IncB/O), IncG = IncU (IncG/U) = IncP-6, IncHI2 = IncS, IncL = IncM (IncL/M), IncP = IncP-1, and IncQ = IncP-4 (Couturier et al., 1988; Haines et al., 2006b; Thomas and Haines, 2004). Thanks to recent studies many new and unclassified plasmid families have emerged. For example, the *repABC*-type and pNL1-like plasmids from α-proteobacteria (see below), and a new group of broad-host-range (BHR) plasmids consisting of pIPO2T (Tauch et al., 2002), pSB102 (Schneiker et al., 2001), and pMOL98 (Gstalder et al., 2003). There are also several classification systems for plasmids from Gram-positive (G^+) bacteria. The plasmids from *Staphylococcus aureus* are categorized into 15 families, Inc1 to Inc15 (Taylor et al., 2004). For the hemolysin (Hly)-encoding plasmids of *Enterococcus feacalis* four groups have been proposed based on different sex pheromone responses (IncHly, IncHlyII, IncHlyIII, and IncHlyIV) (Colmar and Horaud, 1987; Ike and Clewell, 1992). Plasmids from *Bacillus* and *Corynebacterium* can be classified into several groups (Deb and

Nath, 1999; Rasko et al., 2005), but no definite Inc groups have been assigned yet. Thus, the current plasmid classification system is confusing and incomplete, and reclassification and rearrangement of the extant plasmid groups is necessary to better understand the relationship and evolutionary history of plasmids.

HOST RANGE OF PLASMIDS

Despite the important contribution of plasmids to HGT and to the acute drug resistance health crisis, very little is understood about the range of bacterial hosts in which these plasmids can replicate and persist in the absence of selection (e.g. antibiotics). A plasmid's host range is usually defined as the range of hosts in which the plasmid can replicate. This range is typically narrower than the range of hosts into which the plasmid can transfer by conjugation (the transfer range), but wider than the range of hosts in which it is stably maintained without selection (De Gelder et al., 2007; Guiney, 1982). The host range is known to be highly variable among plasmids. While so-called narrow-host-range (NHR) plasmids replicate only in a limited number of closely related species, BHR plasmids can transfer to and replicate in a wide range of hosts. Whereas some researchers consider plasmids that transfer between different genera as having a BHR, we have defined BHR plasmids in Proteobacteria those that can transfer and replicate in bacterial species from at least two subgroups (e.g. between α- and β-Proteobacteria) (Szpirer et al., 1999). Since both NHR and BHR plasmids are found in various bacterial populations today, they each must have developed different strategies to persist as extrachromosomal MGEs. While NHR plasmids succeeded by specializing to a limited range of hosts, BHR plasmids seem to be generalists whose replication mechanism is more flexible. Recent insights in the molecular basis of host-range, and future perspectives on plasmid host-range shifts over evolutionary time are described in the section Future Trends.

BHR plasmids may well be the single most important mediators of HGT between distantly related bacterial hosts (Mazodier and Davies, 1991). In G^- bacteria, IncQ plasmids have the broadest host-range of all, followed by those of the IncP-1 group, and IncA/C, IncN, IncW, and the recently described plasmids pIPO2T, pMOL98, and pSB102 are also promiscuous to various degrees (Gstalder et al., 2003; Krishnan and Iyer, 1988; Llanes et al., 1996; Schneiker et al., 2001; Szpirer et al., 1999; Tauch et al., 2002; Thomas and Smith, 1987). Since the transfer range of some BHR plasmids is even much broader than their replication range, these plasmids can spread genes across even wider taxonomic ranges by conjugative transfer followed by integration of plasmid parts into the recipient chromosome, or by mobilization of non-conjugative vectors with a wider host range, including G^+ bacteria, cyanobacteria, and Eukaryotes such as yeasts and plants (Kado, 2000; Mazodier and Davies, 1991). They can also capture genes or non-self-transmissible plasmids from hosts in which they cannot replicate, a phenomenon called retromobilization or retrotransfer (Szpirer et al., 1999).

Well-known examples of G$^-$ NHR plasmids are the F-like plasmids of the incompatibility group IncFI/IncFII. The F plasmid of *Escherichia coli* has been characterized in most detail, but a large variety of plasmids that encode drug resistance and virulence in human pathogens also belong to this group (Jacoby, 1986; Silva et al., 1988). Although a few exceptions have been reported, their host range is generally limited to the *Enterobacteriaceae*, such as *Shigella*, *Salmonella*, *Klebsiella* etc. (Datta and Hedges, 1972; Guiney, 1982; Mergeay and Gerits, 1978).

Plasmids from G$^+$ bacteria also show NHR or BHR features. Many plasmids from *Staphylococcus aureus*, except those in the Inc1 and Inc2 groups, can replicate in various G$^+$ bacteria, including streptococci and bacilli (Novick, 1989). Two self-transmissible plasmids, pIP501 (Horodniceanu et al., 1976; Thompson and Collins, 2003) and pAMβ1 (LeBlanc et al., 1978) isolated from streptococci, can be transferred to a wide variety of G$^+$ bacteria, including lactobacilli, lactococci, bacilli, clostridia, staphylococci, enterococci, *Listeria* spp., and *Streptomyces lividans* (Kurenbach et al., 2003). The replication and stable maintenance of pIP501 in a G$^-$ *E. coli* strain was confirmed (Kurenbach et al., 2003). The fact that nearly identical plasmids such as the pUB110/pBC16 and pE12/pIM13 pairs were found in staphylococci and bacilli indicates a common source of plasmids in G$^+$ bacteria (Novick, 1989; Polak and Novick, 1982; Projan et al., 1987).

Interestingly, small non-self-transmissible but mobilizable plasmids, irrespective of the origins (from G$^+$ or G$^-$), tend to have a wider host range than large self-transmissible plasmids (see below for mobilizable and self-transmissible plasmids). For example, the IncQ plasmid RSF1010 from *E. coli* (Guerry et al., 1974), pMV158 from *Streptococcus agalactiae* (Burdett, 1980), and pWV01 from *Lactococcus lactis* (Otto et al., 1982) are capable of replicating in both G$^+$ and G$^-$ bacteria and mobilizable by a variety of self-transmissible plasmids (del Solar et al., 1993; Kok et al., 1984; Rawlings and Tietze, 2001). This might reflect specific ecological strategies of these mobilizable plasmids: They may require a broad host range to persist in bacterial communities, since they cannot transfer themselves but rely on the transfer machineries of several kinds of self-transmissible plasmids, which transfer within different sets of hosts.

PLASMIDS HAVE MOSAIC STRUCTURES

Plasmids are not at all static genomes, but instead are plastic genetic mosaics that have evolved over time through the iterative acquisition of various transposons, integrons, gene cassettes, genomic islands, and insertion sequences, which encode the 'accessory' genes (Merlin et al., 2000; Osborn and Boltner, 2002). Those elements are described in detail below. Consequently, plasmids can be thought of as consisting of two regions. The 'plasmid backbone' encodes genes involved in the replication, maintenance, and transfer of the plasmid, and the 'accessory' genes confer specific phenotypic characteristics to the host organism, such as resistance to antibiotics, heavy metals, and UV light, metabolic functions such as those required for the degradation of various man-made pollutants, production of toxins, and bacterial and bacteria-host interactions (Thomas, 2000a; Top et al., 2000). However, the vast majority of traits encoded by plasmids remain "cryptic", i.e., none of the open reading frames code for known phenotypes. As examples of plasmid structures, schematics showing the organization of IncP-1 (self-transmissible) and IncQ (mobilizable) replicons, the best-studied groups of BHR plasmids, are depicted in Fig. 1. The accessory genes in IncP-1 plasmids are typically inserted in the plasmid backbone between two regions involved in plasmid transfer [that encode mating-pair formation (Trb) and plasmid transfer (Tra)], and between the origin of replication (*oriV*) and the gene for plasmid replication (carried in the Rep region in Fig. 1). In the cases of IncQ plasmids, the accessory genes are located between *oriV* and the Rep or Mob region. During the last years, at least 20 complete sequences of IncP-1 plasmids have become available, and detailed analysis and comparison of the sequences have demonstrated that plasmids with almost identical backbones contain entirely different accessory genes on transposons and integrons: Genes that encode degradation of chlorinated herbicides [examples are pADP-1 (Martinez

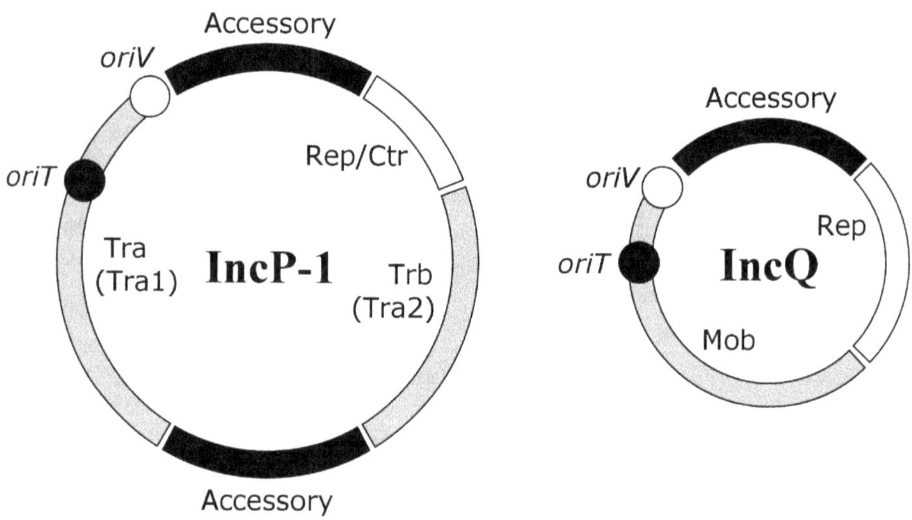

Fig. 1. Schematic structures of IncP-1 (self-transmissible) and IncQ (mobilizable) plasmids. The white, gray, and black arcs show the regions for plasmid replication (and control), plasmid transfer (or mobilization), and accessory genes, respectively. The white and black circles represent the origins of vegetative replication (*oriV*) and plasmid transfer (*oriT*), respectively. Gene products from the Tra and Trb regions are required for DNA transfer and mating-pair formation. Mob proteins in mobilizable plasmids are necessary to convert the plasmids into the transferable form. See "conjugative transfer of plasmids" section for details.

et al., 2001), pUO1 (Sota et al., 2003), and pJP4 (Trefault et al., 2004)] versus antibiotic resistance determinants [the type plasmid R751 (Thorsted et al., 1998), pB3 (Heuer et al., 2004), pB4 (Tauch et al., 2003), pB8 (Schlüter et al., 2005), and pB10 (Schlüter et al., 2003)]. Moreover, two IncP-1 plasmids were recently described that do not carry any transposable elements (Harada et al., 2006; Kamachi et al., 2006). These findings suggest that these backbones are acting as shuttles for genes from diverse sources, thereby facilitating genetic exchange within and among microbial communities (Schlüter et al., 2007).

The horizontal transfer of DNA between plasmids is also evident in the 'plasmid backbones' that appear to have undergone recombination within homologous regions of closely related plasmids. For example, the genome sequence of the BHR IncP-1β plasmid pB10 suggests that recombination occurred between various IncP-1β plasmids (Schlüter et al., 2003). Sequence analysis of the two new BHR plasmids, pSB102 and pIPO2, also shows mosaic plasmid backbones (Schneiker et al., 2001; Tauch et al., 2002). Since 'plasmid backbones' encode the replication, maintenance, and transfer functions of plasmids, these recombination events may play a key role in expanding or contracting the host range of these plasmids. This could be important not only because it is crucial in determining which species can acquire new genetic traits through plasmid-mediated HGT, but also because it determines the genetic backgrounds from which plasmids can acquire novel traits.

CONJUGATIVE TRANSFER OF PLASMIDS

Conjugation is a DNA transfer system by which a single strand of DNA is transferred from a donor into a recipient cell through cell-cell contact. Plasmids and plasmid-like molecules [conjugative transposons (CTns)] use this system to transfer DNA molecules. To date, genes for conjugation are identified on plasmids from G^- and G^+ Bacteria and Archaea (Clewell and Francia, 2004; Garrett et al., 2004; Grohmann et al., 2003; Lawley et al., 2004; Zechner et al., 2000). Most of the plasmids can be categorized into two types, self-transmissible or mobilizable plasmids. While self-transmissible (or conjugative) plasmids can transfer from a donor to a recipient cell using their own transfer machinery, the mobilizable plasmids can only be transferred with the help of conjugative plasmids. Despite much diversity in details of the conjugation systems, there are several common features (except for the system in *Streptomyces*, see below): (i) cell-cell contact established by pili or cells, whose components are encoded by plasmids or CTns, (ii) presence of the origin of transfer (*oriT*), where plasmid transfer is initiated, and (iii) indispensability of the coupling proteins known as TraG-like proteins, which act as intermediates between the DNA processing functions and the DNA transport functions. In all systems, DNA relaxases (Mob proteins) are the key enzymes to initiate the DNA transfer. Since the detailed structures and functions of genes and their products for the conjugation systems are reviewed elsewhere (Clewell and

Francia, 2004; Grohmann et al., 2003; Lawley et al., 2004; Zechner et al., 2000), we will introduce just an overview of conjugation by plasmids.

SELF-TRANSMISSIBLE PLASMIDS

Self-transmissible plasmids carry all the genes necessary for their own transfer by conjugation, and are ubiquitous in bacterial populations. Self-transmissible plasmids from G^- bacteria wherein the conjugation systems are partially or extensively characterized are those in Inc groups IncFI, IncI1, IncHI, IncN, IncP-1, IncW, IncX, and tumor-inducing (Ti) plasmids from *Agrobacterium tumefaciens*. The conjugation systems of G^- bacteria are related to the type IV secretion system (T4SS), a bacterial macromolecular transport system (Christie and Vogel, 2000; Gerlach and Hensel, 2007). Usually the genes, more than 20 in some plasmids, are organized in one or two operons, and the expression is properly regulated by one or several regulatory proteins. The tight regulation of the operons is thought to help plasmids minimize the energy cost to maintain the plasmids in the host cells (Thomas, 2004). In the best characterized case of F plasmid, the transfer (*tra*) genes are composed of 35 genes with a total size of 34 kb, organized in one operon, and transcribed from the P_Y promoter upstream of the *traY* gene (Frost et al., 1994; Frost et al., 2005). TraJ protein, whose gene is located upstream of the P_Y promoter, is responsible for the initiation of high-level transcription of the operon from the promoter (Frost et al., 1994; Willetts, 1977).

Self-transmissible plasmids with a broad host-range have been shown to belong to groups such as IncP-1, IncW and to a lesser extent IncA/C, IncN, and to a new Inc group that includes pIPO2T, pSB102, and pMOL98 (Gstalder et al., 2003; Szpirer et al., 1999). IncP-1 plasmids transfer at very high frequencies on solid medium and are thought to have the broadest range of replication in self-transmissible plasmids so far identified (Adamczyk and Jagura-Burdzy, 2003; Schlüter et al., 2007; Thomas and Smith, 1987). They can not only transfer and replicate within different subgroups of the Proteobacteria, but also mobilize natural non-self-transmissible (mobilizable) plasmids, and so-called 'shuttle vectors' to G^+ bacteria (Mazodier et al., 1989), cyanobacteria (Kreps et al., 1990), or even Eukaryotes such as yeast (Heinemann and Sprague, 1989) (see below). In IncP-1 plasmids, the genes for conjugation are split into two region, *tra* (Tra1) and *trb* (Tra2) (Adamczyk and Jagura-Burdzy, 2003; Schlüter et al., 2007; Thomas and Smith, 1987), usually due to insertion of Tn*402*-like transposons (Sota et al., 2007). The *tra* region contains two transcriptional units, from *traC* to *traJ* and *traKLM*. The function of these products are relaxosome formation, DNA-processing for transfer, initiation of transfer replication, and guidance of the relaxosome complex to the secretion channel. The *oriT* sequence is situated between *traJ* and *traK*, and the TraHIJK proteins assemble the relaxosome complex at the *oriT* sequence. The *trb* region carries the genes for pilus protein synthesis and pilus processing (TrbC and TrbF)

(Kalkum et al., 2004; Kalkum et al., 2002), and for the envelope-spanning secretion apparatus (TrbBDEFGHIJL) (Grahn et al., 2000). It also carries *trbA*, which codes for the regulation of the *tra* and *trb* operons (TrbA) (Jagura-Burdzy et al., 1992; Zatyka et al., 1994).

The smallest conjugation system is the one encoded by the IncW plasmid R388 (Ward and Grinsted, 1978), which contains 14 *trw* genes (*trwA* to *trwN*), *korA*, and *eex* (Fernandez-Lopez et al., 2006). The TrwABC proteins are involved in DNA processing during conjugation and the TrwDEFGHIJKLMN proteins are used for mating-pair formation. KorA and KorB are regulatory proteins that modulate expression of the *trwD-trwN* genes (Fernandez-Lopez et al., 2006; Seubert et al., 2003). The conjugation gene organization of this IncW plasmid R388 is quite similar to that of the IncN plasmid pKM101, suggesting that they evolved from a common ancestor (Fernandez-Lopez et al., 2006).

The Ti plasmids from *A. tumefaciens* have been of great interests for many researchers because they transfer specific DNA sectors of the plasmids, called T-DNAs (transferred DNAs), into plant cells (Christie, 2004; Kado, 2002). This DNA transfer system mediated by the *vir* genes has been used to transform plant cells. Naturally infection of T-DNAs into plants is induced by phenolic compounds and monosaccharide sugars secreted by plants (Cangelosi et al., 1990; Christie, 2004). The T-DNAs carry the genes to (i) generate tumors on plant roots and (ii) synthesize opines (amino acid-sugar conjugates). Plants infected by T-DNAs produce and secrete high level of opines, and thus provide the *Agrobacterium* species with a nutrient rich environment. The genes for the metabolism of opines are also carried on the Ti plasmids. The Ti plasmids also have a second DNA transfer system (*tra*/*trb* genes) that is used for conjugation between agrobacterial species. It is known that environmental stress factors deleterious to *Agrobacterium* species promote the conjugative transfer of the Ti plasmids into recipient bacteria, leading to the survival of the plasmids (Kado, 2002).

Conjugative transfer of plasmids has been studied in many G^+ bacteria, but its mechanism is less understood compared to that of plasmids from G^- bacteria. It has been shown that most plasmids from streptococci and enterococci replicate in a broad range of hosts, such as the representatives pIP501 and pRE25, respectively (Evans and Macrina, 1983; Schwarz et al., 2001). Detailed analysis of the plasmid sequences has revealed the transfer regions of these plasmids, and all of them include several genes related with those of T4SS. A recent biochemical study by Abajy et al. (2007) has lead to the first proposal of an architectural model of a DNA secretion complex for the T4SS-like conjugative transfer of G^+ bacteria. Some plasmids from enterococci have a unique conjugation mechanism, called pheromone-responding conjugation (Kozlowicz et al., 2006), in which plasmid transfer is induced by hydrophobic peptides secreted by the recipient. This is the only known system of bacterial conjugation for which the factor produced by bacteria to initiate conjugation is identified. The peptides known as sex pheromones consist of 7 or 8 amino acids, and their precursors are parts of signal sequences for surface lipoproteins (Clewell et al., 2000). The pheromones are sensed by the receptors in the plasmid donor, e.g., the products of *traC* of pAD1 (Tanimoto et al., 1993) and *prgZ* of pCF10 (Ruhfel et al., 1993), which are lipoproteins binding to the exogenous pheromones at the cell surface. The donor and recipient of a pheromone-inducing plasmid form aggregates in which efficient plasmid transfer occurs without any role of pili (Clewell and Weaver, 1989; Dunny, 1990; Dunny et al., 1978). The pheromone-responding plasmids also produce inhibitor peptides that inhibit production of the corresponding pheromones, blocking self-induction of plasmid transfer of the donor cells and desensitizing donors to pheromones secreted by recipients too far away from the donors (Clewell, 1993; Grohmann et al., 2003).

The conjugation system in *Streptomyces* is quite different from other known DNA transfer systems in bacteria. A model for conjugative transfer in *Streptomyces* has been proposed based on experimental evidence as well as homology of proteins to others (Grohmann et al., 2003). First, these strains do not require any system to mediate cell-cell contact between the donor and recipient probably because of their mycelium, the branching of which allows the establishment of cell contact. Second, the *Streptomyces* system does not require any DNA-processing activities, thus DNA relaxases are absent. This suggests that double-stranded plasmids are transferred from the donor to recipient. Third, a single protein, Tra is essential for the conjugative transfer of plasmids and could be responsible for the mycelial fusion process (Brolle et al., 1993; Hagege et al., 1993; Pettis and Cohen, 1994). It represents a host killing function (Kendall and Cohen, 1987), but the mechanism is unknown. The features described above are common in all different types of *Streptomyces* plasmids (linear, circular, integrative, etc.), which suggests that a unique transfer system evolved in this genus.

Conjugative transfer of archaeal plasmids has been detected in several studies (Erauso et al., 2006; Greve et al., 2004; Prangishvili et al., 1998; Schleper et al., 1995; Stedman et al., 2000) and is described in more detail in the chapter "Archaeal plasmids".

MOBILIZABLE PLASMIDS

Mobilizable plasmids are transferred from donor to recipient cells by means of the transfer machineries provided by co-resident self-transmissible plasmids. These two plasmids do not have to be stably maintained together in a cell but simply co-reside long enough for mobilization to occur. In general, they encode proteins required to produce the mobilizable form of the plasmids (so called Mob proteins or DNA relaxases) (Francia et al., 2004). Mobilizable plasmids are generally small (<15 kb) because of the lack of their own transfer system. An exception is the recently characterized 57-kb IncP-6 plasmid, Rms149, 79% of which consist of multiple mobile elements (Haines et al., 2005). The plasmids in the IncQ group, RSF1010 (Guerry et al., 1974), R1162 (Bryan et al., 1972), and R300B (Barth and Grinter, 1974) are the best-characterized examples of mobilizable plasmids. The IncQ plasmids are mobilized efficiently by the IncP-1 plasmids but less so by other self-transmissible plasmids (Cabezon et al., 1994; Cabezon et al., 1997). They can be transferred into plant and animal cells (Buchanan-Wollaston et al., 1987; Yoshida et al., 1997). A remarkable feature of this group is an extreme BHR, i.e., the IncQ plasmids can replicate in G^- and G^+ bacteria and in cyanobacteria (Rawlings and Tietze, 2001). Another example of mobilizable plasmids is plasmid pBBR1 from *Bordetella bronchiseptica*, derivatives of which are now widely used as BHR mobilizable vectors for genetic manipulation (Antoine and Locht, 1992; Kovach et al., 1995; Kovach et al., 1994). Plasmid pBBR1 codes for only two proteins, the replication initiation protein, Rep, and the mobilization protein, Mob. The plasmid is mobilized by the IncP-1 transfer machinery, can replicate in a wide variety of Proteobacteria (Kovach et al., 1995), but does not belong to the known BHR groups (Antoine and Locht, 1992). Its replication initiation protein shows similarity with that of pM3 from the IncP-9 group, but since they are compatible in a bacterial cell (Greated et al., 2000), the incompatibility group of pBBR1 is not known yet. Thus there is greater diversity of mobilizable plasmids in G^- bacteria than currently understood.

Many mobilizable plasmids have been identified in *Bacillus*, *Lactobacillus*, *Rhodococcus*, *Staphylococcus*, *Streptococcus*, etc. An archetype of such plasmids is pMV158 from *Streptococcus agalactiae* (Burdett, 1980), the causative agent of group B streptococcal infections. This plasmid can be established in both G^- and G^+ bacteria and can also be mobilized by self-transmissible plasmids from both groups (del Solar et al., 1993; Farias and Espinosa, 2000). Its mobilization protein, MobM, was purified and characterized in detail (Guzman and Espinosa, 1997). The MobM protein is a DNA relaxase that cleaves supercoiled DNA at the pMV158 *oriT* sequence. Plasmid pMA67 released from a honeybee pathogen *Paenibacillus larvae* is a relative of pMV158 with a similar size and gene organization (Murray et al., 2007). Two mobilizable plasmids, pB264 (Lessard et al., 2004) and pAN12 (Yang et al., 2007) from *Rhodococcus* strains have been shown to be mobilized in the presence of co-resident megaplasmids.

The accumulating sequence data of mobilizable plasmids from G^- and G^+ bacteria has allowed a classification based on the relaxase similarities. The proposed classes are the MOB_Q (IncQ-like) superfamily, the ColE1-superfamily, the pMV158-

superfamily, and the ColDF13-superfamily (Clewell and Francia, 2004; Francia et al., 2004). It should be noted that this classification is not replacing the traditional classification of plasmids based on their incompatibility, because plasmids within each of the four groups have different replication machineries. As of the time of writing this chapter, no intrinsically mobilizable plasmid has been found in the Archaea, although a few deletion variants of archaeal self-transferable plasmids can be mobilized by the parent plasmids (Stedman et al., 2000).

MOBILE GENETIC ELEMENTS CARRIED ON PLASMIDS

Whereas plasmids themselves are a group of MGEs that mediate gene transfer between cells, even distantly related bacteria, they carry other MGEs such as insertion sequences (ISs), transposons (Tns), and integrons (Ins), which are responsible for transfer of genetic material within the cell (Chandler and Mahillon, 2002; Mahillon and Chandler, 1998; Mazel, 2006; Rowe-Magnus and Mazel, 2001; Tsuda et al., 1999). Through their ability to move between chromosome and plasmids, these elements contribute to HGT as well as to bacterial genome evolution. Plasmids are known to be vehicles that shuttle these elements among bacterial populations. Considering the close association of ISs, Tns, and Ins with plasmids and their significant roles in HGT, we briefly summarize their features here.

INSERTION SEQUENCES (ISs)

ISs, also known as bacterial class I transposons, are the most ubiquitous MGEs in bacterial genomes. Their structure is usually quite simple because most ISs carry only one gene for their own transposition (transposase gene) between short (<50 bp) inverted repeat sequences (IRs) (Fig. 2). Transposition of ISs is usually carried out by the mechanism called "simple insertion", "conservative transposition", or "cut-and-paste" (Fig. 3A): An IS element is physically separated from its donor molecule by double-stranded cleavage at the end of the element (cut), then inserted into the target molecule (paste). Thus no duplication of the element occurs. Most ISs generate short (<15 bp) directly repeated sequences (DRs) of the target DNA upon transposition, resulting in IS elements flanked by DRs. Therefore a DR is evidence that an IS element has transposed into that location. Usually insertion of ISs occurs at random, but some transpose target-specifically. Examples are those in the IS*1111* family, known to be inserted preferentially at specific positions in the 38-bp terminal IRs of Tn*3*-like (class II) transposons without duplication of the target sequence (see below for the Tn*3*-like transposons) (Partridge and Hall, 2003; Yano et al., 2007). Many host factors such as IHF (integration host factor) and Fis (factor for inversion stimulation) and environmental factors such as temperature are involved in transposition of IS elements (Ohtsubo et al., 2005; Signon and Kleckner, 1995; Weinreich and Reznikoff, 1992).

A DNA segment bracketed by two copies of (nearly) identical IS elements is called a composite transposon, since this segment and two ISs can act as a single

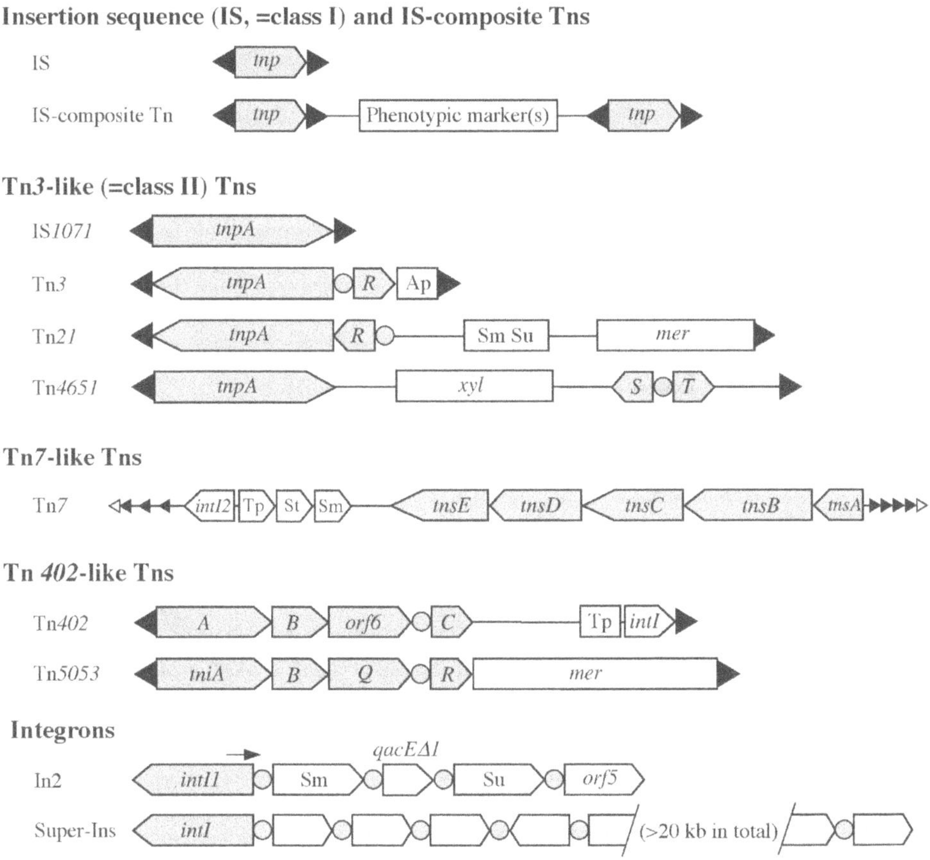

Fig. 2. Schematic structure of MGEs found on plasmids. Black arrowheads except those in Tn*7* indicate the terminal IRs of Tns. In Tn*7*, the 30-bp terminal IR consists of an 8-bp sequence (white arrowhead) followed by a 22-bp sequence (black arrowhead). Pentagons indicate the genes with their direction. Gray genes are those involved in transposition/movement of the elements. The circles show the site-specific recombination sites where DNA resolvases or integrases mediate recombination reactions. Abbreviations: *tnp*, transposase gene; R, S, and T in Tn*3*-like Tns, genes required for resolution; *mer*, genes encoding resistance to mercury; *xyl*, genes coding for degradation of toluene and xylene; Ap, Sm, St, Su, and Tp, genes encoding resistance to respectively ampicillin, streptomycin, streptothricin, sulfonamide, and trimethoprim; and *int*, integrase gene. Details for these MGEs are described in the section of "Mobile genetic elements carried on plasmids".

transposable unit. Tn*5* and Tn*10* are the representatives of the composite Tns, and carry respectively a kanamycin and tetracycline resistance gene (Berg et al., 1984; Kleckner et al., 1996). Large catabolic gene clusters, such as those for benzene, chlorobenzene, toluene, and nylon oligomers are also found to form composite Tn-like structures (Kato et al., 1994; Reddy et al., 1994; Tan, 1999; van der Meer et al., 1991), but only a few of them have been shown to be mobile. This might be explained by the fact that the IS-encoded transposases act at the cognate IS

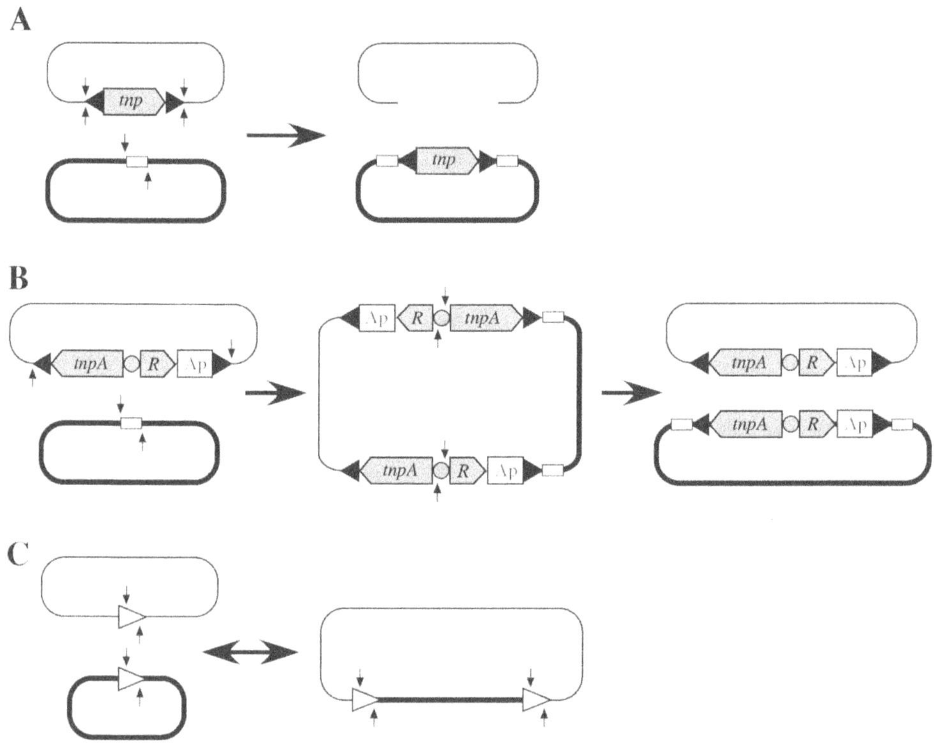

Fig. 3. DNA recombination mechanisms mediated by Tns and integrase-based elements. Vertical arrows indicate cleavages by DNA recombinases (transposase, integrase, or resolvase), and the white rectangles represent the target sequences. (A) Cut-and-paste transposition used by ISs and Tn*7*-like Tns. The model shows the transposition of an IS element. The element is excised from the donor molecule (thin line) by means of double-stranded cleavages at both ends of the IS, and then integrated into the recipient molecule (thick line) with a staggered cleavage. The element is thus conserved through the reaction and generates a target duplication upon transposition. (B) Copy-and-paste transposition used by Tn*3*-like and Tn*402*-like Tns. The transposition model of Tn*3* is depicted. The 3'-ends of the Tn are cleaved by TnpA and jointed with the staggered ends in the recipient molecule. The Tn is copied in the reaction, and two copies of the Tn connect the donor and recipient molecules to form a cointegrate. The cointegrate resolves at the resolution sites (circle) by the function of site-specific recombinases encoded on the Tn. As a result, each of the donor and recipient carries a copy of the Tn. (C) Integration and excision used by integrase-based elements such as Ins and conjugative Tns. Integrases mediate site-specific recombination between two attachment sites (arrowhead), resulting in integration (right to left) or excision (left to right) of a DNA fragment.

ends close to their point of synthesis (Chandler and Mahillon, 2002), resulting in high frequency transposition of each IS element prior to the transposition as a composite Tn.

The growing number of IS elements has created a need for a good classification and nomenclature system. Mahillon and Chandler (1998) have classified the IS elements based on the following criteria: (i) similarities of genetic organization; (ii) similarities in transposases; (iii) similarities of IRs; and (iv) determined length of DRs. The following rule is now widely used for the nomenclature of newly found elements: IS + "first letter of genus" + "first two letters of species" + "number". For instance, IS elements found in *Pseudomonas putida* have been named as IS*Ppu1* to IS*Ppu21*. General information on IS elements is available at the IS Finder web site (http://www-is.biotoul.fr/).

TN*3*-LIKE (CLASS II) TRANSPOSONS

Tn*3*-like Tns, also known as bacterial class II Tns, are basically composed of three parts: a transposase gene (*tnpA*), a resolvase gene (*tnpR*), and IRs (Fig. 2) (Grindley, 2002; Sherratt, 1989). Most of them additionally encode phenotypic traits such as antibiotic resistance and xenobiotic degradation, which have been incorporated during the evolutionary history of the Tns (Fig. 2). The transposition is usually accompanied by duplication of the transposon, which is called the "replicative transposition" or "copy-and-paste" mechanism (Fig. 3B). In the first step, the TnpA protein acts at the IRs to generate a cointegrate of the donor and target molecules connected by two directly repeated copies of the transposon, one at each junction. In the second step, the cointegrate resolves at the resolution (*res*) sites by means of the *tnpR* product (resolvase), and a 4- to 6-bp duplication of the target sequence is generated (Grindley, 2002; Sherratt, 1989; Shintani et al., 2005b; Tsuda et al., 1999; Yano et al., 2007). A detailed molecular model of the transposition is described elsewhere (Grinsted et al., 1990; Shapiro, 1979; Sherratt, 1989). The TnpAs of the Tn*3*-like Tns are able to catalyze their transposition even when the *tnpA* gene and cognate IRs are located on separate molecules, while the transposases of IS elements predominantly act on IRs in *cis* (Chandler and Mahillon, 2002; Grindley and Reed, 1985). This feature allows efficient transposition of Tn*3*-like Tns with very large sizes (>50 kb), e.g., the toluene-catabolic Tns Tn*4651* (56 kb), Tn*4653* (70 kb), and Tn*4657* (86 kb) (Tsuda et al., 1989; Yano et al., 2007) and the carbazole-catabolic Tn Tn*4676* (73 kb) (Maeda et al., 2003; Shintani et al., 2005b). The *tnpA* genes (ca. 3 kb) are larger than the transposase genes from the IS elements (<2 kb), probably because they confer transposition immunity, by which presence of a Tn (more exactly, a single IR) reduces the insertion frequency of the cognate element into the same replicon (Kans and Casadaban, 1989).

Several Tn*3*-like Tns encode for integrases instead of resolvases to resolve the cointegrates. These enzymes also play important roles in the movement of various MGEs (Burrus et al., 2002; Mazel, 2006; Nash, 1996). While the resolvases, most of which belong to the serine recombinases, catalyze only the intramolecular recombination between two copies of the specific *res* sites on the same replicon (Smith and Thorpe, 2002), the integrases, most of which are classified into tyrosine recombinases, catalyze both the intra- and intermolecular recombination between two attachment (*att*) sites (Fig. 3C) (Grainge and Jayaram, 1999; Nash, 1996). Well-characterized examples of integrases encoded on Tn*3*-like Tns are the TnpS protein from the toluene-catabolic Tn Tn*4651*, the TnpI protein from the naphthalene-catabolic Tn Tn*4655*, and the TnpI protein from the cryptic Tn Tn*4430* (Genka et al., 2002; Mahillon and Lereclus, 1988; Sota et al., 2006b). Comparison of the phylogenetic tree of transposases with that of resolution proteins (resolvases and integrases) clearly indicates that the resolution genes have been incorporated into Tn*3*-like Tns from other sources (Grindley, 2002). At present, it is not known how Tn*3*-like Tns acquired the resolution genes.

IS*1071* is a unique IS elements because it is the only one known to belong to the Tn*3*-like (class II) Tns, based on the sequences of its transposase gene and IRs (Nakatsu et al., 1991). It is the only Tn*3*-like Tn that does not encode a resolution function (Fig. 2). This element was first found as a component of the chlorobenzoate-catabolic composite Tn, Tn*5271* (Nakatsu et al., 1991). The IRs of IS*1071* are much longer than those of other known Tn*3*-like Tns, and the TnpA protein show only ca. 20% identity with other Tn*3*-like TnpAs. Since Tn*3*-like Tns are considered to have evolved by acquisition of resolution function as described above, IS*1071* could be an ancestral form of Tn*3*-like Tns (Grindley, 2002; Sota et al., 2006a; Wyndham et al., 1994). Curiously, most IS*1071* elements found so far are flanking genes that encode degradation of man-made chemicals on IncP-1 plasmids. Moreover, IS*1071* is predominantly distributed in β-Proteobacteria, which is consistent with the finding that it is only functional in these strains (Sota et al., 2006a). These observations strongly suggest the importance of IS*1071* together with IncP-1 plasmids in the dissemination of xenobiotic degradation genes, and the close relationship of this element with these plasmids as well as with the β-Proteobacteria.

Tn*3*-like Tns are categorized into four major groups (IS*1071*, Tn*3*-, Tn*21*-, and Tn*4651*-family) based on their structural organization and functional exchangeability (Fig. 2) (Grindley, 2002; Grinsted et al., 1990; Sota et al., 2006a; Tsuda et al., 1989). Tn*21*-family Tns can be further divided into two subfamilies, Tn*21*- and Tn*1722*-subfamily Tns, according to the interchangeability of TnpA proteins (Grinsted et al., 1990). Tn*21*-family Tns are shown to be the predominant Tn*3*-like Tns in natural bacterial communities (Dahlberg and Hermansson, 1995). A novel fifth group, Tn*5393*-family, should be added to Tn*3*-like Tns, because (i) Tn*5393* and its derivatives (Tn*5393*a, Tn*5393*b, Tn*5393*c, and Tn*5393*d) code for TnpAs with at most 40% amino acid identity to those of the other Tns and (ii) their IRs contain 81 bp, which is not found in the other four groups (Chiou and Jones, 1993; L'Abee-Lund and Sorum, 2000; Mantengoli and Rossolini, 2005; Sundin et al., 2004). The Tns and their groups described above were identified in G^- bacterial strains, but Tn*3*-like Tns found in G^+ bacterial strains such as the cryptic Tn Tn*4430* from *Bacillus thuringiensis* (Mahillon and Lereclus, 1988) and the vancomycin Tn

Tn*1546* from *Enterococcus feacium* (Arthur et al., 1993) would have to be added to the classification. In all groups, the 5'-GGGG sequences at the 5' outermost position of the IRs are conserved. Mutations in the four nucleotides, especially at the second position, reduce the transposition frequency of Tns (Sota et al., 2002; Sota et al., 2003; Wiater and Grindley, 1991). Of the four known families, only two TnpAs from the Tn*3*-family [Tn*3* and Tn*1000* (= γδ)] have been investigated biochemically to reveal their enzymatic features (Ichikawa et al., 1987; Wiater and Grindley, 1991), even though the TnpAs of Tn*3*-like Tns are divergent. Such characterization of other TnpA proteins will be necessary to gain insight into their detailed molecular characteristics as well as the evolutionary history of the Tns.

TN7-LIKE TRANSPOSONS

Tn*7* is a 14-kb Tn that was first found on the IncIα plasmid R483 from a bacterium with resistance to trimethoprim and streptomycin isolated from a calf (Barth et al., 1976). To date, Tn*7* derivatives are found in *P. aeruginosa* (Haines et al., 2007) (DQ176869), *Acinetobacter baumannii* (DQ176450, DQ176451), *Enterobacter cloacae* (EF042190, DQ268533), *Raoultella terrigena* (DQ275532), *Shigella sonnei* (Yang et al., 2005), *Thiobacillus ferrooxidans* (Oppon et al., 1998), and *E. coli* (AP002527). All elements carry antibiotic resistance genes. Whereas most Tns carry one or two genes for transposition, Tn*7* carries five genes to mediate its mobility (*tnsABCDE*) (Fig. 2) (Craig, 2002). The terminal IRs of Tn*7* are composed of 30-bp sequences, but the presence of only the IRs is not sufficient for transposition. Seven copies of a 22-bp sequence, which is the 3' part of the IR and known as the TnsB binding site, are located at the ends of Tn*7* (Fig. 2). Several, but not all, copies of the 22-bp sequence are necessary for Tn*7* transposition (Arciszewska et al., 1989). Tn*7* usually moves by the cut-and-paste mechanism (Fig. 3A) and generates a 5-bp duplication upon transposition. A noteworthy feature of Tn*7* is its target specificity, which is determined by two subsets of transposition proteins, TnsABC + TnsD or TnsE. TnsABC + TnsD mediate Tn*7* insertion into the specific site, called *attTn7* site, in bacterial chromosomes, while the TnsABC + TnsE transposition does not recognize any particular target sequences (Waddell and Craig, 1988). The site-specific transposability has provided a useful gene cloning and expression system using mini-Tn*7* (Choi et al., 2005), since no host genes are damaged by the insertion. Sophisticated genetic and biochemical studies have been done for Tn*7* and reviewed in many references (Craig, 1991, 1996, 2002; Peters and Craig, 2001). Although all Tn*7* derivatives characterized so far encoded antibiotic resistance, a recent study has shown that they also form genomic islands with various traits such as heavy metal resistance, restriction modification system, and DNA repair (Parks and Peters, 2007). Whole genome determination of *Vibrio parahaemolyticus* (Makino et al., 2003) has also revealed that a pathogenicity island is carried on a Tn*7*-like element (T. Sugiyama and T. Iida, personal communication).

TN*402*-LIKE (TN*5053*-LIKE) TRANSPOSONS

Tn*402*-like transposons were first found as "classic" integrons that were defined as DNA segments carrying an integrase and antibiotic resistance genes (gene cassettes) between terminal 25-bp IRs (Stokes and Hall, 1989). A few yeas later, Kholodii et al. (1993) reported the mobility and structure of a DNA segment, designated Tn*5053*, found in the chromosome of a mercury-resistant *Xanthomonas* strain. This element does not have an integrase. After the report, detailed analysis of Tn*5053* clearly showed that the four genes, *tniABQR*, and the 25-bp terminal IRs of Tn*5053* are required for transposition (Fig. 2) (Kholodii et al., 1995). In 1996, several integrons (Ins) were redefined as defective Tn*402*-like elements (Brown et al., 1996), and thus the novel family of Tns had been established. Tn*402*-like transposons move by the copy-and-paste mechanism (Fig. 3B) and generate a 5-bp duplication after transposition. Defect in either one of TniABQ proteins completely disrupts transposition (Kholodii et al., 1995). Tn*402*-like Tns have been categorized into the same family as Mu and Tn*7* (Craig, 2002; Radstrom et al., 1994), but we think that they form a separate group since they require a resolution reaction catalyzed by TniR, which is not encoded on Tn*7*.

Several Tn*402*-like Tns (Tn*5053*, Tn*4671*, and Tn*4672*) carry mercury resistance genes and a 38-bp IR sequence that are closely related to those of the mercury resistance Tn, Tn*21*, suggesting that these genes have been incorporated into Tn*21* elements in the past (Kholodii et al., 1993; Sota et al., 2003). Antibiotic resistance determinants in Tn*402*-like Tns seem to have been incorporated by Ins as gene cassettes with the help of integrases, because most of the antibiotic resistance Tn*402*-like elements carry integrases or their remnants (Schlüter et al., 2007). Interestingly, Tn*402*-like Tns region-specifically transpose to the *res* site of resolvases (Minakhina et al., 1999) and therefore are identified near resolvase genes in various kinds of plasmids. This region-specific transposition into the *res* sites for the ParA resolvase genes in IncP-1 plasmids has contributed to the striking structural similarity of these plasmids (Sota et al., 2007).

INTEGRONS (INS)

In the latest definition, an integron is a DNA segment that consists of an integrase gene (*intI*) and accompanying ORFs, so-called gene cassettes, which are excised and incorporated by the IntI integrase through site-specific recombination reaction at the *att* recombination sites (Fig. 2) (Fluit and Schmitz, 2004; Mazel, 2006; Recchia and Hall, 1995). The *attI* site is the primary recombination site situated just upstream of the *intI* gene, and the *attC* site (or previously called 59-bp element) is a 57- to 147-bp sequence carried at the 3' end of a gene cassette. Since the *att* sites are left after excision or integration of a gene cassette, Ins can keep accepting new gene cassettes (Fig. 3C). The ORFs in the gene cassettes are usually transcribed from a promoter that is located in the *intI* gene (Fig. 2). The gene cassettes encode resistance to various antibiotics: β-lactams, aminoglycosides, chloramphenicol, rifampicin, trimethoprim, etc. (Fluit and Schmitz, 2004; Recchia and Hall, 1995). Although Ins themselves were first described as mobile elements (Stokes and Hall, 1989), they cannot move by themselves: only the gene cassettes move between DNA replicons. Details of the mechanism of gene cassette recombination have been described recently by MacDonald et al. (2006).

To date, there are two superfamilies of Ins: mobile (or resistance) Ins and super-Ins. Mobile Ins, although not mobile *sensu stricto* but located on mobile Tn*402*-like Tns (see above) or other MGEs such as Tn*7*, have been subdivided in five classes based on the sequence of the integrases (Mazel, 2006). Class 1 Ins are the predominant group of the mobile Ins and usually found in the Tn*402*-like transposons or their remnants. They are found extensively in clinical bacterial isolates with a wide variety of known antibiotic resistance (Rowe-Magnus and Mazel, 2002). A few examples have been reported in each of the four remaining mobile integron classes (Arakawa et al., 1995; Collis et al., 2002; Hochhut et al., 2001; Sundstrom et al., 1991). Super-Ins have a very large number of gene cassettes. The first super-In from *Vibrio cholerae* contains 216 ORFs, most of which encode unknown functions (Mazel et al., 1998). In contrast to the mobile Ins, super-Ins are located on the bacterial chromosomes and are not linked with MGEs. Other than *Vibrio*, super-Ins have been identified in various species, e.g., *P. alcaligenes* (Vaisvila et al., 2001), *Shewanella oneidensis* (Drouin et al., 2002), and *Xanthomonas campestris* (Rowe-Magnus et al., 2001). The fact that (i) some gene cassettes of mobile Ins are identical to those of super-Ins and (ii) the IntI1 integrase of a class 1 In can recruit gene cassettes of super-In in *V. cholerae* into the *attI1* site, suggests that the gene cassettes in mobile Ins are possibly derived from super-Ins (Fluit and Schmitz, 2004; Mazel, 2006; Rowe-Magnus et al., 2002; Rowe-Magnus et al., 2001). Although the exchange of gene cassettes between different Ins has hardly been investigated, the extant diversity of Ins is probably due to cassette exchange. This is consistent with the fact that the IntI1 and IntI2 from class 1 and class 2 Ins, respectively, can excise each other's gene cassettes at their flanking *attC* sites (Hansson et al., 2002).

PHENOTYPIC TRAITS CARRIED ON PLASMIDS

A wide variety of phenotypes has been found to be encoded on plasmids since the discovery of antibiotic resistance plasmids (so-called R plasmids). These phenotypic traits benefit the host in specific environments, allowing them to expand their niche or simply to survive. As described above, these genetic determinants can be exchanged among bacterial communities and thus further evolve by mutations and/or rearrangements of the genes in new hosts, leading to the generation of novel phenotypes (Springael and Top, 2004). Since covering all phenotypic traits is beyond the scope of this chapter, we focus on a few examples that were recently identified. Most other traits have been described elsewhere (Funnel and Phillips, 2004; Kado, 2000; Thomas, 2000b).

ANTIBIOTIC RESISTANCE

An important mechanism by which bacterial pathogens acquire antibiotic resistance is through conjugative transfer of plasmids that encode various antibiotic resistance determinants. This presents a serious clinical and public health problem world-wide because numerous pathogens have developed resistance to multiple antibiotics, and in some instances treatment options are becoming limited (Gibbons, 1992; McGowan, 2006; Paterson, 2006). Transfer of multiple antibiotic resistance by plasmids between *Enterobacteriaceae* was first confirmed in the late 1950s and early 1960s, and this phenomenon is now widely recognized in almost all bacterial species, especially those that include clinical isolates such as *P. aeruginosa*, *S. enterica*, *E. feacalis* and *S. aureus*. Resistance plasmids are also often found in bacterial communities of wastewater treatment plants. Since these plants form interfaces between different bacterial habitats such as hospitals and surface water, they are considered as important reservoirs from where antibiotic resistance genes can spread further (Schlüter et al., 2007). Antibiotic resistance genes can also be identified in faecal enterococci of wild animals (Poeta et al., 2005), although their presence has shown not to be a universal property of enterobacterial populations, and is possibly a result of the human use of antibiotics (Osterblad et al., 2001; Skurnik et al., 2006). Since drug-sensitive bacteria are rapidly outcompeted by their resistant counterparts in the presence of the drug, resistance plasmids are under strong selection. Thus, bacteria resistant to a certain antibiotic thanks to the acquisition of a mobile resistance gene can appear soon after the drug became available to humans.

Antibiotic resistance genes have been found on almost all Inc groups of plasmids from G^+ and G^- bacteria, especially those isolated from clinical environments, and genes for resistance to almost all major groups of antibiotics have been reported: β-lactams (ampicillin, amoxicillin, cephalosporin, cephamycin, carbapenem etc.), aminoglycosides (kanamycin, streptomycin, gentamycin etc.), chloramphenicol, tetracycline, macrolides (azithromycin, erythromycin, clarithromycin etc.), glycopeptides (vancomycin and teicoplanin), quinolones (ciprofloxacin) etc. There are three mechanisms of plasmid-mediated resistance to antibiotics: (i) destruction or modification of the antibiotics, (ii) altered uptake of the antibiotics (e.g., multi-drug efflux system), and (iii) modification of the target of the antibiotics. We refer to the recent review article by Taylor et al. (2004) for the details of the mechanisms as well as many examples of resistance plasmids from various bacterial sources.

The BHR plasmids of the IncP-1 group are one of the best-studied archetypes of resistance plasmids. The first ones were isolated in 1969 from multi-resistant clinical isolates of the species *P. aeruginosa* and *Enterobacter aerogenes* (formerly *Klebsiella aerogenes*) (Datta et al., 1971; Jobanputra and Datta, 1974; Thomas and Smith, 1987). Plasmid RK2 (synonymously R18, R68, RP1, and RP4, so-called Birmingham IncP-1α plasmids) was shown to carry genes for resistance to ampicillin, kanamycin, and tetracycline. Since then, many IncP-1 or IncP-1-like plasmids with drug resistance determinants and other phenotypic markers have been isolated. The antibiotics they code resistance to are very diverse and include aminoglycosides, β-lactams, macrolides, sulfonamides, tetracycline, trimethoprim, and chloramphenicol. The resistance genes are usually carried on MGEs, especially Tn*3*- and Tn*402*-like Tns (Schlüter et al., 2007; Sota et al., 2007). Since these Tns can transpose from IncP-1 resistance plasmids onto other replicons in the same cell, the resistance genes can be shuttled onto other plasmids, and pooled or scattered in bacterial genomes even when the plasmids disappear within a few bacterial generations. IncP-1 plasmids are found in different subgroups of the Proteobacteria (α, β, γ) and able to transfer (but not replicate) into G^+ bacteria (Mazodier et al., 1989), yeast (Heinemann and Sprague, 1989), and mammalian cells (Waters, 2001). Thus IncP-1 plasmids must have evolved by passing through diverse bacterial strains, and meanwhile acquiring accessory elements. The presence of Ins mostly carried on Tn*402*-like Tns could enable the plasmids to incorporate novel resistance genes from bacterial gene pools such as genomic islands including super-Ins. On the macrolide-resistance IncP-1β plasmid pRSB111, isolated from the final effluent of a wastewater treatment plant, the accessory element may have replaced the Tn*402*-like element present on the sister plasmid pB3, a nice example of IncP-1 plasmid evolution. Since IncP-1 plasmids not only transfer at high rates within a wide host-range, but also mobilizable drug reistance plasmids such as RSF1010 (IncQ) (Rawlings and Tietze, 2001) and pBBR1 (unknown Inc group) (Antoine and Locht, 1992) at similar rates (Adamczyk and Jagura-Burdzy, 2003; Schlüter et al., 2007; Thomas and Smith, 1987), they must have contributed to the worldwide spread of antibiotic resistance determinants.

One of greatest concerns in the current crisis of increasing antibiotic resistance is the appearance of vancomycin-resistant (Van^r) bacteria. Therefore we will focus on this particular example of horizontal transfer of drug resistance. Vancomycin, a glycopeptide antibiotic, has provided effective therapy against serious infections by G^+ bacteria since its introduction in the mid 1980s. It blocks the synthesis of the bacterial cell wall by binding to the D-alanyl-D-alanine terminus of the peptidoglycan precursors at the cell surface (Reynolds, 1989). In 1988, however, a clinical isolate of *E. feacium* was first reported to be resistant to vancomycin (Leclercq et al., 1988). Since then, Van^r enterococci (VRE) have been identified and cause serious problems in hospitals worldwide. Because the Van^r phenotype was transferred between enterococcal species as well as from enteroccoci to *S. aureus* (Leclercq et al., 1989; Noble et al., 1992), the emergence of Van^r *S. aureus* (VRSA) in clinical circumstances was anticipated. In 1997, the first *S. aureus* isolate with reduced susceptibility to vancomycin (<8 mg/l) was reported (Hiramatsu et al., 1997), and these type of *S. aureus* strains - now isolated worldwide - are designated as vancomycin intermediate-susceptible *S. aureus* (VISA) (Gemmell, 2004). Finally, a high-level VRSA (1024 mg/l) was isolated from a dialysis patient in Michigan (Chang et al., 2003).

Vancomycin resistance is often encoded on plasmids, and the prototype of the Van^r plasmids is the 34-kb mobilizable plasmid pIP816 from *E. faecium* BM4147 (Leclercq et al., 1988). This plasmid has a cluster of seven genes, *vanRSHAXYZ*, which is known as the *vanA*-type, one of seven categories (*vanA* to *vanG*) based on the resulting phenotypes of the host strain. The VanR and VanS proteins encode for a two-component regulatory system that activates transcription of *vanH*, *vanA*, and *vanX* in response to the presence of vancomycin (Arthur et al., 1992). The VanH, VanA, and VanX proteins are necessary for replacing the carboxy terminal D-alanyl-D-alanine of the cell wall precursor with a depsipeptide D-alanyl-D-lactate (Bugg et al., 1991), which shows a more than 1000-fold reduced affinity to vancomycin than the normal cell wall precursor (Bugg et al., 1991). The VanY and VanZ proteins are not strictly required for vancomycin resistance but contribute to resistance to glycopeptides (Arthur et al., 1995; Arthur et al., 1994). All seven genes on plasmid pIP816 are carried on a Tn*3*-like Tn, Tn*1546*, which has only a transposase gene (ORF1) and a resolvase gene (ORF2) in addition to the *van* genes (Arthur et al., 1993). The mobility of the Tn has been demonstrated in *E. coli* and *S. aureus* (Arthur et al., 1993; Weigel et al., 2003).

Dissemination of the *vanA*-type Van^r genes is closely associated with Tn*1546*-like elements, irrespective of the replicons that carry them. A plasmid from *E. faecium* CVM1869 isolated from a dog with a urinary tract infection in Michigan carried a Tn*1546* derivative with a 889-bp deletion in ORF1, an IS*1216V* insertion at its 5'-end, and an IS*1251* insertion between *vanS* and *vanH* (Simjee et al., 2002). Three Van^r plasmids, pHTα, pHTβ, and pHTγ were isolated from three different *E. faecium* strains in Japan (Tomita and Ike, 2005; Tomita et al., 2003). Plasmid pHTα (65.9 kb) and pHTγ (66.5 kb) are derivatives of pHTβ (63.7 kb) with insertions of a 2.2-kb IS*232*-like element and a 2.8-kb group II intron. Plasmid pHTβ was revealed to be a pheromone-independent conjugative plasmid with an insertion of a Tn*1546*-like element. A strain of VRE was also identified in France, and harbored a 200-kb self-transmissible plasmid with a Tn*1546*-like Tn (Naas et al., 2005). This Tn had an insertion of IS*16* in the *vanY* gene, which may explain the particular *vanD*-type resistance (low-level resistance to vancomycin and mostly susceptible to teicoplanin). Other VRE and VRSA isolates from Korea (Lim et al., 2006), Poland (Kawalec et al., 2007), Norway (Johnsen et al., 2005), and New York (Perichon and Courvalin, 2006) also harbor plasmids carrying Tn*1546*-like elements. Chromosomally located Tn*1546*-like elements with the *vanA* cluster were found in *E. gallinarum* isolates in Italy (Foglia et al., 2003). This worldwide distribution of Tn*1546* and its derivatives among pathogenic enterococci and staphylococci is really striking and a threat to human health. The dissemination of Tn*1546*-like elements among bacterial strains must have been promoted by conjugative plasmids, because most of them are carried on such plasmids, even though the first Van^r plasmid described, pIP816, is not self-transmissible.

In addition to plasmids with their transposons, conjugative Tns (CTns), also have been found to carry vanccomycin resistance determinants, with the archetype being Tn*1549* from a clinical *Enterococcus* isolate (Garnier et al., 2000). Since these MGEs are not the topic of this book, they are not discussed further.

HEAVY METAL RESISTANCE

Bacteria show resistance to toxic heavy metal ions including silver (Ag^+), arsenite (AsO_2^-), arsenate (AsO_4^{3-}), cadmium (Cd^{2+}), cobalt (Co^{2+}), chromate (CrO_4^{2-}), copper (Cu^{2+}), mercury (Hg^{2+}), nickel (Ni^{2+}), lead (Pb^{2+}), antimony (Sb^{3+}), tellurite (TeO_3^{2-}), thallium (Tl^+), and zinc (Zn^{2+}). Most of the determinants for resistance to these toxic ions have been identified on plasmids, and their function is usually based on energy-dependent efflux of toxic ions from the cell by membrane proteins that function either as ATPases or as chemiosmotic cation/proton antiporters (Silver, 1996, 2003; Silver and Phung, 1996). Analogous to resistance to antibiotics, acquisition of resistance to heavy metals allows bacteria to outcompete their metal-sensitive counterparts in a metal-polluted environment. For example, *E. coli* strains resistant to copper were isolated from the effluent of a piggery where pigs were fed a diet supplemented with copper sulfate, and bacteria with silver resistance were reported from hospitals and burn wounds where silver compounds were used as microbicidal agents (Bridges et al., 1979; Tetaz and Luke, 1983). Some resistance determinants are shown to be loaded on Tns, further suggesting their importance in rapid bacterial adaptation to new environmental conditions (Kholodii et al., 1993; Tuffin et al., 2005). Examples of plasmids with heavy metal resistance are described below.

Bacterial mercury resistance (Hg^r) is the only heavy metal resistance system that transforms the metal ion (Hg^{2+}) into a volatile and uncharged form (Hg^0). Plasmids that confer resistance to inorganic and organic mercury have been identified from both G$^+$ and G$^-$ bacteria. Although there are several variations on the structure and organization of Hg^r operons (*mer*), the mercury reductase gene, *merA*, the membrane-bound protein gene, *merT*, and the regulator gene, *merR*, are the minimum components of the operons. MerA is the key enzyme of Hg^r, which catalyzes the conversion of Hg^{2+} to Hg. MerT is required for uptake of Hg^{2+} into the cell, and MerR is a metal-specific activator-repressor of the *mer* operon that controls expression of the *mer* genes. Other Mer proteins are MerB, which cleaves the Hg-carbon bond in organomercurial compounds, MerC, which can also be involved in Hg^{2+} uptake, MerD, which allows continued expression of the structural genes even after Hg^{2+} is eliminated from the medium, MerE, with unknown function, MerF, which facilitates volatilization of Hg^{2+}, MerG, which prevents organomercuric uptake, and MerP, responsible for the transport of Hg^{2+} from the periplasm to the inner-membrane-bounded MerT. It should be noted that not all the Mer proteins shown above are necessary components of the resistance module, and that there are many combinations of *mer* genes. The molecular aspects of Hg^r are described in detail in a recent review (Barkay et al., 2003).

Resistance to organomercurial compounds is more common in G^+ than G^- bacterial strains. Plasmid pI258 from *S. aureus* confers resistance to inorganic mercury (Hg^{2+}) and phenylmercuric acetate (Weiss et al., 1977). The phenylmercury portion of the compound is hydrolyzed enzymatically to benzene and mercuric ion, which is then also enzymatically reduced to metallic mercury. The *mer* genes on pI258 cannot volatilize mercury from other organomercurial compounds such as methyl-, ethyl-, and p-hydroxymercuribenzoate. The genes are also functional in *B. subtilis* but not in *E. coli* (Chu et al., 1992; Laddaga et al., 1987). More recently, two *Streptomyces* linear plasmids, pRJ3l (322 kb) and pRJ28 (330 kb) were shown to carry *mer* genes homologous to those found on the chromosome of a different *S. lividans* strain, 66 (Ravel et al., 1998). Both plasmids could be transferred to a *S. lividans* 66 derivative that lacked the genes for mercury resistance, again confirming the high mobility of Hg^R determinants (Ravel et al., 2000). In G^- bacteria, plasmid pMR26 from *Pseudomonas* sp. K-62 encodes resistance to phenyl-, methyl-, and ethylmercury (Kiyono et al., 1995). A specific feature of the pMR62 *mer* genes is the presence of *merG*, deletion of which had no effect on resistance to Hg^{2+}, but rendered the host more sensitive to phenylmercury (Kiyono and Pan-Hou, 1999). The MerG protein, located in the periplasm, is thought to reduce cell permeability to phenylmercury (Kiyono and Pan-Hou, 1999). The *mer* operon (*merRTPABD*) of plasmid pDU1358 from *Serratia marccens* is highly homologous to those of Tn*21* (*merRTPCAD*) and Tn*501* (*merRTPAD*) but in addition carries the *merB* gene required for cleavage of the C-Hg bond of phenylmercury (Griffin et al., 1987).

Resistance determinants to inorganic mercury ion have been identified on various plasmids such as the *S. flexneri* virulence plasmid pWR501 (Venkatesan et al., 2001) and the *P. stutzeri* mercury resistance plasmid pPB (Reniero et al., 1998). Well-studied examples are pVS1 from *P. aeruginosa* (Stanisich, 1974) and R100 (= NR1 or R222) from *S. flexneri* (de la Cruz and Grinsted, 1982), which carry respectively the Tn*3*-like Hg^r Tns Tn*501* and Tn*21*. These transposons are apparently contributing to the dissemination of the *mer* genes because Tn*501*- or Tn*21*-like sequences are often found near the genes. The Tn*402*-like Hg^r transposon Tn*5053* (Kholodii et al., 1993) is also associated with the distribution of *mer* genes. Among the Hg^r plasmids, the IncP-1 plasmids are known to have a very close relationship with the Hg^r Tns (Schlüter et al., 2007; Thomas and Smith, 1987). For example, the multidrug-resistance plasmid pB10 (Schlüter et al., 2003) carries a truncated Hg^r Tn whose *mer* cluster is nearly identical to that on Tn*501*, and the closely related 2,4-dichlorophenoxyacetic acid degradative plasmid pJP4 carries a Tn*501* remnant (Trefault et al., 2004). The atrazine-catabolic plasmid pADP-1 (Martinez et al., 2001), the haloacetate-catabolic plasmid pUO1 (Sota et al., 2003), and the trimethoprim-resistance plasmid R751 (Thorsted et al., 1998) also possess full or partial Hg^r Tns. Plasmid pTP6 is the first example of an IncP-1 plasmid that carries only a Hg^r Tn and no other MGEs (Smalla et al., 2006). All these IncP-1 plasmids belong to the subgroup β. Four of the nine Hg^r plasmids, pQKH6, pQKH33, pQKH54, and pQM719, isolated from river epilithon were identified as IncP-1 (Haines et al., 2006a). Since Hg is very common in the Earth's crust, and Hg^r Tns were detected in *Pseudomonas* strains isolated from permafrost samples aged 15,000 to 40,000 years (Mindlin et al., 2005), it is not so surprising that Hg^r plasmids and Tns are ubiquitous and have spread widely among bacterial populations.

Arsenic, a well-known poisonous metalloid, and its compounds are used as pesticides, herbicides, insecticides and various alloys. A few examples of plasmids that encode resistance to arsenical ions (As^r; AsO_2^- and AsO_4^{3-}) are R478 from *S. marcescens* (Gilmour et al., 2004), R46 and R773 from *E. coli* (Hedges and Baumberg, 1973), pI258 from *S. aureus* (Silver et al., 1981), and pSX267 from *S. xylosus* (Götz et al., 1983). The As^r phenotype is generally conferred by either of the three *ars* operons, *arsRBC*, *arsRBCH*, and *arsRDABC*. The key components of As^r are ArsR, ArsB and ArsC. The ArsR protein is the regulator that controls the expression of the operons (Wu and Rosen, 1991); ArsC is an arsenate reductase that converts arsenate (AsO_4^{3-}) to arsenite (AsO_2^-), which can be pumped out through the transmembrane efflux channel protein, ArsB (Ji and Silver, 1992). Two novel ars genes, *arsO* and *arsT*, have recently been identified on the linear plasmid pHZ227 from *Streptomyces* sp. strain FR-008 (Wang et al., 2006). The ArsO and ArsT proteins appeared to be a flavin-binding monooxygenase and thioredoxin reductase. Recently an As^r operon was shown to be carried on a Tn*3*-like Tn, Tn*AtcArs*, located on the chromosome of the moderately thermophilic, acidphilic, sulfur-oxidizing bacterium *Acidithiobacillus caldus* (Kotze et al., 2006; Tuffin et al., 2005). Although it is unclear how the As^r operon was incorporated into a Tn*3*-like Tn, Tn*AtcArs* could be an intriguing example of evolution of heavy metal resistance Tns.

Silver compounds are used widely as antimicrobial agents in clinics (silver sulfadiazine), as disinfectants, etc., because silver ions (Ag^+) are highly toxic to microbes but generally have no adverse effects to humans (Silver, 2003). Molecular studies on bacterial silver resistance (Ag^r) have recently revealed the genes and proteins and mechanism of resistance (Gupta et al., 1999; Gupta et al., 2001; Silver, 2003). Plasmid pMG101 isolated from *S. typhimurium* encodes resistance to Ag^+, Hg^{2+}, TeO_3^{2-}, and several antibiotics (McHugh et al., 1975). The silver resistance involves nine ORFs (*silP*, ORF105, *silA*, *silB*, ORF96, *silC*, *silR*, *silS*, and *silE* in this order) arranged in three transcriptional units. Seven ORFs are named *sil* genes based on homologies to known proteins for other metal resistance. SilA is an inner membrane proton/cation anbtiporter, SilB is a membrane fusion protein that spans the inner and outer membranes of G$^-$ bacteria, and SilC is an outer membrane protein (Gupta et al., 1999). These proteins presumably form a three-polypeptide complex for a chemiosmotic cation/proton antiporter. The SilR and SilS proteins are thought to function as a two-component transductional regulator. SilE is a periplasmic Ag^+-binding protein, and SilP is predicted to be a P-type ATPase, a member of heavy metal resistance ATPases (Rensing et al., 1999). Comparison of the pMG101 *sil* genes (*silA* through *silS*) with those in the *E. coli* chromosomes (*agr* genes of K12 and O157:H7) indicate a close relationship in gene order as well as nucleotide identity (Gupta et al., 1999; Silver, 2003). Interestingly, the five IncH plasmids with Ag^r determinants isolated from *E. coli* all belong to the IncHI group (Gupta et al., 2001). Sequence analysis of these Ag^r IncHI plasmids will provide more genetic information on the association between these determinants and the IncHI plasmids.

A well-studied and intriguing bacterial strain that is highly resistant to at least nine heavy metals is *Cupriavidus* (formerly *Ralstonia*) *metallidurans* CH34. Two large plasmids, pMOL28 and pMOL30, confer resistance to Ag, Cd, Co, Cr, Cu, Hg, Ni, Pb, and Zn (Mergeay et al., 2003; Monchy et al., 2007). Plasmid pMOL28 carries the genes for resistance, to Co, Cr, Hg, and Ni, and its replication machinery is found to be similar to those of the symbiotic plasmid pSym of *C. taiwanensis* and pHG1 of *C. necator*, suggesting that these three plasmids form a new Inc group of plasmids (Monchy et al., 2007). Plasmid pMOL30 encodes resistance to Ag, Cd, Co, Cu, Hg, Pb, and Zn. The Hg^r determinants on pMOL28 and pMOL30 are identical and carried on Tn*3*-like Tns, Tn*4378* and Tn*4380*, respectively (Mergeay et al., 2003). The resistance genes for each metal are well organized and located on one and two genomic islands in pMOL28 and pMOL30, respectively. Both plasmids appear not to be self-transmissible, and it is not clear at present if they are mobilizable or not, because previously demonstrated transfer with IncP-1 plasmids seemed to have been mediated by the cointegrate formation of the two Hg^r Tns (Mergeay et al., 2003). Since the resistance genes for heavy metals are similar to those found on other bacterial plasmids and chromosomes and also associated with various MGEs, the present architecture of these large plasmids is a clear example of adaptive plasmid evolution, allowing the host to thrive in harsh environments contaminated with high concentrations of toxic metals.

For information on resistance to other heavy metals, we refer the reader to the followings recent reports of heavy metal resistance plasmids: Cu (Gilmour et al., 2004; Gilotra and Srivastava, 1997), Cd (Bruins et al., 2003; Udo et al., 2000), Co

(Alonso et al., 2002; Park et al., 2003), Zn (Alonso et al., 2002; Mahapatra et al., 2002), Cr (Juhnke et al., 2002; Tauch et al., 2003), Te (Chen et al., 2004; Vavrova et al., 2006), and Pb (Alonso et al., 2002; Chen et al., 2004). The mechanisms of resistance to these metals are described in several reviews, although some are yet to be elucidated (Nies, 2003; Silver, 1996; Silver and Phung le, 2005; Silver and Phung, 1996).

VIRULENCE

The involvement of plasmids in the virulence of pathogenic bacteria has long been known, but it was not until the late 1990's when entire sequences of large virulence plasmids were first published (Burland et al., 1998; Lindler et al., 1998). Such virulence plasmids, whose targets cover all living creatures, were reported from enterobacteria (*E. coli*, *S. enterica*, *S. flexneri*, *Y. pestis*, etc.), sporulating and non-sporulating G^+ bacteria (*C. botulinum*, *B. anthracis*, *S. aureus*, etc.), phytopatogenic bacteria (*A. tumefaciens*, *P. syringae*, *X. campestris*, etc.) and many more (Christie, 2004; Cornelis et al., 1998; Lindler, 2004; Pillar and Gilmore, 2004; Rood, 2004; Rychlik et al., 2006; Sundin, 2007; Venkatesan and Burland, 2004). These plasmids are generally low copy number and large in size (>50 kb). Most plasmids encoding virulence to plants such as the Ti plasmid from *A. tumefaciens* (Christie, 2004) and pXcB from *X. citri* (Marques et al., 2001) are transferable. In stark contrast, virulence plasmids from animal and humans pathogens such as pO157 from *E. coli* O157:H7 (Burland et al., 1998; Makino et al., 1998) and pCD1 from *Y. pestis* KIM5 (Lindler et al., 1998), have lost most of the genes required for the transfer machinery, and are therefore not transferable by conjugation. Since the virulence plasmids are known to have sequences in common with other non-virulence conjugative plasmids, the loss of transfer function in the mammalian virulence plasmids is very intriguing. It is tempting to speculate that loss of genes required for horizontal transfer is an example of so-called pathoadaptive mutations, i.e., mutations that allow pathogens to adapt to new hosts (Day et al., 2001; Sokurenko et al., 1999). If conjugative plasmid transfer is not required for persistence of a plasmid in natural populations, or even negatively interferes with the function of the bacterial pathogen in its new host, the genes may accumulate deleterious mutations and/or be deleted form the plasmid genome over evolutionary time.

Virulence plasmids in *Salmonella* have been identified in eight out of over 2500 serovars, *S. enterica* serovars Abortusoivis, Abortusequi, Choloraesuis, Dublin, Enteritidis, Gallinarum-Pollorum, Sendai, and Typhimurium (Chu and Chiu, 2006). It is very interesting that most of the serovars of *S. enterica* subspecies *enterica* do not harbor any plasmids, but those associated with infections of humans and animals, i.e., the eight serovars shown above, do (Rychlik et al., 2006). This might be explained by the need for these strains to rapidly adapt to changing environment such as increasing presence of antibiotics. The sizes of the serovar-specific virulence plasmids are from 50 to 100 kb, and all of the plasmids so far investigated carry the *Salmonella* plasmid virulence (*spv*) genes (*spvRABCD*). The Spv proteins are necessary for transfer of bacterial cells through host membranes, systemic spread to other tissues, and intraceller survival in macrophages (Venkatesan and Burland, 2004), but the detailed molecular mechanism for the virulence remains unclear. Only the SpvB protein has been determined to be a mono ADP-ribosyl transferase that regulates the conversion of G-actin to F-actin to destabilize the cytoskeleton of the macrophage (Otto et al., 2000; Tezcan-Merdol et al., 2001). The SpvR protein is a positive transcriptional regulator of the *spvABCD* genes. Expression of the *spv* genes is dependent on the growth condition such as carbon starvation and low pH. The *Salmonella* virulence plasmids are known to belong to F-like plasmids, i.e., they have the IncFIB and/or IncFIIA replicon (Rychlik et al., 2006). Most of the *Salmonella* virulence plasmids have lost their self-transferability, but a few are still conjugative (Ahmer et al., 1999) as well as mobilizable by the F plasmid (Barrow and Lovell, 1989).

Plasmid pO157 is found in all strains of enterohemorrhagic *E. coli* (EHEC) O157: H7, which causes sporadic cases of hemorrhagic colitis. While most *E. coli* strains are harmless and normally present in the gastrointestinal tract of mammals, this strain causes severe food-borne illness, sometimes resulting from eating raw or undercooked beef. Two pO157 sequences have been deposited in databases by an American group (from strain EDL933, 92,077 bp, AF074613) and Japanese group (from strain Sakai, 92,721 bp, AB011547), respectively (Burland et al., 1998; Makino et al., 1998). The difference between the two plasmids is that the latter has a 644-bp additional insertion with unknown function. The genes associated to virulence are (i) a cluster with 13 genes (*etpC-O*) that are highly homologous to the genes for the type II secretion system of G^- bacteria (Schmidt et al., 1997), (ii) a *hlyCABD* cluster that encodes a hemolysin (HlyA) and its secretion system (HlyB, HlyC, and HlyD) (Schmidt et al., 1995), (iii) *katP*, which encodes a bifunctional catalase-peroxidase (Brunder et al., 1996), (iv) *espP*, which codes for an extracelluler serine protease capable of cleaving pepsin and human coagulation factor V (Brunder et al., 1997), (v) *toxB*, whose product show limited homology (20%) to the Toxin B of *Clostridium difficile* (Barroso et al., 1990), and (vi) *tagA* whose product is similar to TagA of *Vibrio cholerae* and promotes aggregation of human T cells (Harkey et al., 1995; Venkatesan and Burland, 2004). Although the functions of these gene products are known, the mechanism of disease causation is still unclear. Recently, the third sequence of a plasmid from *E. coli* O157 (O157: H^- strain 3072/96) was published (Brunder et al., 2006). The plasmid, pSFO157 has a size of 121,239 bp, ca. 29 kb larger than the two other O157 plasmids. This plasmid does not possess the *katP*, *espP*, and *toxB* genes that are present on the two pO157 but instead has the *sfp* gene cluster (*sfpAHCDJG*) and F-like transfer genes. The Sfp proteins are responsible for mannose-resistant hemagglutination and the expression of *sfp* fimbriae (Brunder et al., 2001). Twenty-six *tra-* and *trb*-related genes of the F-like transfer system were identified in pSFO157, but lack of several genes lead to the conclusion that this plasmid is not conjugative at present. These three *E. coli* plasmids, the *Salmonella* virulence plasmids, and plasmid F apparently evolved from a similar ancestor since they seem to have similar replication machineries.

B. anthracis is the etiologic agent of anthrax, an acute fatal disease among mammals. Unfortunately, it was used as a bioterrorism agent in 2001 in the United States, resulting in five deaths. Three forms of the disease have been recognized: cutaneous (through skin), gastrointestinal (through alimerntary tract), and pulmonary (by inhalation of spores) (Bhatnagar and Batra, 2001). The major virulence factors of *B. anthracis* are a poly-D-glutamic acid capsule and a three-component protein exotoxin. Host-related signals affecting transcription of the toxin and capsule genes include temperature (37°C) and CO_2 concentration. The genes coding for the toxin and the enzymes responsible for capsule production are carried on plasmid pXO1 and pXO2, respectively. Both plasmids are essential for virulence. The most recognizable feature of the plasmid pXO1 is a pathogenicity island defined by a 44.8-kb region that is bordered by inverted IS*1627* elements at each end, indicating the incorporation of the island as an IS-composite Tn (Okinaka et al., 1999). Genetic analysis has revealed that the genes for the anthrax toxin proteins, *pagA*, *lef*, and *cya*, are located in distinct loci on pXO1, and the biosynthetic genes for capsule synthesis, *capABC*, and a gene associated with depolymerization of capsule, *dep* are arranged in an apparent operon on pXO2. The pathogenicity island contains a Tn*3*-like Tn, Tn*XO1*, which carries the *gerX* operon involved in the germination of *B. anthracis* spores within phagocytic cells (Van der Auwera and Mahillon, 2005). The pXO1-encoded gene *atxA* positively controls expression of the toxin and capsule genes as well as numerous other genes on both plasmids and the chromosome, therefore *atxA* mutants are avirulent in mice (Koehler, 2002). The pXO2-encoded regulator, *acpA*, which shares sequence similarity with *atxA*, appears to positively control capsule gene expression, although the *acpA* expression is positively affected by *atxA* (Bourgogne et al., 2003; Koehler, 2002). The chromosomal gene *abrB* also controls growth-phase-specific transcription of the toxin genes. Thus, the expression of virulence of *B. anthracis* is tightly regulated by many regulatory proteins. Although pXO1 and pXO2 are not self-transmissible, they can be mobilized by other conjugative plasmids. This may explain the recent detection of a pXO1-like plasmid, pBCXO1, in *B. cereus* strain G9241, with 99.6% nucleotide identity to pXO1 (Hoffmaster et al., 2004). This *B. cereus* strain does not bear a pXO2-like plasmid, but another plasmid, pBC218, which encode a polysaccharide capsule (Hoffmaster et al., 2004). Comparison of two disease-causing *B. cereus* plasmids, pPER272 from strain AH820 and pCER270 from AH187, with pXO1 revealed that these plasmids have a highly conserved region in their backbone sequences, suggesting a common evolutionary history (Rasko et al., 2007). Screening of a population of *B. cereus* group isolates revealed that pXO1-like plasmids are more often associated with clinical isolates, demonstrating that these plasmids may define pathogenic *B. cereus* isolates in the same way *B. anthracis* is conceptually defined as a species with the pXO1 and pXO2 plasmids (Helgason et al., 2000; Rasko et al., 2005).

SYMBIOSIS

Rhizobia are soil bacteria that live in a symbiotic relationship with legumes. Usually the rhizobia supply ammonia or amino acids to the plant and in return receive organic acids as a carbon and energy source, a good example of mutualism. Since rhizobia are also able to grow in soil outside the plant, this symbiotic function is not strictly required for the bacteria, but provides them with an additional niche. Symbiotic plasmids or symbiotic islands in bacterial chromosomes are the carriers of the genes required for symbiosis. They have been shown to be transferable (Romero and Brom, 2004; Sullivan and Ronson, 1998), suggesting that the symbiotic property of rhizobia was acquired by HGT. This was shown by the fact that the genes for nodulation (*nod*) of several β-proteobacterial strains (*Burkholderia phymatum*, *B. caribensis*, *B. tuberum*, and *C. taiwanensis;* nitrogen-fixing symbionts of the legume *Mimosa)* are very similar to those of the rhizobia, which belong to the α-Proteobacteria (Chen et al., 2005; Chen et al., 2003; Elliott et al., 2007; Moulin et al., 2001).

Symbiotic nitrogen fixation requires about 60 genes. Those genes are *nod* (nodulation), *nol* (nodulation loci), *noe* (nodulation efficiency), *nif* and *fix* (nitrogen fixation), etc. The *nod* genes are activated by signal compounds, flavonoid or isoflavonoids, which are commonly present in legume root exudates. The gene products play roles in the synthesis and secretion of the nodulation factors. The *nodABC* genes are essential for production of the nodulation factor. The nodulation factors allow rhizobia to invade the cytoplasm of root cells. In response to the nodule environment where O_2 is limited, transcription of the *nif* and *fix* genes (more than 20 genes) is initiated (Fischer, 1994, 1996). The gene products include a nitrogenase (*nifHDK* gene product) and proteins that provide energy for the nitrogen fixation (Romero and Brom, 2004). An interesting feature of the symbiotic plasmids is the scattering of the genes for symbiosis (nodulation and nitrogen fixation) on one or sometimes several plasmids (Freiberg et al., 1997; Barnett et al., 2001; Finan et al., 2001). An extreme example is *R. etli* CFN42, where five of six plasmids (p42a to p42f) are associated with symbiosis (Brom et al., 2000). Although p42a is not involved in symbiosis, it is indispensable for the transfer of the symbiotic plasmid p42d (Brom et al., 2004; Gonzalez et al., 2003); it is self-transmissible and can form a cointegrate with p42d possibly by an integrase-mediated recombination between two *att* sites (Fig. 3C), one on each plasmid (Brom et al., 2004). The mosaic and divergent structure of symbiotic plasmids as well as the scattered gene positioning may reflect frequent rearrangement events in and between these plasmids and genes during their evolutionary history.

Plasmids with the genes for symbiosis are usually very large (>300 kb). For example, plasmid pNGR234a from *Rhizobium* sp. NGR234, the first symbiotic plasmid to be completely sequenced (Freiberg et al., 1997), is 536 kb, and two symbiotic plasmids, pSymA and pSymB from *S. meliloti* 1021 are 1,354 and 1,683 kb in size, respectively (Barnett et al., 2001; Finan et al., 2001). Like other plasmids from *Rhizobiaceae*, replication of these enormous plasmids is controlled by the *repABC* replication/partitioning system. The RepA and RepB proteins are homologues of the ParAB family of replicon partitioning proteins (Bignell and Thomas, 2001), and RepC is the replication-initiation protein (Ramirez-Romero et al., 2000). The *repABC*-type replicons cannot replicate in *E. coli* (Bartosik et al., 2001). Interestingly, several *repABC*-type replicons are compatible as they can be stably maintained in the same cell. The regions responsible for the incompatibility of the *repABC*-type replicons are the *incα* and *incβ* sequences, which are located between *repB* and *repC* and downstream of *repC*, respectively (Ramirez-Romero et al., 2000). The *incα* region is essential for correct plasmid segregation, indicating that it encodes a centromere-like sequence, and the *incβ* region is essential for plasmid replication-initiation, thus probably represents an *oriV*-like sequence. It is noteworthy that the *repABC*-type replicons do not seem to possess direct repeats considered as iterons. A recent study has revealed the presence of a gene, namely *incA*, which is carried between *repB* and *repC* but transcribed antisense to the *repABC* genes (MacLellan et al., 2005). The gene product is an approximately 56-nucleotide untranslated RNA molecule that regulates plasmid replication by negatively regulating the synthesis of the replication-initiation protein, RepC. Therefore, the expression of *incA* mediates incompatibility for its cognate plasmids. Since the sequences of the *incA*-coding region are diverse among the *repABC*-type replicons (MacLellan et al., 2005), replicons that do not show mutual *incA*-mediated interference may be able to stably replicate in the same cell.

DEGRADATION OF ORGANIC COMPOUNDS

Plasmids encoding degradation of man-made (xenobiotics) and/or natural organic compounds allow the bacterial host to utilize these molecules as a source of carbon and energy and/or nitrogen. Such plasmids have been found in environments contaminated with herbicides, pesticides, petroleum, etc. Catabolic plasmids are usually large in size (>50 kb), because (i) typically several enzymes are required to degrade an organic pollutant, (ii) in most cases the genes for these catabolic cascades are organized as one or a few operons on the plasmid, and (iii) most of these plasmids are self-transmissible. Usually the genes and operons are regulated by regulatory proteins encoded by genes near or in the cascade genes. To the best of our knowledge, in G$^-$ bacteria catabolic genes are found only on six groups of plasmids, i.e., IncP-1, IncP-2, IncP-7, and IncP-9 plasmids mostly from *Pseudomonas* and β-Proteobacteria and unclassified plasmids from the α-Proteobacteria (mainly sphingomonads, see below) and *Flavobacterium* strains (Nojiri et al., 2004; Ogawa et al., 2004; Top et al., 2000). Catabolic plasmids of G$^+$ bacteria are mainly found in *Arthrobacter* and *Rhodococcus* (Ogawa et al., 2004). Interestingly, genes for degradation of naturally occurring compounds such as naphthalene and toluene are carried on IncP-2, IncP-7, and IncP-9 plasmids, while those encoding degradation of man-made compounds such as herbicides, and other chlorinated organics are found on IncP-1 and α-proteobacterial plasmids (Nojiri et al., 2004; Ogawa et al., 2004; Top et al., 2000). Since the catabolic gene clusters are usually associated with transposable elements, especially with Tn*3*-like or IS-composite Tns, most of which are thought to transpose randomly between replicons, the biased association of the degradation genes with this limited set of plasmid groups is fascinating. In this section we will focus on our recent and related studies on several catabolic plasmids belonging to the IncP-1, IncP-7, IncP-9, on α-proteobacterial plasmids, as well as the functions of catabolic Tns residing on the analyzed plasmids.

The IncP-1β plasmid pUO1 was detected in a soil bacterium *Delftia acidovorans* strain B isolated from industrial wastewater in Japan (Kawasaki et al., 1981b). The plasmid encodes two monohaloacetate dehalogenases: H-1, acting predominantly on fluoroacetate, and H-2, acting on chloro-, bromo-, and iodoacetate (Kawasaki et al., 1981a). It also confers resistance to mercury (Kawasaki et al., 1981b), and the mercury resistance genes are located on two Tn*402*-like Tns, Tn*4671* and Tn*4672*, which share one end (Sota et al., 2003). The genes for the dehalogenases, *dehH1* and *dehH2*, respectively, are associated with Tns. The *dehH2* gene is flanked by two intact and identical copies of IS*1071*, therefore leading to formation of an IS-composite Tn named Tn*Had1* (Sota et al., 2002). The *dehH1* gene is also bracketed by two copies of IS*1071* (one is shared with Tn*Had1*), but one of them is truncated (Sota et al., 2002). The two dehalogenase genes and three copies of IS*1071* are carried on a Tn*3*-like Tn, Tn*Had2*, which has lost its transposition genes but is still transposable when the Tn*21* transposase is supplied in *trans*. Tn*Had2* is located between the replication-initiation protein gene, *trfA*, and its *cis*-acting site, *oriV*, one of the two typical insertion sites for mobile elements in IncP-1 plasmids (Schlüter et al., 2007). This suggests that Tn*Had2* transposition into an ancestral form of IncP-1β plasmids generated pUO1. We have recently shown that the sequences around *oriV* and the *parA* resolvase gene, respectively, are hot spots for specific insertion of Tn*3*-like and Tn*402*-like Tns (Sota et al., 2007). It should be noted that all catabolic IncP-1 plasmids so far reported have IS*1071* sequences in the vicinity of the catabolic gene clusters and that IS*1071* has not been identified on any other group of plasmids from G$^-$ bacteria except IncP-1, indicating the special partnership between degadation genes, IS*1071* and IncP-1 plasmids. An exception is pTC1 from a G^{+} strain *A. aurescens* TC1 (Sajjaphan et al., 2004). This plasmid might have acquired IS*1071* from IncP-1 plasmids, since *Arthrobacter* sp. strains were shown to be good recipients for IncP-1 plasmids, although replication of these plasmids in the strains has not been confirmed (Musovic et al., 2006; Sørensen et al., 2005). Again, since IncP-1β plasmids and IS*1071* are predominantly found in β-proteobacteria (Top et al., 2002), where the IS element is also actively transposing (Sota et al., 2006a), we postulate that catabolic IncP-1 plasmids have evolved in the β-proteobacteria.

The IncP-9 plasmids NAH7 and pWW0 from *P. putida* G7 (Dunn and Gunsalus, 1973; Sota et al., 2006b) and mt-2 (Benson and Shapiro, 1978; Greated et al., 2002), respectively, have been extensively characterized with respect to the naphthalene and toluene/xylene degradation genes (*nah* and *xyl*, respectively). In both plasmids, the catabolic pathways consist of an upper and lower pathway. In NAH7, the gene products encoded by the upper pathway operon convert naphthalene to salicylate, and those encoded by the lower *nah* operon convert salicylate to pyruvate and acetaldehyde via the catechol *meta*-cleavage pathway (Yen and Serdar, 1988). The gene products encoded on pWW0 degrade toluene/xylene to benzoic acid (toluene) or toluic acid (xylene) in the upper pathway, and the lower pathway products metabolize them also to pyruvate and acetaldehyde (or propionaldehyde for *p*-xylene) via the *meta*-cleavage pathway (Harayama and Rekik, 1990). On both plasmids the genes for the *meta*-cleavage pathways, show high nucleotide identity and similar gene organization to each other as well as to those that encode aerobic degradation of several other aromatic compounds (Ono et al., 2007; Sota et al., 2006b). This strongly suggests that these *meta*-cleavage pathways have been distributed by HGT (Williams and Sayers, 1994). The complete sets of *nah* and *xyl* genes are carried on Tn*3*-like Tns, Tn*4655* and Tn*4653*, respectively (Sota et al., 2006b; Tsuda and Iino, 1988, 1990; Tsuda et al., 1989). Tn*4653* harbors another Tn*3*-like element, Tn*4651*, which contains the *xyl* genes (Tsuda et al., 1989). The *xyl* genes are flanked by two copies of IS*1246*, thus forming an IS-composite Tn (Reddy et al., 1994). However, incorporation of the *xyl* genes into Tn*4651* is thought to have occurred not by transposition of the composite Tn, but by homologous recombination between two copies of IS*1246*, one on the pWW0 (or Tn*4651*) precursor and the other with the *xyl* genes on another replicon (Williams et al., 2004). The mechanism by which the *nah* genes on NAH7 were introduced into Tn*4655* is still not clear. Recently, detailed comparison of three IncP-9 plasmids, NAH7, pWW0, and pDTG1 (Dennis and Zylstra, 2004) as well as that of three naphthalene-catabolic plasmids, NAH7, pDTG1, and pND6-1 (IncP-7) (Li et al., 2004) allowed defining the IncP-9 backbone (Dennis, 2005; Sota et al., 2006b). The comparison also led to the proposal that an integrase, TnpI, which is responsible for the cointegrate resolution reaction of Tn*4655*, may be involved in the integration of the *nah* genes into these plasmids for the following reasons. First, the TnpI-like integrases are found near catabolic genes in IncP-7 and IncP-9 plasmids (Maeda et al., 2003). Second, the two *nah* operons and their intervening sequences on pDTG1 (IncP-9) and pND6-1 (IncP-7) are almost identical to each other, although they are not carried on any Tns but inserted in different plasmid backbones. Third, the segment containing all the *nah* genes on pDTG1 is bracketed by two copies of the *tnpI*-like gene (*intA*), although one is truncated. These observations suggest that the *nah* genes on pDTG1 and pND6-1 may have been incorporated by TnpI-mediated site-specific recombination between two attachment sites on two different molecules. Thus, catabolic IncP-9 plasmids have been established by compilation of various MGEs and their subsequent rearrangement on the backbones.

The IncP-7 plasmid pWW53 is a non-conjugative toluene/xylene-catabolic plasmid that was identified in *P. putida* MT53 from North Wales (Keil et al., 1985; Yano et al., 2007). The transfer machinery is thought to have been lost over evolutionary time, since the related carbazole-catabolic IncP-7 plasmid pCAR1 carries the complete set of transfer genes (Maeda et al., 2003). Plasmid pWW53 carries a single copy of the upper *xyl* operon and two homologous but functionally distinguishable lower *xyl* (*meta*-cleavage) operons (*meta* 1 and *meta* 2). pWW53 also carries two distinct transposase-resolvase gene clusters (*tnpAR* modules), five short terminal IRs, and three *res* sites, which are all typical of Tn*3*-like Tns. pWW53 includes four possible transposable regions, Tn*4657* to Tn*4660*. The largest 86-kb region (Tn*4657*) spans the three other Tns, and contains all of the 36 *xyl* genes. Three Tns, Tn*4657*, Tn*4658* (5.7 kb), and Tn*4659* (3.7 kb) still actively transpose. Plasmid pWW53 also contains another 37-kb toluene/xylene-catabolic transposon, Tn*4656*, which carries two separated and inversely oriented segments of pWW53. This Tn is formed by the Tn*4658* resolvase-mediated site-specific inversion of a large pWW53 segment between the two inversely oriented *res* sites. The Tn*4658* transposase is able to mediate the transposition of Tn*4658*, Tn*4657*, and Tn*4656*, while the Tn*4659* transposase catalyzes only the transposition of Tn*4659*. These findings clearly indicate that multiple copies of transposition-related genes and sites on one plasmid and their recombination activities contribute greatly to the diversification of plasmid structures. As in pWW53 and the catabolic IncP-9 plasmids, the carbazole-catabolic genes on the IncP-7 palsmid pCAR1 are also carried on a very large, active Tn*3*-like Tn, Tn*4676* (73 kb) (Maeda et al., 2003; Shintani et al., 2005b). The large catabolic Tns and *Pseudomonas*-specific IncP-7 and IncP-9 plasmids (Greated et al., 2000; Shintani et al., 2005a; Shintani et al., 2006) have apparently contributed to wide dissemination of evolutionary common gene clusters in *Pseudomonas* strains.

Recent sequence analyses of catabolic plasmids from α-proteobacterial strains have provided new insight into the distribution of catabolic gene clusters. The self-transmissible plasmid pNL1 from *Novoshingobium* (formerly *Sphingomonas*) *aromaticivorans* F199 is the first entirely sequenced catabolic plasmid of α-proteobacterial strains (Romine et al., 1999). The plasmid carries the genes for degradation of biphenyl, naphthalene, *m*-xylene, and *p*-crezol, which are distributed among 15 gene clusters. The complete sequence of another pNL1-like plasmid, pCAR3 from *Sphingomonas* sp. KA1, was published in 2007. The plasmid encodes for degradation of carbazole, and possible degradation of anthranilate, catechol, dibenzofuran, protocatechuate, etc (Shintani et al., 2007). Detailed comparison and analysis of the genes and their products in the two plasmids revealed that (i) the replication of these plasmids seems to be controlled by a single protein, RepAb (pNL1) or RepA (pCAR3), with the target iteron sequences and (ii) these plasmids are clustered into a new Inc group phylogenetically distant from other groups of catabolic plasmids (Shintani et al., 2007). It has been shown that plasmids from sphingomonads encoding degradation of various xenobiotics and polycyclic aromatic compounds have replication machineries similar to that of pNL1 (Basta et al., 2005), suggesting an important role of the pNL1-like plasmids in the dissemination of catabolic genes among *Sphingomonadaceae*.

A new family of catabolic plasmids from *Sphingomonadaceae* has recently been reported. *S. japonicum* (formerly *S. paucimobilis*) UT26 utilizes a highly chlorinated pesticide, γ-hexachlorocyclohexane (γ-HCH = lindane) as a sole source of carbon and energy. It is known that the six structural genes (*linA* to *linF*) and one regulatory gene (*linR*) for the γ-HCH degradation to β-ketoadipate are distributed on three replicons: the chromosomes I (*linABC*) and II (*linF*) and the 185-kb plasmid pCHQ1 (*linREF*) (Nagata et al., 2006). A *linB* UT26 mutant was used to capture plasmids from soil contaminated with HCH isomers by complementation of the gene. The 66-kb self-transmissible plasmid pLB1 thus obtained carries an IS*6100*-composite Tn containing two copies of *linB*, and putative genes coding for replication, maintenance, and conjugative transfer with low levels of similarity to known genes (Miyazaki et al., 2006). A 2-kb region containing a predicted *repA* gene and its upstream region supports autonomous replication of the plasmid. The *repA* gene is unique because only three homologues with similarity at a low level were found in the databases: RepA of *A. radiobacter* K84 plasmid pAgK84, Rep of *A. tumefaciens* 1D1422 plasmid pTAR, and a hypothetical protein of a *Zymomonas mobilis* ZM4 plasmid (33%, 31%, and 27% identity respectively). Plasmid pLB1 can conjugally transfer to α-proteobacterial strains but not to any β- or γ-proteobacteria so far tested, suggesting a rather narrow host-range. Since HCH-degrading sphingomonads have been isolated from Japan, India, France, Germany and Spain, and the *lin* genes are closely linked with IS*6100* (Lal et al., 2006), the pLB1-like plasmids and IS*6100* may together have distributed the genes among sphingomonads worldwide. It is of great interest that the *lin* genes are not clustered as one or a few transcriptional units but scattered over several replicons. Such scattered positioning of genes or gene clusters that work cooperatively is also found among the catabolic genes of pNL1 and pCAR3 as well as the symbiotic p42 plasmids (see above). The "partitioned" carriage of genes might be a successful way for α-proteobacterial genomes to evolve a new, or further improve an existing function by replacing a small portion of extant DNA regions with new corresponding alleles.

THE ECOLOGY OF PLASMID TRANSFER AND PERSISTENCE

PLASMID MEDIATED GENE TRANSFER IN MICROCOSMS AND IN THE FIELD

As described above, plasmids can acquire, rearrange, and distribute genetic material encoding various phenotypic traits between distantly related bacterial populations through HGT. However, for many of these plasmids the evidence that they actually mediate horizontal transfer of diverse genes is mostly based on plasmid transfer experiments under laboratory conditions and on the retrospective observation that similar genes/proteins are found in phylogenetically distinct bacteria. Obviously laboratory conditions with high nutrient concentration, high cell density, and mostly pure cultures of donor and recipient strains used to investigate plasmid transfer do not at all reflect the actual movement of the plasmids in natural ecosystems. Thus microcosm and field experiments have been performed for some plasmids to determine under what conditions genes are spread in various habitats by plasmid-mediated HGT. Since most of these studies are not done very recently, we only summarize some findings and cite some of the most recent studies.

There are several mechanisms or combinations thereof by which microbial populations can evolve and adapt to changes in their environment, and one of them is the acquisition of genetic information from either related or phylogenetically distinct populations in the community by HGT (Top and Springael, 2003). Studies done in laboratory microcosms and in the field have shown that HGT readily occurs in various habitats between distinct bacterial lineages, and results in the wholesale acquisition of traits that provide a selective advantage or are sometimes essential such as degradation of organic compounds and resistance to heavy metals and antibiotics (De Rore et al., 1994; Dejonghe et al., 2000; Dröge et al., 1999; Götz and Smalla, 1997; Hill and Top, 1998; Smalla et al., 2006; Top et al., 1990; Top and Springael, 2003). It became obvious that some habitats and environmental conditions are more conducive to conjugative transfer than others (see below). However, the most important factor that determines if a plasmid will become dominant due to horizontal and vertical transfer seems to be selection for one or more traits it encodes [for a more detailed review, see (Top and Springael, 2003; Top et al., 2002)]. For example, the presence of heavy metals or organic recalcitrant pollutants more significantly increased the number of transconjugants in various microcosm studies (De Rore et al., 1994; Dejonghe et al., 2000; Newby and Pepper, 2002; Smalla et al., 2006), and mercury from dental fillings or low levels of antibiotics can have positive effects on spread and abundance of drug resistance plasmids in the intestinal tract (Licht et al., 2003; Summers et al., 1993). These are clear examples of how plasmids can allow bacterial populations to rapidly adapt to specific environmental conditions by providing them with the necessary genetic information. As a result, these populations can outcompete others and become more abundant, thereby also increasing the plasmid population density. Thus if selection for a horizontally transferred gene or gene cluster is very strong, even low rates of gene transfer can result in amplification of the transferred gene in the community.

Since a detailed overview of environmental factors that affect plasmid transfer efficiencies has been has been published a few years ago (van Elsas and Bailey, 2002), we will provide only a very brief summary here. In most natural environments microbial growth and activity is limited by one or more nutrients (C, N, P, …), and this is no different for conjugative plasmid transfer. Several studies have shown that the addition of nutrients to soil or other microcosms enhances the number of transconjugants detected after addition of a donor and recipient, as compared to unamended microcosms. Moreover a positive effect of manure on plasmid transfer has been demonstrated in field studies as well (Götz and Smalla, 1997; Heuer and Smalla, 2007). For reasons of high nutrient availability as well as high cell densities, specific sites within habitats are often conducive to conjugation, and are therefore regarded as "hot" spots for conjugative plasmid transfer. Examples are the rhizosphere of plants, decaying organic material of animal or plant origin, and guts of soil animals like earthworms (Daane et al., 1996; Hoffmann et al., 1998; Lilley et al., 2003; Mølbak et al., 2007). We do not know however if plasmid-mediated HGT is the most significant of all mechanisms of gene spread in these hot-spots, and how it shapes the bacterial community structure.

Biofilms and microcolonies are common forms of bacterial growth in natural and clinical habitats. They are found in marine and fresh water habitats, soils, sediments, sewage sludge of wastewater treatment plants, the gastrointestinal tract, on medical implants, as dental plaque and in the lungs of cystic fibrosis patients (Hoiby et al., 2001; Singh et al., 2000). Bacterial biofilms are surface-attached, matrix-enclosed bacterial communities that can cause serious health and economic problems because of biofilm-associated phenotypes such as increased resistance against antimicrobial agents and disinfectants (Costerton et al., 1994; Lawrence et al., 1991). The drastic changes in phenotypes as well as in gene expression patterns that have recently been shown between bacteria grown in biofilm versus planktonic cultures demonstrate that there is a major lifestyle difference between these two growth forms (Beloin et al., 2004). The fate of plasmids due to gene transfer can be fundamentally different in these spatially structured populations than in completely mixed liquids for several reasons. First, close contact between cells on a substrate can positively affect conjugation, which requires cell-cell contact. This is especially true for plasmids like those of the IncP-1, IncW, IncN groups, which have fragile rigid short pili and transfer at ca. 1,000 times higher efficiencies on solid medium than in stirred liquids (Bradley et al., 1980). Recent publications and our own unpublished data show that conjugative transfer of IncP-1 plasmids on solid surfaces, and in the gastrointestinal tract of mice, significantly contributes to plasmid persistence because the high transfer efficiency compensate for plasmid loss and cost (see further) (Bahl et al., 2007a; Bahl et al., 2007b). Moreover, because of the spatial organization different dynamics take place than in completely mixed systems, influencing for example the effect of initial cell densities on the fate of a plasmid, as clearly shown in experimental studies and mathematical simulations (Krone et al., 2007; Simonsen, 1990). Another important factor affecting plasmid persistence and spread in structured versus non-structured populations is local versus global competition, as cells only have to compete with their immediate neighbors when fixed in a matrix. Finally, plasmid pili themselves may positively affect cell attachments and thus biofilm formation. This may be beneficial to the plasmid host in habitats where attachment to surfaces is important for growth and/or survival (Ghigo, 2001; Reisner et al., 2003; Tolker-Nielsen and Molin, 2004).

During the last 10 years, several demonstrations of plasmid transfer in natural and laboratory-based biofilms have been published. For example, Christensen et al. (1998) showed that the TOL plasmid, marked with the green fluorescent protein gene *gfp*, was readily transferred among strains in a three-member biofilm community grown in laboratory flow cells. Using those same flow cells but different strains and/or plasmids, the same group showed that plasmid transfer occurred readily in biofilm communities consisting of a plasmid-free host (recipient) and a plasmid-bearing 'donor' strain (Haagensen et al., 2002; Licht et al., 1999). Successful conjugative transfer or triparental mobilization of antibiotic resistance genes was also observed in different biofilm systems or reactors, including one derived from human saliva, simulating dental plaque (Beaudoin et al., 1998; Haagensen et al., 2002; Hausner and Wuertz, 1999; Roberts et al., 1999; Wuertz et al., 2001). In spite of these studies, we do not really know at this time if plasmids in general, and IncP-1 plasmids in particular, contribute more to HGT in biofilms or microcolonies than in completely mixed liquid environments.

ARE PLASMIDS SELFISH, PARASITIC ELEMENTS?

An additional critical question in plasmid ecology that remains unanswered today is: What are the factors that allow plasmids to persist in bacterial communities? First, in the presence of selection for plasmid-encoded traits, it is easy to understand that the plasmid genes encoding these traits are essential for the host and therefore maintained. However, their eventual integration into the host chromosome could be a better solution for the individual cell and would make plasmids dispensable

(Bergstrom et al., 2000). Second, in the absence of selection, plasmids can be lost upon cell division at a certain frequency, and often confer a cost that reduces the host's growth rate (fitness). However, by virtue of their ability to transfer horizontally, they can also 'infect' or 'reinfect' plasmid-free cells by conjugative transfer. As pointed out 30 years ago by Stewart and Levin (1977), in order for plasmids to persist in the absence of selection for the traits they encode, high transfer rates are required to completely compensate for occasional plasmid loss and cost. The significance of conjugative transfer in overall plasmid stability has also been empirically shown in liquid broth and on surfaces (Bahl et al., 2007b; Easter et al., 1997; Sia et al., 1995). However, Stewart and Levin (1977) and later studies (Bergstrom et al., 2000; Lundquist and Levin, 1986) reported that the transfer rates required for plasmid R1, measured in chemostats, were not high enough for the plasmid to persist by infection alone, i.e. be parasitic. The question as to whether or not a high horizontal transfer rate can ensure the persistence of plasmids as parasitic genetic elements without ever benefiting their host, remains unanswered. Current plasmid knowledge suggest that different plasmid groups have found different solutions to survive: from controlling low and high copy numbers, a very efficient plasmid partitioning system, post-segregational killing of plasmid-free segregants, and multimer resolution, all processes that directly affect stable plasmid inheritance, to minimizing their cost to the cell, and maximizing their transfer rate to spread fast enough to overcome loss and host fitness cost (Sørensen et al., 2005). In addition, all plasmids improve their survival rate when they acquire a mobile elements that benefits their host under the existing environmental conditions, because improved host survival guarantees improved plasmid persistence as long as the beneficial gene does not move into the chromosome (Kado, 1998).

Recent studies in other laboratories and ours have tried to improve our insight into the factors that determine persistence and invasion of the BHR IncP-1 plasmids. These plasmids transfer very efficiently on surfaces such as agar and filters on top of agar, even between phylogenetically distinct hosts (Bradley et al., 1980; Schlüter et al., 2005). Thus bacterial microcolonies or biofilms, common forms of bacterial growth in nature and in clinical environments, may very well promote the persistence and spread of these plasmids in environmental and clinical habitats. The IncP-1 plasmid pKJK5 could completely invade a plasmid-free population of *E. coli* and *Kluyvera* sp. on agar surfaces in the absence of selection when the plasmid was initially present in 10% of the population, whereas a transfer-negative mutant was not able to invade (Bahl et al., 2007b). The same authors also showed that conjugative transfer significantly contributed to plasmid persistence in the gastrointestinal tract of germfree rats (Bahl et al., 2007a). Our own experiments with the IncP-1β plasmid pB10 on agar surface in the absence of antibiotics suggest that it can invade an *E. coli* population under certain conditions even when the initial fraction of plasmid-bearing cells is extremely low (Fox et al., unpublished data). Since these plasmids do not confer an obvious fitness benefit to these hosts under those conditions, these studies suggest parasitism of IncP-1 plasmids due to their efficient conjugative transfer in combination with high stability and low cost. Detection of the first 'cryptic' IncP-1 plasmids without any obvious host-beneficial genes or any traces of their presence in the past may support the notion that IncP-1 plasmids can be purely parasitic for prolonged time periods (Harada et al., 2006; Kamachi et al., 2006).

Thus, although we have a long way to go, we have begun to disclose the ecology of plasmids in their natural habitats: We have gained great insight in various factors that affect plasmid spread in natural and clinical ecosystems, and we have begun to address the role of conjugative transfer, stable inheritance and plasmid cost in the persistence of plasmids in their natural environments. However, even though there is no doubt that plasmids transfer occurs across taxonomic barriers in natural and simulated ecosystems, we still cannot quantify the relative contribution of plasmid mediated gene transfer to the evolution and adaptation of various natural, clinical and engineered bacterial communities, as compared to other gene transfer and intragenomic evolutionary mechanisms. As described below under Future Trends, we need better tools to assess rates of plasmid transfer and abundance of different plasmids in natural communities, and better theoretical models to help us understand the relative importance of the various factors that affect plasmid mediated gene transfer.

FUTURE TRENDS

If we are to further improve our understanding of the role of plasmids in HGT and bacterial evolution in general, we need to address at least the following questions. This is not at all meant to be an exhaustive overview of future relevant studies, but a few examples of avenues that we find most important and interesting to pursue.

MOLECULAR MECHANISMS AND ECOLOGICAL CONSEQUENCES OF PLASMID-HOST INTERACTIONS

To thoroughly understand HGT by plasmids, future studies need to further focus on plasmid-host interactions. We have a rather good understanding of the replication, conjugative transfer and maintenance functions, encoded by the plasmid backbone, and have unraveled some of the proteins encoded by the accessory genes. We have also determined their plasmid-encoded regulators and unraveled some of the complex plasmid-encoded regulatory circuits (Thomas, 2006). However, even though plasmid are often defined as 'autonomously replicating' DNA elements, they depend on their host for many of their functions and regulation, and may also affect their host in ways that are not yet understood. Examples of recent progress and future needs is this area are summarized below.

Recent progress in microarray technology has enabled comprehensive transcriptional analyses of plasmid-host interactions. This is beginning to shed new light on when and how plasmids affect expression of host genes and *vice versa*. For example, in the anthrax agent, *B. anthracis*, expression of plasmid-encoded anthrax toxin genes was shown to be regulated by chromosomally encoded genes (Koehler, 2002). Very recently, using a custom-designed microarray, Miyakoshi et al. (2007) investigated the effect of different growth substrates on the transcriptional pattern of strain *P. putida* KT2440 with and without the carbazole degradation plasmid pCAR1. First, they found that when KT2440 (pCAR1) was grown with carbazole, plasmid-encoded *car* and *ant* operons (degradation for carbazole and the intermediate anthranilate) and chromosomal genes *cat* and *pca* (for catechol degradation) were induced, and the regulatory gene *antR*, specific for the *car* and *ant* operons was upregulated. Importantly, the *antR* promoter is an RpoN-dependent promoter, suggesting that successful carbazole degradation depends on the host containing the specific RpoN-dependent activator. Interestingly, they also showed that the presence of pCAR1, and specificially its *parA* gene (involved in plasmid partitioning), caused significant induction of a chromosomal orphan *parA* homologue, named *parI*, whose product also regulated itself. Future studies with different plasmid-host combinations will help to understand the differential expression of plasmid-encoded proteins in different hosts, and disclose factors that work together to maintain the plasmid in its host.

Of the three main plasmid backbone functions, plasmid replication has long been known to require cooperation between host- and plasmid-encoded proteins. Detailed studies of interactions of host proteins with the plasmid-encoded replication initiation protein Rep at the origin of replication *oriV* have elucidated at least in part the molecular basis of the replication range of some plasmids (see next paragraph on host-range). Many low-copy number plasmids also employ various strategies to ensure their stable inheritance in diverse species, such as active centromere-like partitioning systems, multimer resolution systems, post-segregational killing systems, and plasmid encoded restriction-modification systems. Although it has been suggested that not all systems are functional in a wide host range, very little is known so far about the effect of the host on the successful functioning of plasmid maintenance systems (Ebersbach and Gerdes, 2005; Gerdes et al., 2000). We have recently shown that long-term stability can vary greatly between hosts within the replication range of the IncP-1 plasmid pB10, even at the species level, but the molecular basis of these differences was not examined (De Gelder et al., 2007). Similarly, progress has been made in our understanding of the regulation of the conjugative transfer process by host factors for plasmid F (Will and Frost, 2006), but for most other plasmids host-specific regulation of plasmid transfer has not been studied at the molecular level. Plasmid transfer studies have however shown that the type of donor can affect the range of hosts into which a plasmid transfers, not only in laboratory media (Shintani et al., 2005a), but also in an activated sludge sample from a wastewater treatment plant (De Gelder et al., 2005). Earlier, Dionisio et al. demonstrated a large diversity in 'donor ability' for an F-type plasmid. (Dionisio *et al.*, 2002). These studies together reveal our lack of understanding about the effect of the hosts on the rate and range of conjugative plasmid transfer. In conclusion, we should not forget that the host is the plasmid's immediate environment and future studies should focus on the molecular biology, ecology and evolution of host-plasmid interactions.

MOLECULAR BASIS AND EVOLUTION OF PLASMID HOST RANGE

To better understand HGT mediated by plasmids, it will also be indispensable to disentangle the question of plasmid host range, a long-standing enigma in plasmid biology. In spite of the important contribution of transmissible plasmids to the rapid adaptation of bacteria to new environments, our current understanding of factors that determine plasmid host range is largely based on limited studies of plasmid replication mechanisms mostly using the NHR and BHR plasmids F (IncFI) and RK2 (IncP-1).

BHR plasmids have evolved various strategies that enable them to replicate as extrachromosomal elements in a wide range of hosts (Chattoraj, 2000; Espinosa et al., 2000; Krishnapillai, 1988; Sherzinger et al., 1991). For instance, the promiscuity of the BHR IncP-1 plasmids is dependent on meeting the requirements of the replication machinery of various hosts at the origin of plasmid replication. These plasmids encode a long (TrfA1) and short (TrfA2) version of the replication initiation protein TrfA that are differentially used depending on the host (Doran et al., 1999; Durland and Helinski, 1987; Fang and Helinski, 1991; Shingler and Thomas, 1989; Smith and Thomas, 1984). The host range of a plasmid may thus be limited by the inability of a plasmid replication system to productively interact with host proteins, or by the poor expression of essential plasmid-encoded proteins. For example, the failure of the F plasmid to replicate in *Pseudomonas* hosts is due to the inability of the replication protein (Rep) to form a stable complex with the helicase (DnaB) of these hosts (Zhong et al., 2005). Similarly, the restricted host range of pPS10 from *P. syringae* is due to unproductive interactions of its Rep protein with DnaA, and a single base-pair mutation in *repA* was sufficient to expand the plasmid's host range (Fernandez-Tresguerres et al., 1995; Maestro et al., 2003; Maestro et al., 2002). These findings suggest that small genetic changes may allow many NHR plasmids to shift or expand their host range (Sevastsyanovich et al., 2005). To better understand the molecular basis of host range differences between plasmids, similar studies should be performed in the future using a larger range of hosts and many different replicons that have not yet been characterized at this level.

While we have gained some understanding of the factors that determine plasmid host range, we still know little if anything about the potential and mechanisms of plasmid host-range shifts over evolutionary time. In our attempts to assess a plasmid's host-range, we should remember that a plasmid backbone sequence can evolve over time through various mutation mechanisms as well as recombination with other plasmids (Boyd et al., 1996; Schlüter et al., 2003; Schlüter et al., 2007; Thomas, 2004). Thus the range of hosts in which plasmids can persist in the absence of selection (antibiotics) can expand or contract over time. Whereas a few studies, like those mentioned in the paragraph above, have shown that very small genetic changes can be sufficient to expand a plasmid's host-range, it is not known if these laboratory-generated plasmid mutants would be competitive in their natural populations. Since there may be trade-offs such as less stable replication of these plasmids in other hosts or decreased transferability, we need to complement these *in vitro* mutagenesis studies with population biology and ecology studies to understand the true evolvability of a plasmid's host-range.

We postulate that the natural host range of plasmids can evolve, much like that of parasites, to become wider, narrower, or different over evolutionary time. The approach we and others have chosen to test this hypothesis is a combination of experimental evolution studies and mathematical modeling of plasmid-host adaptation. Several experimental evolution studies have demonstrated that the fitness cost a plasmid confers to its host can decrease over evolutionary time under selection for plasmid maintenance. These clear examples of plasmid-host adaptation could be explained by either plasmid- or host- encoded genetic changes, or both. However, plasmid host range shifts over evolutionary time were not examined, except for one study. Dionisio et al (2005) showed that after adaptation of plasmid R1 to *E. coli* K12, some of the evolved plasmids conferred increased fitness not only on the ancestral *E. coli*, but also on *Salmonella enterica* cells that had not carried the plasmid before (Dionisio et al., 2005). These observations suggest that once drug resistance plasmids and their host have adapted to each other, the plasmids can be retained in the bacterial population even when the antibiotic treatment is discontinued. Using an experimental evolution approach, we have recently demonstrated for two different IncP-1 plasmids that a single genetic change can allow IncP-1 plasmids to further expand the range of hosts in which they are stably maintained (De Gelder et al., 2008). Given that many BHR plasmids are involved in the rapid spread of multiple antibiotic resistance determinants, there is a need to further investigate if and how these plasmids shift or expand their host-range in natural and clinical ecosystems.

THE ECOLOGY OF PLASMID PERSISTENCE AND SPREAD

If we want to understand and control the impact of plasmids on spread of unwanted and wanted phenotypic traits (e.g. drug resistance and pollutant degradation), we need to be able to monitor, understand and predict the dynamics of plasmid spread between indigenous bacteria in undisturbed microbial communities. An important aspect of correctly assessing plasmid mediated HGT in natural populations is the availability of adequate tools to detect the plasmids and their hosts. The biggest problem of most studies until recently is that they typically relied on cultivation-based techniques, i.e. enumeration of donor, recipient and transconjugant colonies after plating samples on agar media. Therefore, the actual transfer frequency and host range could have been underestimated. Introduction of new marker genes, has allowed detection and enumeration of the total number of the recipients of plasmids without cultivation (Sørensen et al., 2005). Of these genes, those encoding fluorescent proteins [green (*gfp*), red (*rfp*), cyan (*cfp*), yellow (*yfp*), etc.] are especially useful for monitoring in situ plasmid transfer because the proteins do not require any chemical substrates to generate the fluorescence. For example, Sørensen et al. (2003) constructed a *gfp*-labeled plasmid, whose *gfp* expression is under the control of a *lac* promoter, and donor strains with the *lac* repressor gene, *lacIq*, in their chromosomes. The green fluorescence is expressed when the plasmid was transferred to recipients, thus detectable with flow cytometry or confocal laser microscopy. A dual-labeling system has also been developed, where a plasmid and either donor or recipient are labeled with the *gfp* or *rfp* (Nancharaiah et al., 2003). With these techniques, in situ transfer, localization, and stability of plasmids have been estimated in natural or model environments, such as the rhizosphere (Musovic et al., 2006), plant sprouts and roots (Mølbak et al., 2003; Mølbak et al., 2007), activated sludge (Geisenberger et al., 1999), biofilms (Tolker-Nielsen and Molin, 2004), the animal intestinal tract (Bahl et al., 2007a), etc. Some of these studies showed that the actual transfer frequency of plasmids was 20- to 1000-fold higher than that measured by the standard culture-based methods (Hausner and Wuertz, 1999; Sørensen et al., 2005). The marker-gene assay system combined with flow cytometry can also be used to identify the host range of plasmids in natural communities, since one can separate the cells that acquired the marked plasmid and determine their 16S rRNA gene sequences. For example, Musovic et al. (2006) showed that the gfp gene on the IncP-1 plasmid pKJK10 transferred into the G+ *Arthrobacter* sp. strains, even though IncP-1 plasmids are thought not to replicate in G+ bacteria. Since the authors did not demonstrate the physical presence of the plasmid, integration of the plasmid or the *gfp* gene may have occurred. Nevertheless the fluorescent marker genes facilitate detection of such gene transfer across the Gram barrier, and future research may reveal that it occurs more often than we realize. Thus, these approaches can begin the address questions about plasmid persistence and horizontal transfer rates in ways that are not possible with the traditional cultivation-dependent methods.

Even with the appropriate detection tools available, we still have a long ways to go before we will be able to quantitatively assess the role of plasmids in HGT in natural microbial communities. A large number of studies have documented the effect of abiotic and biotic factors on plasmid spread, and some have assessed the range of hosts that acquired the plasmid (see Ecology section). However, this is usually done by introducing a donor strain that is grown in the laboratory into a microcosm, which is a rather artificial system. The rate and range of plasmid transfer from such a laboratory-grown host may be very different than from indigenous hosts. One particular aspect of plasmid mediated HGT that needs more attention is the pathway that plasmids follow during their horizontal spread through the complex maze that microbial communities form. As described above, the plasmid host affects the rate and range of conjugative plasmid and thus may very well determine the pathways of plasmid spread. Donor-specific effects on plasmid host range could be due to restriction-modification (R-M) systems, biophysical interactions between cells, host physiology determining cell location in spatially structured habitats, etc. In the future we will need to reveal the flow of plasmid-encoded genes in the complex networks of natural and clinical microbial communities, and the main factors that control it. A combination of molecular, biochemical and environmental studies, and mathematical models and computer simulations will be required to reach this goal.

Another missing link in our attempts to understand and predict plasmid-mediated HGT is a mathematical model that is suitable for spatially structured populations such as those in soils, sediments and many forms of biofilms. There are very few quantitative studies that enable predictions about the rate at which plasmids may spread in these populations horizontally and vertically in the presence and absence of selection. Strategies for preventing the spread of antibiotic resistance in clinical settings or for promoting the spread of catabolic genes in polluted habitats or in bioreactors will depend to a large extent on our ability to predict the dynamics of microbial communities in a quantitative way. For ca. 30 years, mathematical modeling of horizontal gene transfer has been based on mass-action differential equations (Stewart and Levin, 1977). This model assumes random encounters between members of the plasmid-bearing and plasmid-free populations at a frequency that is jointly proportional to the densities of the respective populations. The model and numerous variations thereof can successfully simulate conjugative plasmid dynamics in mixed liquids (chemostats and batch cultures) (Bergstrom et al., 2000; Levin, 1981; Levin and Rice, 1980; Levin et al., 1979; Simonsen, 1991; Simonsen et al., 1990). However, because of the implicit assumption of random encounters in a homogenous medium, they do not, and cannot, account for spatial structure. It has been demonstrated mathematically that ignoring spatial structure can lead to inaccurate predictions that are sometimes dramatic (Bascompte and Sole, 1997; Dieckmann et al., 2000; Durrett and Levin, 1994b; Hanski and Gilpin, 1997; Tilman and Kareiva, 1997; Wei and Krone, 2005). There is thus clearly a need to develop a new class of spatially explicit models for the study of plasmid-bacteria interactions in spatially structured populations. Surprisingly, very few attempts to model plasmid transfer on surfaces have been made, and they either make assumption that have not been tested, or have other limitations (Beaudoin et al., 1998; Lagido et al., 2003). Since there was an obvious need for realistic, spatially explicit models for describing and predicting the vertical and horizontal spread of plasmids in bacterial communities, we developed an individual-based lattice model that provides a predictive framework for understanding the dynamics of plasmid-bacteria interactions in spatially structured populations. This model belongs to the family of discrete-space continuous-time stochastic models known as *interacting particle systems* (IPS) or asynchronously updated *stochastic cellular automata* (CA) (Krone et al., 2007). While other, even more sophisticated models for simulation of bacterial growth in biofilms and colonies have been developed (Durrett and Levin, 1994a, b; Kreft and Bonhoeffer, 2005; Kreft et al., 1998; Picioreanu et al., 2004; Picioreanu et al., 1998a; Picioreanu et al., 1998b; Wimpenny and Colasanti, 1997), so far as we know an experimentally validated plasmid dynamics model has not yet been published. Briefly, in our 2-dimensional square lattice (up to 1,000 x 1,000 sites), every site can be vacant, or contain nutrients, or one or maximum 2 cells (donor, recipient or transconjugant), and cells are assumed to divide, transfer or receive a plasmid, or lose a plasmid. Some of the model parameters, like relative growth rate of plasmid-bearing and plasmid-free cells, were estimated based on growth experiments, and the model was validated with qualitative and quantitative experimental observations of plasmid loss and transfer on agar medium (Krone et al., 2007). In the future, expansion and validation of this and/or other 2-D and 3-D individual based models for microcolonies and biofilms will help us to better understand and predict the fate of plasmids and their contribution to HGT in many spatially structured populations.

PLASMID DIVERSITY AND FUNCTION

Future research on the ecology of plasmids should also include characterization of the untapped diversity of plasmids by expanding the paltry collection of well-annotated plasmid genome sequences. Most of our knowledge is derived from studies of plasmids that were chosen because of their relevance to human or animal health, or to industrial or agricultural processes, or because they could be readily isolated from cultivable organisms. It is likely that the information from these studies yields a distorted understanding of the diversity found in plasmid 'backbones' and 'accessory' genes. Recently completed sequences of BHR plasmids isolated from various soil environments support the notion that the diversity of plasmids and the genes they encode in natural habitats is much greater than previously envisioned (Gstalder et al., 2003; Rhodes et al., 2004; Schneiker et al., 2001; Tauch et al., 2002; Tett et al., 2007). These plasmids represent sources of unknown biological diversity, and may be reservoirs of unwanted drug resistance genes, or code for novel functions that could benefit society. Thus we need to focus on plasmids that are not found in 'popular', easily culturable strains, and use plasmid isolation methods that circumvent cultivation of the host. Such methods include exogenous plasmid isolation and direct plasmid DNA isolation from bacterial communities (Schlüter et al., 2007; Smalla and Sobecky, 2002).

Even if a better representation of the extant diversity of plasmids were available in the DNA sequence databases, our understanding of the ecological role of most plasmids would still be limited by our inability to assign proper functions to the large diversity of accessory plasmid genes. At the time this chapter was written, more than 1,100 plasmids with complete sequences have been deposited in databases. However, if we want to better understand the ecological role that plasmids play in bacterial communities, a major challenge is to determine the function of the countless 'hypothetical' and unknown plasmid-encoded proteins. One of the reasons of the poor progress in annotation is that many plasmid sequences are byproducts of whole genome sequencing projects, and are not being properly annotated let alone functionally analyzed. For example, among the plasmid proteins in the ACLAME database (Leplae et al., 2006), many are categorized as 'unknown function' because they lack sequence similarity to functionally known proteins. Some other proteins are referred to as 'ORFans' because they lack similarity to any protein sequences in the current databases. Even when the plasmid genome is the focus of the study, and special efforts are made to assign functions to the plasmids genes, we fail to ascribe functions to the majority of genes when the plasmids are from non-clinical or non-polluted habitats. For example, complete sequencing of the megaplasmid pQBR103, the largest self-transmissible plasmid yet sequenced from the phytosphere, revealed an extremely high degree of genetic novelty, with 80% of the 478 predicted coding sequences with no known functions, and 60% ORFans (Tett et al., 2007). These findings underscore our poor understanding of the diversity of functions encoded on these important genetic elements of the horizontal gene pool (Frost et al., 2005). Future plasmid annotation studies and experimental testing will be required to better understand when, how and why plasmids are selected in the host's natural environment, and if and when plasmid encoded function are absolutely essential for the competitive success of their host.

CONCLUSIONS

Since the discovery of resistance plasmids from the late 1950s to the early 1960s, an enormous number of plasmids encoding a wide variety of phenotypic traits have been reported. The examples we have described above represent just the tip of a huge iceberg for transmissible plasmids. The fact that genetic information can flow among different species, even between kingdoms of living creatures, is now generally recognized, and plasmids are obviously playing a leading role in this genetic flow, called HGT. They are vehicles for genetic information of a multitude of traits that move around the bacterial world. These accessory genes are sometimes lost, modified, or replaced, and constantly being exchanged among distantly related bacteria as a kind of virtual genome shared by many. As described previously, the activities of these plasmids and their MGEs have been, and still are, contributing to their own evolution as well as that of their hosts. Evolution in the genetic structure and composition of plasmids is very well illustrated by the continuing emergence of plasmids that code for previously unnoticed resistances to antibiotics and disinfectants (Schlüter et al., 2007). However, even though recent studies have led to enormous progress in our understanding of plasmid biology, ecology and evolution, we still know very little about how, when and where plasmids acquire and assemble new genes and gene clusters, at what rate and in which range they exchange these between populations in various ecosystems, and what the true diversity is of their backbones and the phenotypes they encode. Future research should aim at systematically addressing these questions, with a greater emphasis on quantitative studies.

WEB RESOURCES

Several web pages that may be useful for studies on plasmids are listed in the chapter "Sequence analysis of plasmids, bioinformatic tools". A few other sources that may be useful are as follows: IBEST (Initiative for Bioinformatics and Evolutionary Studies), http://www.ibest.uidaho.edu/ibest/index.html; International Society for Plasmid Biology, http://www.ispb.org/; Eva Top's laboratory, http://people.ibest.uidaho.edu/~etop/index.html; and Eva Top's Virtual Genome Project, http://people.ibest.uidaho.edu/~etop/vgp/index.html (under development).

ACKNOWLEDGEMENTS

We are grateful for Dr. Patty Sobecky for her very useful comments on this chapter. M.S. and E.M.T. were supported by NIH Grant # P20 RR16448 from the COBRE Program of the National Center for Research Resources. E.M.T. was also supported by the NIH-NIGMS (National Institute of General Medical Sciences) Grant # R01 GM073821.

REFERENCES

Abajy, M.Y., Kopec, J., Schiwon, K., Burzynski, M., Doring, M., Bohn, C., and Grohmann, E. (2007). A type IV-secretion-like system is required for conjugative DNA transport of broad-host-range plasmid pIP501 in gram-positive bacteria. J. Bacteriol. *189*, 2487-2496.

Abby, S., and Daubin, V. (2007). Comparative genomics and the evolution of prokaryotes. Trends Microbiol. *15*, 135-141.

Adamczyk, M., and Jagura-Burdzy, G. (2003). Spread and survival of promiscuous IncP-1 plasmids. Acta Biochim. Pol. *50*, 425-453.

Ahmer, B.M., Tran, M., and Heffron, F. (1999). The virulence plasmid of *Salmonella typhimurium* is self-transmissible. J. Bacteriol. *181*, 1364-1368.

Alonso, G., Vilchez, G., Bruzual, I., and Rodriguez-Lemoine, V. (2002). Characterization of plasmid MIP233 (IncHI3) of the H complex. Res. Microbiol. *153*, 149-153.

Antoine, R., and Locht, C. (1992). Isolation and molecular characterization of a novel broad-host-range plasmid from *Bordetella bronchiseptica* with sequence similarities to plasmids from gram-positive organisms. Mol. Microbiol. *6*, 1785-1799.

Arakawa, Y., Murakami, M., Suzuki, K., Ito, H., Wacharotayankun, R., Ohsuka, S., Kato, N., and Ohta, M. (1995). A novel integron-like element carrying the metallo-β-lactamase gene bla_{IMP}. Antimicrob. Agents Chemother. *39*, 1612-1615.

Arciszewska, L.K., Drake, D., and Craig, N.L. (1989). Transposon Tn*7*. *cis*-Acting sequences in transposition and transposition immunity. J. Mol. Biol. *207*, 35-52.

Arthur, M., Depardieu, F., Molinas, C., Reynolds, P., and Courvalin, P. (1995). The *vanZ* gene of Tn*1546* from *Enterococcus faecium* BM4147 confers resistance to teicoplanin. Gene *154*, 87-92.

Arthur, M., Depardieu, F., Snaith, H.A., Reynolds, P.E., and Courvalin, P. (1994). Contribution of VanY D,D-carboxypeptidase to glycopeptide resistance in *Enterococcus faecalis* by hydrolysis of peptidoglycan precursors. Antimicrob. Agents Chemother. *38*, 1899-1903.

Arthur, M., Molinas, C., and Courvalin, P. (1992). The VanS-VanR two-component regulatory system controls synthesis of depsipeptide peptidoglycan precursors in *Enterococcus faecium* BM4147. J. Bacteriol. *174*, 2582-2591.

Arthur, M., Molinas, C., Depardieu, F., and Courvalin, P. (1993). Characterization of Tn*1546*, a Tn*3*-related transposon conferring glycopeptide resistance by synthesis of depsipeptide peptidoglycan precursors in *Enterococcus faecium* BM4147. J. Bacteriol. *175*, 117-127.

Bahl, M.I., Hansen, L.H., Licht, T.R., and Sorensen, S.J. (2007a). Conjugative transfer facilitates stable maintenance of IncP-1 plasmid pKJK5 in *Escherichia coli* cells colonizing the gastrointestinal tract of the germfree rat. Appl. Environ. Microbiol. *73*, 341-343.

Bahl, M.I., Hansen, L.H., and Sørensen, S.J. (2007b). Impact of conjugal transfer on the stability of IncP-1 plasmid pKJK5 in bacterial populations. FEMS Microbiol. Lett. *266*, 250-256.

Barkay, T., Miller, S.M., and Summers, A.O. (2003). Bacterial mercury resistance from atoms to ecosystems. FEMS Microbiol. Rev. *27*, 355-384.

Barnett, M.J., Fisher, R.F., Jones, T., Komp, C., Abola, A.P., Barloy-Hubler, F., Bowser, L., Capela, D., Galibert, F., Gouzy, J., *et al.* (2001). Nucleotide sequence and predicted functions of the entire *Sinorhizobium meliloti* pSymA megaplasmid. Proc. Natl. Acad. Sci. U. S. A. *98*, 9883-9888.

Barroso, L.A., Wang, S.Z., Phelps, C.J., Johnson, J.L., and Wilkins, T.D. (1990). Nucleotide sequence of *Clostridium difficile* toxin B gene. Nucleic Acids Res *18*, 4004.

Barrow, P.A., and Lovell, M.A. (1989). Functional homology of virulence plasmids in *Salmonella gallinarum*, *S. pullorum*, and *S. typhimurium*. Infect. Immun. *57*, 3136-3141.

Barth, P.T., Datta, N., Hedges, R.W., and Grinter, N.J. (1976). Transposition of a deoxyribonucleic acid sequence encoding trimethoprim and streptomycin resistances from R483 to other replicons. J. Bacteriol. *125*, 800-810.

Barth, P.T., and Grinter, N.J. (1974). Comparison of the deoxyribonucleic acid molecular weights and homologies of plasmids conferring linked resistance to streptomycin and sulfonamides. J. Bacteriol. *120*, 618-630.

Bartosik, D., Szymanik, M., and Wysocka, E. (2001). Identification of the partitioning site within the *repABC*-type replicon of the composite *Paracoccus versutus* plasmid pTAV1. J. Bacteriol. *183*, 6234-6243.

Bascompte, J., and Sole, R. (1997). Modelling Spatiotemporal Dynamics in Ecology (Landes Bioscience).

Basta, T., Buerger, S., and Stolz, A. (2005). Structural and replicative diversity of large plasmids from sphingomonads that degrade polycyclic aromatic compounds and xenobiotics. Microbiology *151*, 2025-2037.

Beaudoin, D.L., Bryers, J.D., Cunningham, A.B., and Peretti, S.W. (1998). Mobilization of broad host range plasmid from *Pseudomonas putida* to established biofilm of *Bacillus azotoformans*. I. Experiments. Biotech. Bioeng. *57*, 272-279.

Beloin, C., Valle, J., Latour-Lambert, P., Faure, P., Kzreminski, M., Balestrino, D., Haagensen, J.A.J., Molin, S., Prensier, G., Arbeille, B., *et al.* (2004). Global impact of mature biofilm lifestyle on *Escherichia coli* K-12 gene expression. Mol. Microbiol. *51*, 659-674.

Benson, S., and Shapiro, J. (1978). TOL is a broad-host-range plasmid. J. Bacteriol. *135*, 278-280.

Berg, D.E., Berg, C.M., and Sasakawa, C. (1984). Bacterial transposon Tn*5*: evolutionary inferences. Mol. Biol. Evol. *1*, 411-422.

Bergstrom, C.T., Lipsitch, M., and Levin, B.R. (2000). Natural selection, infectious transfer and the existence conditions for bacterial plasmids. Genetics *155*, 1505-1519.

Bhatnagar, R., and Batra, S. (2001). Anthrax toxin. Crit. Rev. Microbiol. *27*, 167-200.

Bignell, C., and Thomas, C.M. (2001). The bacterial ParA-ParB partitioning proteins. J. Biotechnol. *91*, 1-34.

Bourgogne, A., Drysdale, M., Hilsenbeck, S.G., Peterson, S.N., and Koehler, T.M. (2003). Global effects of virulence gene regulators in a *Bacillus anthracis* strain with both virulence plasmids. Infect. Immun. *71*, 2736-2743.

Boyd, E.F., Hill, C.W., Rich, S.M., and Hartl, D.L. (1996). Mosaic structure of plasmids from natural populations of *Escherichia coli*. Genetics *143*, 1091-1100.

Bradley, D.E., Taylor, D.E., and Cohen, D.R. (1980). Specification of surface mating systems among conjugative drug resistance plasmids in *Escherichia coli* K-12. J. Bacteriol. *143*, 1466-1470.

Bridges, K., Kidson, A., Lowbury, E.J., and Wilkins, M.D. (1979). Gentamicin- and silver-resistant *Pseudomonas* in a burns unit. Br. Med. J. *1*, 446-449.

Brolle, D.F., Pape, H., Hopwood, D.A., and Kieser, T. (1993). Analysis of the transfer region of the *Streptomyces* plasmid SCP2. Mol. Microbiol. *10*, 157-170.

Brom, S., Garcia-de los Santos, A., Cervantes, L., Palacios, R., and Romero, D. (2000). In *Rhizobium etli* symbiotic plasmid transfer, nodulation competitivity and cellular growth require interaction among different replicons. Plasmid *44*, 34-43.

Brom, S., Girard, L., Tun-Garrido, C., Garcia-de los Santos, A., Bustos, P., Gonzalez, V., and Romero, D. (2004). Transfer of the symbiotic plasmid of *Rhizobium etli* CFN42 requires cointegration with p42a, which may be mediated by site-specific recombination. J. Bacteriol. *186*, 7538-7548.

Brown, H.J., Stokes, H.W., and Hall, R.M. (1996). The integrons In0, In2, and In5 are defective transposon derivatives. J. Bacteriol. *178*, 4429-4437.

Bruins, M.R., Kapil, S., and Oehme, F.W. (2003). Characterization of a small plasmid (pMBCP) from bovine *Pseudomonas pickettii* that confers cadmium resistance. Ecotoxicol. Environ. Saf. *54*, 241-248.

Brunder, W., Karch, H., and Schmidt, H. (2006). Complete sequence of the large virulence plasmid pSFO157 of the sorbitol-fermenting enterohemorrhagic *Escherichia coli* O157:H$^-$ strain 3072/96. Int. J. Med. Microbiol. *296*, 467-474.

Brunder, W., Khan, A.S., Hacker, J., and Karch, H. (2001). Novel type of fimbriae encoded by the large plasmid of sorbitol-fermenting enterohemorrhagic *Escherichia coli* O157:H($^-$). Infect. Immun. *69*, 4447-4457.

Brunder, W., Schmidt, H., and Karch, H. (1996). KatP, a novel catalase-peroxidase encoded by the large plasmid of enterohaemorrhagic *Escherichia coli* O157:H7. Microbiology *142*, 3305-3315.

Brunder, W., Schmidt, H., and Karch, H. (1997). EspP, a novel extracellular serine protease of enterohaemorrhagic *Escherichia coli* O157:H7 cleaves human coagulation factor V. Mol. Microbiol. *24*, 767-778.

Bryan, L.E., Van Den Elzen, H.M., and Tseng, J.T. (1972). Transferable drug resistance in *Pseudomonas aeruginosa*. Antimicrob. Agents Chemother. *1*, 22-29.

Buchanan-Wollaston, V., Passiatore, J.E., and Cannon, F. (1987). The *mob* and *oriT* mobilization functions of a bacterial plasmid promote its transfer to plants. Nature *328*, 172-175.

Bugg, T.D., Wright, G.D., Dutka-Malen, S., Arthur, M., Courvalin, P., and Walsh, C.T. (1991). Molecular basis for vancomycin resistance in *Enterococcus faecium* BM4147: biosynthesis of a depsipeptide peptidoglycan precursor by vancomycin resistance proteins VanH and VanA. Biochemistry (Mosc). *30*, 10408-10415.

Burdett, V. (1980). Identification of tetracycline-resistant R-plasmids in *Streptococcus agalactiae* (group B). Antimicrob. Agents Chemother. *18*, 753-760.

Burland, V., Shao, Y., Perna, N.T., Plunkett, G., Sofia, H.J., and Blattner, F.R. (1998). The complete DNA sequence and analysis of the large virulence plasmid of *Escherichia coli* O157:H7. Nucleic Acids Res. *26*, 4196-4204.

Burrus, V., Pavlovic, G., Decaris, B., and Guedon, G. (2002). Conjugative transposons: the tip of the iceberg. Mol. Microbiol. *46*, 601-610.

Cabezon, E., Lanka, E., and de la Cruz, F. (1994). Requirements for mobilization of plasmids RSF1010 and ColE1 by the IncW plasmid R388: *trwB* and RP4 *traG* are interchangeable. J. Bacteriol. *176*, 4455-4458.

Cabezon, E., Sastre, J.I., and de la Cruz, F. (1997). Genetic evidence of a coupling role for the TraG protein family in bacterial conjugation. Mol. Gen. Genet. *254*, 400-406.

Cangelosi, G.A., Ankenbauer, R.G., and Nester, E.W. (1990). Sugers induce the *Agrobacterium* virulence genes through a periplasmic binding protein and a transmembrane signal protein. Proc. Natl. Acad. Sci. U. S. A. *87*, 6708-6712.

Chandler, M., and Mahillon, J. (2002). Insertion sequences revisited. In Mobile DNA II, N.L. Craig, R. Craigie, M. Gellert, and A.M. Lambowitz, eds. (Washington, D. C., ASM Press), pp. 305-366.

Chang, S., Sievert, D.M., Hageman, J.C., Boulton, M.L., Tenover, F.C., Downes, F.P., Shah, S., Rudrik, J.T., Pupp, G.R., Brown, W.J., *et al.* (2003). Infection with vancomycin-resistant *Staphylococcus aureus* containing the *vanA* resistance gene. N. Engl. J. Med. *348*, 1342-1347.

Chattoraj, D.K. (2000). Control of plasmid DNA replication by iterons: no longer paradoxal. Mol. Microbiol. *37*, 467-476.

Chen, W.M., de Faria, S.M., Straliotto, R., Pitard, R.M., Simoes-Araujo, J.L., Chou, J.H., Chou, Y.J., Barrios, E., Prescott, A.R., Elliott, G.N., *et al.* (2005). Proof that *Burkholderia strains* form effective symbioses with legumes: a study of novel Mimosa-nodulating strains from South America. Appl. Environ. Microbiol. *71*, 7461-7471.

Chen, W.M., Moulin, L., Bontemps, C., Vandamme, P., Bena, G., and Boivin-Masson, C. (2003). Legume symbiotic nitrogen fixation by b-Proteobacteria is widespread in nature. J. Bacteriol. *185*, 7266-7272.

Chen, Y.T., Chang, H.Y., Lai, Y.C., Pan, C.C., Tsai, S.F., and Peng, H.L. (2004). Sequencing and analysis of the large virulence plasmid pLVPK of *Klebsiella pneumoniae* CG43. Gene *337*, 189-198.

Chiou, C.S., and Jones, A.L. (1993). Nucleotide sequence analysis of a transposon (Tn*5393*) carrying streptomycin resistance genes in *Erwinia amylovora* and other gram-negative bacteria. J. Bacteriol. *175*, 732-740.

Chiu, C.M., and Thomas, C.M. (2004). Evidence for past integration of IncP-1 plasmids into bacterial chromosomes. FEMS Microbiol. Lett. *241*, 163-169.

Choi, K.H., Gaynor, J.B., White, K.G., Lopez, C., Bosio, C.M., Karkhoff-Schweizer, R.R., and Schweizer, H.P. (2005). A Tn7-based broad-range bacterial cloning and expression system. Nat. Methods *2*, 443-448.

Christensen, B.B., Sternberg, C., Andersen, J.B., Eberl, L., Moller, S., Givskov, M., and Molin, S (1998). Establishment of new genetic traits in a microbial biofilm community. Appl. Environ. Microbiol. *64*, 2247-2255.

Christie, P.J. (2004). The *Agrobacterium* Ti plasmids. In Plasmid biology, B.E. Funnell, and G.J. Phillips, eds. (Washington, D. C., ASM Press), pp. 455-472.

Christie, P.J., and Vogel, J.P. (2000). Bacterial type IV secretion: conjugation systems adapted to deliver effector molecules to host cells. Trends Microbiol. *8*, 354-360.

Chu, C., and Chiu, C.H. (2006). Evolution of the virulence plasmids of non-typhoid *Salmonella* and its association with antimicrobial resistance. Microbes and infection / Institut Pasteur *8*, 1931-1936.

Chu, L., Mukhopadhyay, D., Yu, H., Kim, K.S., and Misra, T.K. (1992). Regulation of the *Staphylococcus aureus* plasmid pI258 mercury resistance operon. J. Bacteriol. *174*, 7044-7047.

Clewell, D.B. (1993). Bacterial sex pheromone-induced plasmid transfer. Cell *73*, 9-12.

Clewell, D.B., An, F.Y., Flannagan, S.E., Antiporta, M., and Dunny, G.M. (2000). Enterococcal sex pheromone precursors are part of signal sequences for surface lipoproteins. Mol. Microbiol. *35*, 246-247.

Clewell, D.B., and Francia, M.V. (2004). Conjugation in Gram-positive bacteria. In Plasmid biology, B.E. Funnell, and G.J. Phillips, eds. (Washington, D. C., ASM Press), pp. 227-256.

Clewell, D.B., and Weaver, K.E. (1989). Sex pheromones and plasmid transfer in *Enterococcus faecalis*. Plasmid *21*, 175-184.

Collis, C.M., Kim, M.J., Partridge, S.R., Stokes, H.W., and Hall, R.M. (2002). Characterization of the class 3 integron and the site-specific recombination system it determines. J. Bacteriol. *184*, 3017-3026.

Colmar, I., and Horaud, T. (1987). *Enterococcus faecalis* hemolysin-bacteriocin plasmids belong to the same incompatibility group. Appl. Environ. Microbiol. *53*, 567-570.

Cornelis, G.R., Boland, A., Boyd, A.P., Geuijen, C., Iriarte, M., Neyt, C., Sory, M.P., and Stainier, I. (1998). The virulence plasmid of *Yersinia*, an antihost genome. Microbiol. Mol. Biol. Rev. *62*, 1315-1352.

Costerton, J.W., Lewandowski, Z., DeBeer, D., Caldwell, D., Korber, D., and James, G. (1994). Biofilms, the customized microniche. J. Bacteriol. *176*, 2137-2142.

Couturier, M., Bex, F., Bergquist, P.L., and Maas, W.K. (1988). Identification and classification of bacterial plasmids. Microbiol. Rev. *52*, 375-395.

Craig, N.L. (1991). Tn*7*: a target site-specific transposon. Mol. Microbiol. *5*, 2569-2573.

Craig, N.L. (1996). Transposon Tn*7*. Curr. Top. Microbiol. Immunol. *204*, 27-48.

Craig, N.L. (2002). Tn*7*. In Mobile DNA II, N.L. Craig, R. Craigie, M. Gellert, and A.M. Lambowitz, eds. (Washington, D. C., U. S. A., ASM Press), pp. 423-456.

Daane, L.L., Molina, J.A., Berry, E.C., and Sadowsky, M.J. (1996). Influence of earthworm activity on gene transfer from *Pseudomonas* fluorescens to indigenous soil bacteria. Appl. Environ. Microbiol. *62*, 515-521.

Dahlberg, C., and Hermansson, M. (1995). Abundance of Tn*3*, Tn*21*, and Tn*501* transposase (tnpA) sequences in bacterial community DNA from marine environments. Appl. Environ. Microbiol. *61*, 3051-3056.

Datta, N., and Hedges, R.W. (1972). Host range of R factors. J. Gen. Microbiol. *70*, 453-460.

Datta, N., Hedges, R.W., Shaw, E.J., Sykes, R.B., and Richmond, M.H. (1971). Properties of an R factor from *Pseudomonas aeruginosa*. J. Bacteriol. *108*, 1244-1249.

Daubin, V., Moran, N.A., and Ochman, H. (2003). Phylogenetics and the cohesion of bacterial genomes. Science *301*, 829-832.

Day, W.A., Jr., Fernandez, R.E., and Maurelli, A.T. (2001). Pathoadaptive mutations that enhance virulence: genetic organization of the *cadA* regions of *Shigella* spp. Infect. Immun. *69*, 7471-7480.

De Gelder, L., Ponciano, J.M., Joyce, P., and Top, E.M. (2007). Stability of a promiscuous plasmid in different hosts: no guarantee for a long-term relationship. Microbiology *153*, 452-463.

De Gelder, L., Vandecasteele, F.P.J., Brown, C.J., Forney, L.J., and Top, E.M. (2005). Plasmid donor affects host range of promiscuous IncP-1β plasmid pB10 in an activated-sludge microbial community. Appl. Environ. Microb. *71*, 5309-5317.

De Gelder, L., Williams, J.J., Ponciano, J., Sota, M., and Top, E.M. (2008). Adaptive plasmid evolution esults in host range expansion f a broad-host-range plasmid. Genetics In press.

de la Cruz, F., and Grinsted, J. (1982). Genetic and molecular characterization of Tn*21*, a multiple resistance transposon from R100.1. J. Bacteriol. *151*, 222-228.

De Rore, H., Top, E.M., Houwen, F., Mergeay, M., and Verstraete, W. (1994). Evolution of heavy metal resistant transconjugants in a soil environment with a concomitant selective pressure. FEMS Microbiol. Lett. *14*, 1263-1274.

Deb, J.K., and Nath, N. (1999). Plasmids of corynebacteria. FEMS Microbiol. Lett. *175*, 11-20.

Dejonghe, W., Goris, J., El Fantroussi, S., Hofte, M., De Vos, P., Verstraete, W., and Top, E.M. (2000). Effect of dissemination of 2,4-dichlorophenoxyacetic acid (2,4-D) degradation plasmids on 2,4-D degradation and on bacterial community structure in two different soil horizons. Appl. Environ. Microbiol. *66*, 3297-3304.

del Solar, G., Moscoso, M., and Espinosa, M. (1993). Rolling circle-replicating plasmids from gram-positive and gram-negative bacteria: a wall falls. Mol. Microbiol. *8*, 789-796.

Dennis, J.J. (2005). The evolution of IncP catabolic plasmids. Curr. Opin. Biotechnol. *16*, 291-298.

Dennis, J.J., and Zylstra, G.J. (2004). Complete sequence and genetic organization of pDTG1, the 83 kilobase naphthalene degradation plasmid from *Pseudomonas putida* strain NCIB 9816-4. J. Mol. Biol. *341*, 753-768.

Dieckmann, U., Law, R., and Metz, J.A.J. (2000). The geometry of ecological interactions: Simplifying spatial complexity (Cambridge, Cambridge University Press).

Dionisio, F., Conceicao, I.C., Marques, A.C.R., Fernandes, L., and Gordo, I. (2005). The evolution of a conjugative plasmid and its ability to increase bacterial fitness. Biology Letters *1*, 250-252.

Dionisio, F., Matic, I., Radman, M., Rodrigues, O.R., and Taddei, F. (2002). Plasmids spread very fast in heterogeneous bacterial communities. Genetics *162*, 1525-1532.

Doran, K.S., Helinski, D.R., and Konieczny, I. (1999). Host-dependent requirement for specific DnaA boxes for plasmid RK2 replication. Mol. Microbiol. *33*, 490-498.

Dröge, M., Pühler, A., and Selbitchka, W. (1999). Horizontal gene transfer among bacteria in terrestrial and aquatic habitats as assessed by microcosm and field studies. Biol. Fertil. Soils *29*, 221-245.

Drouin, F., Melancon, J., and Roy, P.H. (2002). The IntI-like tyrosine recombinase of *Shewanella oneidensis* is active as an integron integrase. J. Bacteriol. *184*, 1811-1815.

Dunn, N.W., and Gunsalus, I.C. (1973). Transmissible plasmid coding early enzymes of naphthalene oxidation in *Pseudomonas putida*. J. Bacteriol. *114*, 974-979.

Dunny, G.M. (1990). Genetic functions and cell-cell interactions in the pheromone-inducible plasmid transfer system of *Enterococcus faecalis*. Mol. Microbiol. *4*, 689-696.

Dunny, G.M., Brown, B.L., and Clewell, D.B. (1978). Induced cell aggregation and mating in *Streptococcus faecalis*: evidence for a bacterial sex pheromone. Proc. Natl. Acad. Sci. U. S. A. *75*, 3479-3483.

Durland, R.H., and Helinski, D.R. (1987). The sequence encoding the 43-kilodalton TrfA protein is required for efficient replication or maintenance of minimal RK2 replicons in *Pseudomonas aeruginosa*. Plasmid *18*, 164-169.

Durrett, R., and Levin, S.A. (1994a). Stochastic spatial models: a user's guide to ecological applications. Phil. Trans. R. Soc. B *343*, 329-350.

Durrett, R., and Levin, S.A. (1994b). The importance of being discrete (and spatial). Theor. Popul. Biol. *46*, 363-394.

Easter, C.L., Sobecky, P.A., and Helinski, D.R. (1997). Contribution of different segments of the *par* region to stable maintenance of the broad-host-range plasmid RK2. J. Bacteriol. *179*, 6472-6479.

Ebersbach, G., and Gerdes, K. (2005). Plasmid segregation mechanisms. Annu. Rev. Genet. *39*, 453-479.

Elliott, G.N., Chen, W.M., Chou, J.H., Wang, H.C., Sheu, S.Y., Perin, L., Reis, V.M., Moulin, L., Simon, M.F., Bontemps, C., *et al.* (2007). *Burkholderia phymatum* is a highly effective nitrogen fixing symbiont of *Mimosa* spp. and fixes nitrogen ex planta. The New Phytologist *173*, 168-180.

Erauso, G., Stedman, K.M., van de Werken, H.J., Zillig, W., and van der Oost, J. (2006). Two novel conjugative plasmids from a single strain of *Sulfolobus*. Microbiology *152*, 1951-1968.

Espinosa, M., Cohen, S.N., Couturier, M., del Solar, G., Diaz-Orejas, R., Giraldo, R., Janniere, L., Miller, C., Osborn, M., and Thomas, C.M. (2000). Plasmid replication and copy number control. In The horizontal gene pool: Bacterial plasmids and gene spread, C.M. Thomas, ed. (Amsterdam, Harwood Academic Publishers), pp. 1-47.

Evans, R.P., Jr., and Macrina, F.L. (1983). Streptococcal R plasmid pIP501: endonuclease site map, resistance determinant location, and construction of novel derivatives. J. Bacteriol. *154*, 1347-1355.

Fang, F.C., and Helinski, D.R. (1991). Broad-host-range properties of plasmid RK2: importance of overlapping genes encoding the plasmid replication initiation protein TrfA. J. Bacteriol. *173*, 5861-5868.

Farias, M.E., and Espinosa, M. (2000). Conjugal transfer of plasmid pMV158: uncoupling of the pMV158 origin of transfer from the mobilization gene *mobM*, and modulation of pMV158 transfer in *Escherichia coli* mediated by IncP plasmids. Microbiology *146*, 2259-2265.

Fernandez-Lopez, R., Garcillan-Barcia, M.P., Revilla, C., Lazaro, M., Vielva, L., and de la Cruz, F. (2006). Dynamics of the IncW genetic backbone imply general trends in conjugative plasmid evolution. FEMS Microbiol. Rev. *30*, 942-966.

Fernandez-Tresguerres, M.E., Martin, M., Garcia de Viedma, D., Giraldo, R., and Diaz-Orejas, R. (1995). Host growth temperature and a conservative amino acid substitution in the replication protein of pPS10 influence plasmid host range. J. Bacteriol. *177*, 4377-4384.

Finan, T.M., Weidner, S., Wong, K., Buhrmester, J., Chain, P., Vorholter, F.J., Hernandez-Lucas, I., Becker, A., Cowie, A., Gouzy, J., *et al.* (2001). The complete sequence of the 1,683-kb pSymB megaplasmid from the N2-fixing endosymbiont *Sinorhizobium meliloti*. Proc. Natl. Acad. Sci. U. S. A. *98*, 9889-9894.

Fischer, H.M. (1994). Genetic regulation of nitrogen fixation in rhizobia. Microbiol. Rev. *58*, 352-386.

Fischer, H.M. (1996). Environmental regulation of rhizobial symbiotic nitrogen fixation genes. Trends Microbiol. *4*, 317-320.

Fluit, A.C., and Schmitz, F.J. (2004). Resistance integrons and super-integrons. Clin. Microbiol. Infect. *10*, 272-288.

Foglia, G., Del Grosso, M., Vignaroli, C., Bagnarelli, P., Varaldo, P.E., Pantosti, A., and Biavasco, F. (2003). Molecular analysis of Tn*1546*-like elements mediating high-level vancomycin resistance in *Enterococcus gallinarum*. J. Antimicrob. Chemother. *52*, 772-775.

Francia, M.V., Varsaki, A., Garcillan-Barcia, M.P., Latorre, A., Drainas, C., and de la Cruz, F. (2004). A classification scheme for mobilization regions of bacterial plasmids. FEMS Microbiol. Rev. *28*, 79-100.

Freiberg, C., Fellay, R., Bairoch, A., Broughton, W.J., Rosenthal, A., and Perret, X. (1997). Molecular basis of symbiosis between *Rhizobium* and legumes. Nature *387*, 394-401.

Frigaard, N.U., Martinez, A., Mincer, T.J., and DeLong, E.F. (2006). Proteorhodopsin lateral gene transfer between marine planktonic Bacteria and Archaea. Nature *439*, 847-850.

Frost, L.S., Ippen-Ihler, K., and Skurray, R.A. (1994). Analysis of the sequence and gene products of the transfer region of the F sex factor. Microbiol. Rev. *58*, 162-210.

Frost, L.S., Leplae, R., Summers, A.O., and Toussaint, A. (2005). Mobile genetic elements: the agents of open source evolution. Nat. Rev. Microbiol. *3*, 722-732.

Funnel, B.E., and Phillips, G.J., eds. (2004). Plasmid Biology (Washington, D. C., ASM Press).

Garnier, F., Taourit, S., Glaser, P., Courvalin, P., and Galimand, M. (2000). Characterization of transposon Tn*1549*, conferring VanB-type resistance in *Enterococcus* spp. Microbiology *146*, 1481-1489.

Garrett, R.A., Redder, P., Greve, B., Brügger, K., Chen, L., and She, Q. (2004). Archaeal plasmids. In Plasmid biology, B.E. Funnell, and G.J. Phillips, eds. (Washington, D. C., ASM Press), pp. 377-392.

Geisenberger, O., Ammendola, A., Christensen, B.B., Molin, S., Schleifer, K.H., and Eberl, L. (1999). Monitoring the conjugal transfer of plasmid RP4 in activated sludge and in situ identification of the transconjugants. FEMS Microbiol. Lett. *174*, 9-17.

Gemmell, C.G. (2004). Glycopeptide resistance in *Staphylococcus aureus*: is it a real threat? J. Infect. Chemother. *10*, 69-75.

Genka, H., Nagata, Y., and Tsuda, M. (2002). Site-specific recombination system encoded by toluene catabolic transposon Tn*4651*. J. Bacteriol. *184*, 4757-4766.

Gerdes, K., Ayora, S., Canosa, I., Ceglowski, P., Diaz-Orejas, R., Franch, T., Gultyaev, A.P., Bugge Jensen, R., Kobayashi, I., Macpherson, C., *et al.* (2000). Plasmid maintenance systems. In The horizontal gene pool: Bacterial plasmids and gene spread, C.M. Thomas, ed. (Amsterdam, Harwood Academic Publishers), pp. 49-85.

Gerlach, R.G., and Hensel, M. (2007). Protein secretion systems and adhesins: The molecular armory of Gram-negative pathogens. Int. J. Med. Microbiol. *297*, 401-415.

Ghigo, J.M. (2001). Natural conjugative plasmids induce bacterial biofilm development. Nature *412*, 442-445.

Gibbons, A. (1992). Exploring new strategies to fight drug-resistant microbes. Science *257*, 1036-1038.

Gilmour, M.W., Thomson, N.R., Sanders, M., Parkhill, J., and Taylor, D.E. (2004). The complete nucleotide sequence of the resistance plasmid R478: defining the backbone components of incompatibility group H conjugative plasmids through comparative genomics. Plasmid *52*, 182-202.

Gilotra, U., and Srivastava, S. (1997). Plasmid-encoded sequestration of copper by *Pseudomonas pickettii* strain US321. Curr. Microbiol. *34*, 378-381.

Gogarten, J.P., Doolittle, W.F., and Lawrence, J.G. (2002). Prokaryotic evolution in light of gene transfer. Mol. Biol. Evol. *19*, 2226-2238.

Gogarten, J.P., and Townsend, J.P. (2005). Horizontal gene transfer, genome innovation and evolution. Nat. Rev. Microbiol. *3*, 679-687.

Gonzalez, V., Bustos, P., Ramirez-Romero, M.A., Medrano-Soto, A., Salgado, H., Hernandez-Gonzalez, I., Hernandez-Celis, J.C., Quintero, V., Moreno-Hagelsieb, G., Girard, L., *et al.* (2003). The mosaic structure of the symbiotic plasmid of *Rhizobium etli* CFN42 and its relation to other symbiotic genome compartments. Genome biology *4*, R36.

Götz, A., and Smalla, K. (1997). Manure enhances plasmid mobilization and survival of *Pseudomonas putida* introduced into field soil. Appl. Environ. Microbiol. *63*, 1980-1986.

Götz, F., Zabielski, J., Philipson, L., and Lindberg, M. (1983). DNA homology between the arsenate resistance plasmid pSX267 from *Staphylococcus xylosus* and the penicillinase plasmid pI258 from *Staphylococcus aureus*. Plasmid *9*, 126-137.

Grahn, A.M., Haase, J., Bamford, D.H., and Lanka, E. (2000). Components of the RP4 conjugative transfer apparatus form an envelope structure bridging inner and outer membranes of donor cells: implications for related macromolecule transport systems. J. Bacteriol. *182*, 1564-1574.

Grainge, I., and Jayaram, M. (1999). The integrase family of recombinase: organization and function of the active site. Mol. Microbiol. *33*, 449-456.

Greated, A., Lambertsen, L., Williams, P.A., and Thomas, C.M. (2002). Complete sequence of the IncP-9 TOL plasmid pWW0 from *Pseudomonas putida*. Environ. Microbiol. *4*, 856-871.

Greated, A., Titok, M., Krasowiak, R., Fairclough, R.J., and Thomas, C.M. (2000). The replication and stable-inheritance functions of IncP-9 plasmid pM3. Microbiology *146*, 2249-2258.

Greve, B., Jensen, S., Brugger, K., Zillig, W., and Garrett, R.A. (2004). Genomic comparison of archaeal conjugative plasmids from *Sulfolobus*. Archaea *1*, 231-239.

Griffin, H.G., Foster, T.J., Silver, S., and Misra, T.K. (1987). Cloning and DNA sequence of the mercuric- and organomercurial-resistance determinants of plasmid pDU1358. Proc. Natl. Acad. Sci. U. S. A. *84*, 3112-3116.

Grindley, N.D., and Reed, R.R. (1985). Transpositional recombination in prokaryotes. Annu. Rev. Biochem. *54*, 863-896.

Grindley, N.D.F. (2002). The movement of Tn*3*-like elements: Transposition and cointegrate resolution. In Mobile DNA II, N.L. Craig, R. Craigie, M. Gellert, and A.M. Lambowitz, eds. (Washington, D. C., ASM Press), pp. 272-302.

Grinsted, J., de la Cruz, F., and Schmitt, R. (1990). The Tn*21* subgroup of bacterial transposable elements. Plasmid *24*, 163-189.

Grohmann, E., Muth, G., and Espinosa, M. (2003). Conjugative plasmid transfer in gram-positive bacteria. Microbiol. Mol. Biol. Rev. *67*, 277-301.

Gstalder, M.E., Faelen, M., Mine, N., Top, E.M., Mergeay, M., and Couturier, M. (2003). Replication functions of new broad host range plasmids isolated from polluted soils. Res. Microbiol. *154*, 499-509.

Guerry, P., van Embden, J., and Falkow, S. (1974). Molecular nature of two nonconjugative plasmids carrying drug resistance genes. J. Bacteriol. *117*, 619-630.

Guiney, D.G. (1982). Host range of conjugation and replication functions of *Escherichia coli* sex plasmid F *lac*: comparison with the broad host range plasmid RK2. J. Mol. Biol. *162*, 699-703.

Gupta, A., Matsui, K., Lo, J.F., and Silver, S. (1999). Molecular basis for resistance to silver cations in *Salmonella*. Nat. Med. *5*, 183-188.

Gupta, A., Phung, L.T., Taylor, D.E., and Silver, S. (2001). Diversity of silver resistance genes in IncH incompatibility group plasmids. Microbiology *147*, 3393-3402.

Guzman, L.M., and Espinosa, M. (1997). The mobilization protein, MobM, of the streptococcal plasmid pMV158 specifically cleaves supercoiled DNA at the plasmid *oriT*. J. Mol. Biol. *266*, 688-702.

Haagensen, J.A.J., Hansen, S.K., Johansen, T., and Molin, S. (2002). In situ detection of horizontal transfer of mobile genetic elements. FEMS Microbiol. Ecol. *42*, 261-268.

Hagege, J., Pernodet, J.L., Sezonov, G., Gerbaud, C., Friedmann, A., and Guerineau, M. (1993). Transfer functions of the conjugative integrating element pSAM2 from *Streptomyces ambofaciens*: characterization of a *kil-kor* system associated with transfer. J. Bacteriol. *175*, 5529-5538.

Haines, A.S., Akhtar, P., Stephens, E.R., Jones, K., Thomas, C.M., Perkins, C.D., Williams, J.R., Day, M.J., and Fry, J.C. (2006a). Plasmids from freshwater environments capable of IncQ retrotransfer are diverse and include pQKH54, a new IncP-1 subgroup archetype. Microbiology *152*, 2689-2701.

Haines, A.S., Cheung, M., and Thomas, C.M. (2006b). Evidence that IncG (IncP-6) and IncU plasmids form a single incompatibility group. Plasmid *55*, 210-215.

Haines, A.S., Jones, K., Batt, S.M., Kosheleva, I.A., and Thomas, C.M. (2007). Sequence of plasmid pBS228 and reconstruction of the IncP-1a phylogeny. Plasmid *58*, 76-83.

Haines, A.S., Jones, K., Cheung, M., and Thomas, C.M. (2005). The IncP-6 plasmid Rms149 consists of a small mobilizable backbone with multiple large insertions. J. Bacteriol. *187*, 4728-4738.

Hanski, I.A., and Gilpin, M.E. (1997). Metapopulation Biology: Ecology, Genetics, and Evolution (Academic Press).

Hansson, K., Sundstrom, L., Pelletier, A., and Roy, P.H. (2002). IntI2 integron integrase in Tn*7*. J. Bacteriol. *184*, 1712-1721.

Harada, K.M., Aso, Y., Hashimoto, W., Mikami, B., and Murata, K. (2006). Sequence and analysis of the 46.6-kb plasmid pA1 from *Sphingomonas* sp. A1 that corresponds to the typical IncP-1β plasmid backbone without any accessory gene. Plasmid *56*, 11-23.

Harayama, S., and Rekik, M. (1990). The *meta* cleavage operon of TOL degradative plasmid pWW0 comprises 13 genes. Mol. Gen. Genet. *221*, 113-120.

Harkey, C.W., Everiss, K.D., and Peterson, K.M. (1995). Isolation and characterization of a *Vibrio cholerae* gene (*tagA*) that encodes a ToxR-regulated lipoprotein. Gene *153*, 81-84.

Hausner, M., and Wuertz, S. (1999). High rates of conjugation in bacterial biofilms as determined by quantitative in situ analysis. Appl. Environ. Microbiol. *65*, 3710-3713.

Hayashi, T., Makino, K., Ohnishi, M., Kurokawa, K., Ishii, K., Yokoyama, K., Han, C.G., Ohtsubo, E., Nakayama, K., Murata, T., *et al.* (2001). Complete genome sequence of enterohemorrhagic *Escherichia coli* O157:H7 and genomic comparison with a laboratory strain K-12. DNA Res. *8*, 11-22.

Hedges, R.W., and Baumberg, S. (1973). Resistance to arsenic compounds conferred by a plasmid transmissible between strains of *Escherichia coli*. J. Bacteriol. *115*, 459-460.

Heinemann, J.A., and Sprague, G.F., Jr. (1989). Bacterial conjugative plasmids mobilize DNA transfer between bacteria and yeast. Nature *340*, 205-209.

Helgason, E., Okstad, O.A., Caugant, D.A., Johansen, H.A., Fouet, A., Mock, M., Hegna, I., and Kolsto, A.B. (2000). *Bacillus anthracis*, *Bacillus cereus*, and *Bacillus thuringiensis*--one species on the basis of genetic evidence. Appl. Environ. Microbiol. *66*, 2627-2630.

Heuer, H., and Smalla, K. (2007). Manure and sulfadiazine synergistically increased bacterial antibiotic resistance in soil over at least two months. Environ. Microbiol. *9*, 657-666.

Heuer, H., Szczepanowski, R., Schneiker, S., Pühler, A., Top, E.M., and Schlüter, A. (2004). The complete sequences of plasmids pB2 and pB3 provide evidence for a recent ancestor of the IncP-1β group without any accessory genes. Microbiology *150*, 3591-3599.

Hill, K.E., and Top, E.M. (1998). Gene transfer in soil systems using microcosms. FEMS Microbiol. Ecol. *25*, 319-329.

Hiramatsu, K., Aritaka, N., Hanaki, H., Kawasaki, S., Hosoda, Y., Hori, S., Fukuchi, Y., and Kobayashi, I. (1997). Dissemination in Japanese hospitals of strains of *Staphylococcus aureus* heterogeneously resistant to vancomycin. Lancet *350*, 1670-1673.

Hochhut, B., Lotfi, Y., Mazel, D., Faruque, S.M., Woodgate, R., and Waldor, M.K. (2001). Molecular analysis of antibiotic resistance gene clusters in *Vibrio cholerae* O139 and O1 SXT constins. Antimicrob. Agents Chemother. *45*, 2991-3000.

Hoffmann, A., Thimm, T., Droge, M., Moore, E.R., Munch, J.C., and Tebbe, C.C. (1998). Intergeneric transfer of conjugative and mobilizable plasmids harbored by *Escherichia coli* in the gut of the soil microarthropod Folsomia candida (Collembola). Appl. Environ. Microbiol. *64*, 2652-2659.

Hoffmaster, A.R., Ravel, J., Rasko, D.A., Chapman, G.D., Chute, M.D., Marston, C.K., De, B.K., Sacchi, C.T., Fitzgerald, C., Mayer, L.W., *et al.* (2004). Identification of anthrax toxin genes in a *Bacillus cereus* associated with an illness resembling inhalation anthrax. Proc. Natl. Acad. Sci. U. S. A. *101*, 8449-8454.

Hoiby, N., Johansen, H.K., Moser, C., Song, Z., Ciofu, O., and Kharazmi, A. (2001). *Pseudomonas aeruginosa* and the *in vitro* and *in vivo* biofilm mode of growth. Microbes Infect. *3*, 23-35.

Horodniceanu, T., Bouanchaud, D.H., Bieth, G., and Chabbert, Y.A. (1976). R plasmids in *Streptococcus agalactiae* (group B). Antimicrob. Agents Chemother. *10*, 795-801.

Ichikawa, H., Ikeda, K., Wishart, W.L., and Ohtsubo, E. (1987). Specific binding of transposase to terminal inverted repeats of transposable element Tn*3*. Proc. Natl. Acad. Sci. U. S. A. *84*, 8220-8224.

Ike, Y., and Clewell, D.B. (1992). Evidence that the hemolysin/bacteriocin phenotype of *Enterococcus faecalis* subsp. *zymogenes* can be determined by plasmids in different incompatibility groups as well as by the chromosome. J. Bacteriol. *174*, 8172-8177.

Jacob, A.E., Shapiro, J.A., Yamamoto, L., Smith, D.I., Cohen, S.N., and Berg, D. (1977). Plasmids studied in *Escherichia coli* and other enteric bacteria. In DNA insertion elements, plasmids and episomes., J.A.S. A.I. Bukhari, and S.L. Adhya, ed. (New York, Cold Spring Harbor Laboratory), pp. 607-638.

Jacoby, G.A. (1986). Resistance plasmids of *Pseudomonas*. In The bacteria: a treatise on structure and function, J.R. Sokatch, ed. (New York, U. S. A., Academic Press), pp. 265-293.

Jagura-Burdzy, G., Khanim, F., Smith, C.A., and Thomas, C.M. (1992). Crosstalk between plasmid vegetative replication and conjugative transfer: repression of the *trfA* operon by *trbA* of broad host range plasmid RK2. Nucleic Acids Res. *20*, 3939-3944.

Jain, R., Rivera, M.C., Moore, J.E., and Lake, J.A. (2002). Horizontal gene transfer in microbial genome evolution. Theor. Popul. Biol. *61*, 489-495.

Jain, R., Rivera, M.C., Moore, J.E., and Lake, J.A. (2003). Horizontal gene transfer accelerates genome innovation and evolution. Mol. Biol. Evol. *20*, 1598-1602.

Ji, G., and Silver, S. (1992). Reduction of arsenate to arsenite by the ArsC protein of the arsenic resistance operon of *Staphylococcus aureus* plasmid pI258. Proc. Natl. Acad. Sci. U. S. A. *89*, 9474-9478.

Jobanputra, R.S., and Datta, N. (1974). Trimethoprim R factors in enterobacteria from clinical specimens. J. Med. Microbiol. *7*, 169-177.

Johnsen, P.J., Osterhus, J.I., Sletvold, H., Sorum, M., Kruse, H., Nielsen, K., Simonsen, G.S., and Sundsfjord, A. (2005). Persistence of animal and human glycopeptide-resistant enterococci on two Norwegian poultry farms formerly exposed to avoparcin is associated with a widespread plasmid-mediated vanA element within a polyclonal *Enterococcus faecium* population. Appl. Environ. Microbiol. *71*, 159-168.

Juhnke, S., Peitzsch, N., Hubener, N., Grosse, C., and Nies, D.H. (2002). New genes involved in chromate resistance in *Ralstonia metallidurans* strain CH34. Arch. Microbiol. *179*, 15-25.

Kado, C.I. (1998). Origin and evolution of plasmids. Antonie Van Leeuwenhoek *73*, 117-126.

Kado, C.I. (2000). Horizontal gene transfer, Second edn (London, Academic Press).

Kado, C.I. (2002). Horizontal transmission of genes by *Agrobacterium* species. In Horizontal gene transfer, M. Syvanen, and C.I. Kado, eds. (London, Academic Press), pp. 45-50.

Kalkum, M., Eisenbrandt, R., and Lanka, E. (2004). Protein circlets as sex pilus subunits. Curr. Protein Pept. Sci. *5*, 417-424.

Kalkum, M., Eisenbrandt, R., Lurz, R., and Lanka, E. (2002). Tying rings for sex. Trends Microbiol. *10*, 382-387.

Kamachi, K., Sota, M., Tamai, Y., Nagata, N., Konda, T., Inoue, T., Top, E.M., and Arakawa, Y. (2006). Plasmid pBP136 from *Bordetella pertussis* represents an ancestral form of IncP-1β plasmids without accessory mobile elements. Microbiology *152*, 3477-3484.

Kans, J.A., and Casadaban, M.J. (1989). Nucleotide sequences required for Tn*3* transposition immunity. J. Bacteriol. *171*, 1904-1914.

Kato, K., Ohtsuki, K., Mitsuda, H., Yomo, T., Negoro, S., and Urabe, I. (1994). Insertion sequence IS*6100* on plasmid pOAD2, which degrades nylon oligomers. J. Bacteriol. *176*, 1197-1200.

Kawalec, M., Kedzierska, J., Gajda, A., Sadowy, E., Wegrzyn, J., Naser, S., Skotnicki, A.B., Gniadkowski, M., and Hryniewicz, W. (2007). Hospital outbreak of vancomycin-resistant enterococci caused by a single clone of *Enterococcus raffinosus* and several clones of *Enterococcus faecium*. Clin. Microbiol. Infect. *13*, 893-901.

Kawasaki, H., Tone, N., and Tonomura, K. (1981a). Plasmid-determined dehalogenation of haloacetates in *Moraxella* species. Agric. Biol. Chem. *45*, 29-34.

Kawasaki, H., Yahara, H., and Tonomura, K. (1981b). Isolation and characterization of plasmid pUO1 mediating dehalogenation of haloacetate and mercury resistance in *Moraxella* sp. B. Agric. Biol. Chem. *45*, 1477-1481.

Keil, H., Keil, S., Pickup, R.W., and Williams, P.A. (1985). Evolutionary conservation of genes coding for *meta* pathway enzymes within TOL plasmids pWW0 and pWW53. J. Bacteriol. *164*, 887-895.

Kendall, K.J., and Cohen, S.N. (1987). Plasmid transfer in *Streptomyces lividans*: identification of a *kil-kor* system associated with the transfer region of pIJ101. J. Bacteriol. *169*, 4177-4183.

Kholodii, G.Y., Mindlin, S.Z., Bass, I.A., Yurieva, O.V., Minakhina, S.V., and Nikiforov, V.G. (1995). Four genes, two ends, and a *res* region are involved in transposition of Tn*5053*: a paradigm for a novel family of transposons carrying either a mer operon or an integron. Mol. Microbiol. *17*, 1189-1200.

Kholodii, G.Y., Yurieva, O.V., Lomovskaya, O.L., Gorlenko, Z., Mindlin, S.Z., and Nikiforov, V.G. (1993). Tn*5053*, a mercury resistance transposon with integron's ends. J. Mol. Biol. *230*, 1103-1107.

Kiyono, M., Omura, T., Fujimori, H., and Pan-Hou, H. (1995). Organomercurial resistance determinants in *Pseudomonas* K-62 are present on two plasmids. Arch. Microbiol. *163*, 242-247.

Kiyono, M., and Pan-Hou, H. (1999). The *merG* gene product is involved in phenylmercury resistance in *Pseudomonas* strain K-62. J. Bacteriol. *181*, 726-730.

Kleckner, N., Chalmers, R.M., Kwon, D., Sakai, J., and Bolland, S. (1996). Tn*10* and IS*10* transposition and chromosome rearrangements: mechanism and regulation in *vivo* and in *vitro*. Curr. Top. Microbiol. Immunol. *204*, 49-82.

Koehler, T.M. (2002). *Bacillus anthracis* genetics and virulence gene regulation. Curr. Top. Microbiol. Immunol. *271*, 143-164.

Kok, J., van der Vossen, J.M., and Venema, G. (1984). Construction of plasmid cloning vectors for lactic streptococci which also replicate in *Bacillus subtilis* and *Escherichia coli*. Appl. Environ. Microbiol. *48*, 726-731.

Kotze, A.A., Tuffin, I.M., Deane, S.M., and Rawlings, D.E. (2006). Cloning and characterization of the chromosomal arsenic resistance genes from *Acidithiobacillus caldus* and enhanced arsenic resistance on conjugal transfer of *ars* genes located on transposon Tn*AtcArs*. Microbiology *152*, 3551-3560.

Kovach, M.E., Elzer, P.H., Hill, D.S., Robertson, G.T., Farris, M.A., Roop II, R.M., and Peterson, K.M. (1995). Four new derivatives of the broad-host-range cloning vector pBBR1MCS, carrying different antibiotic-resistance cassettes. Gene *166*, 175-176.

Kovach, M.E., Phillips, R.W., Elzer, P.H., Roop II, R.M., and Peterson, K.M. (1994). pBBR1MCS: a broad-host-range cloning vector. Biotechniques *16*, 800-802.

Kozlowicz, B.K., Dworkin, M., and Dunny, G.M. (2006). Pheromone-inducible conjugation in *Enterococcus faecalis*: a model for the evolution of biological complexity? Int. J. Med. Microbiol. *296*, 141-147.

Kreft, J.U., and Bonhoeffer, S. (2005). The evolution of groups of cooperating bacteria and the growth rate versus yield trade-off. Microbiology *151*, 637-641.

Kreft, J.U., Booth, G., and Wimpenny, J.W. (1998). BacSim, a simulator for individual-based modelling of bacterial colony growth. Microbiology *144*, 3275-3287.

Kreps, S., Ferino, F., Mosrin, C., Gertis, J., Mergeay, M., and Thuriaux, P. (1990). Conjugative transfer and autonomous replication of a promiscuous IncQ plasmid in the cyanobacterium *Synechocystis* PCC 6803. Mol. Gen. Genet. *221*, 129-133.

Krishnan, B.R., and Iyer, V.N. (1988). Host ranges of the IncN group plasmid pCU1 and its minireplicon in Gram-negative purple bacteria. Appl. Environ. Microbiol. *54*, 2273-2276.

Krishnapillai, V. (1988). Molecular genetic analysis of bacterial plasmid promiscuity. FEMS Microbiol. Rev. *54*, 223-338.

Krone, S.M., Lu, R., Fox, R., Suzuki, H., and Top, E.M. (2007). Modelling the spatial dynamics of plasmid transfer and persistence. Microbiology *153*, 2803-2816.

Kurenbach, B., Bohn, C., Prabhu, J., Abudukerim, M., Szewzyk, U., and Grohmann, E. (2003). Intergeneric transfer of the *Enterococcus faecalis* plasmid pIP501 to *Escherichia coli* and *Streptomyces lividans* and sequence analysis of its tra region. Plasmid *50*, 86-93.

Kurland, C.G., Canback, B., and Berg, O.G. (2003). Horizontal gene transfer: a critical view. Proc. Natl. Acad. Sci. U. S. A. *100*, 9658-9662.

L'Abee-Lund, T.M., and Sorum, H. (2000). Functional Tn*5393*-like transposon in the R plasmid pRAS2 from the fish pathogen *Aeromonas salmonicida* subspecies salmonicida isolated in Norway. Appl. Environ. Microbiol. *66*, 5533-5535.

Laddaga, R.A., Chu, L., Misra, T.K., and Silver, S. (1987). Nucleotide sequence and expression of the mercurial-resistance operon from *Staphylococcus aureus* plasmid pI258. Proc. Natl. Acad. Sci. U. S. A. *84*, 5106-5110.

Lagido, C., Wilson, I.J., Glover, L.A., and Prosser, J.I. (2003). A model for bacterial conjugal gene transfer on solid surfaces. FEMS Microbiol. Ecol. *44*, 67-78.

Lal, R., Dogra, C., Malhotra, S., Sharma, P., and Pal, R. (2006). Diversity, distribution and divergence of *lin* genes in hexachlorocyclohexane-degrading sphingomonads. Trends Biotechnol. *24*, 121-130.

Lavigne, J.P., Vergunst, A.C., Bourg, G., and O'Callaghan, D. (2005). The IncP island in the genome of *Brucella suis* 1330 was acquired by site-specific integration. Infect. Immun. *73*, 7779-7783.

Lawley, T., Wilkins, B.M., and Frost, L.S. (2004). Bacterial conjugation in Gram-negative bacteria. In Plasmid biology, B.E. Funnell, and G.J. Phillips, eds. (Washington, D. C., ASM Press), pp. 203-226.

Lawrence, J.G., and Hendrickson, H. (2003). Lateral gene transfer: when will adolescence end? Mol. Microbiol. *50*, 739-749.

Lawrence, J.G., and Ochman, H. (2002). Reconciling the many faces of lateral gene transfer. Trends Microbiol. *10*, 1-4.

Lawrence, J.R., Korber, D.R., Hyde, B.D., Costerton, J.W., and Caldwell, D.E. (1991). Optical sectioning of microbial biofilms. J. Bacteriol. *173*, 6558-6567.

LeBlanc, D.J., Hawley, R.J., Lee, L.N., and St Martin, E.J. (1978). "Conjugal" transfer of plasmid DNA among oral streptococci. Proc. Natl. Acad. Sci. U. S. A. *75*, 3484-3487.

Leclercq, R., Derlot, E., Duval, J., and Courvalin, P. (1988). Plasmid-mediated resistance to vancomycin and teicoplanin in *Enterococcus faecium*. N. Engl. J. Med. *319*, 157-161.

Leclercq, R., Derlot, E., Weber, M., Duval, J., and Courvalin, P. (1989). Transferable vancomycin and teicoplanin resistance in *Enterococcus faecium*. Antimicrob. Agents Chemother. *33*, 10-15.

Leplae, R., Lima-Mendez, G., and Toussaint, A. (2006). A first global analysis of plasmid encoded proteins in the ACLAME database. FEMS Microbiol. Rev. *30*, 980-994.

Lerat, E., Daubin, V., Ochman, H., and Moran, N.A. (2005). Evolutionary origins of genomic repertoires in bacteria. PLoS Biology *3*, e130.

Lessard, P.A., O'Brien, X.M., Currie, D.H., and Sinskey, A.J. (2004). pB264, a small, mobilizable, temperature sensitive plasmid from *Rhodococcus*. BMC Microbiol *4*, 15.

Levin, B.R. (1981). Periodic selection, infectious gene exchange, and the genetic structure of *E. coli* populations. Genetics *99*, 1-23.

Levin, B.R., and Rice, V.A. (1980). The kinetics of conjugative plasmids by mobilizing conjugative vectors. Genet. Res. *35*, 241-259.

Levin, B.R., Stewart, F.M., and Rice, V.A. (1979). The kinetics of conjugative plasmid transmission: fit of a simple mass action model. Plasmid *2*, 247-260.

Li, W., Shi, J., Wang, X., Han, Y., Tong, W., Ma, L., Liu, B., and Cai, B. (2004). Complete nucleotide sequence and organization of the naphthalene catabolic plasmid pND6-1 from *Pseudomonas* sp. strain ND6. Gene *336*, 231-240.

Licht, T.R., Christensen, B.B., Krogfelt, K.A., and Molin, S. (1999). Plasmid transfer in the animal intestine and other dynamic bacterial populations: the role of community structure and environment. Microbiology *145*, 2615-2622.

Licht, T.R., Struve, C., Christensen, B.B., Poulsen, R.L., Molin, S., and Krogfelt, K.A. (2003). Evidence of increased spread and establishment of plasmid RP4 in the intestine under sub-inhibitory tetracycline concentrations. FEMS Microbiol. Ecol. *44*, 217-223.

Lilley, A.K., Bailey, M.J., Barr, M., Kilshaw, K., Timms-Wilson, T.M., Day, M.J., Norris, S.J., Jones, T.H., and Godfray, H.C. (2003). Population dynamics and gene transfer in genetically modified bacteria in a model microcosm. Mol. Ecol. *12*, 3097-3107.

Lim, S.K., Tanimoto, K., Tomita, H., and Ike, Y. (2006). Pheromone-responsive conjugative vancomycin resistance plasmids in *Enterococcus faecalis* isolates from humans and chicken feces. Appl. Environ. Microbiol. *72*, 6544-6553.

Lindler, L.E. (2004). Virulence plasmids of *Yersinia*: characteristics and comparison. In Plasmid biology, B.E. Funnell, and G.J. Phillips, eds. (Washington, D. C., ASM Press), pp. 423-437.

Lindler, L.E., Plano, G.V., Burland, V., Mayhew, G.F., and Blattner, F.R. (1998). Complete DNA sequence and detailed analysis of the *Yersinia pestis* KIM5 plasmid encoding murine toxin and capsular antigen. Infect. Immun. *66*, 5731-5742.

Llanes, C., Gabant, P., Couturier, M., Bayer, L., and Plesiat, P. (1996). Molecular analysis of the replication elements of the broad-host-range RepA/C replicon. Plasmid *36*, 26-35.

Lundquist, P.D., and Levin, B.R. (1986). Transitory derepression and the maintenance of conjugative plasmids. Genetics *113*, 483-497.

MacDonald, D., Demarre, G., Bouvier, M., Mazel, D., and Gopaul, D.N. (2006). Structural basis for broad DNA-specificity in integron recombination. Nature *440*, 1157-1162.

MacLellan, S.R., Smallbone, L.A., Sibley, C.D., and Finan, T.M. (2005). The expression of a novel antisense gene mediates incompatibility within the large *repABC* family of a-proteobacterial plasmids. Mol. Microbiol. *55*, 611-623.

Maeda, K., Nojiri, H., Shintani, M., Yoshida, T., Habe, H., and Omori, T. (2003). Complete nucleotide sequence of carbazole/dioxin-degrading plasmid pCAR1 in *Pseudomonas resinovorans* strain CA10 indicates its mosaicity and the presence of large catabolic transposon Tn*4676*. J. Mol. Biol. *326*, 21-33.

Maestro, B., Sanz, J.M., Diaz-Orejas, R., and Fernandez-Tresguerres, E. (2003). Modulation of pPS10 host range by plasmid-encoded RepA initiator protein. J. Bacteriol. *185*, 1367-1375.

Maestro, B., Sanz, J.M., Faelen, M., Couturier, M., Diaz-Orejas, R., and Fernandez-Tresguerres, E. (2002). Modulation of pPS10 host range by DnaA. Mol. Microbiol. *46*, 223-234.

Mahapatra, N.R., Ghosh, S., Deb, C., and Banerjee, P.C. (2002). Resistance to cadmium and zinc in *Acidiphilium symbioticum* KM2 is plasmid mediated. Curr. Microbiol. *45*, 180-186.

Mahillon, J., and Chandler, M. (1998). Insertion sequences. Microbiol. Mol. Biol. Rev. *62*, 725-774.

Mahillon, J., and Lereclus, D. (1988). Structural and functional analysis of Tn*4430*: identification of an integrase-like protein involved in the co-integrate-resolution process. EMBO J. *7*, 1515-1526.

Makino, K., Ishii, K., Yasunaga, T., Hattori, M., Yokoyama, K., Yutsudo, C.H., Kubota, Y., Yamaichi, Y., Iida, T., Yamamoto, K., *et al.* (1998). Complete nucleotide sequences of 93-kb and 3.3-kb plasmids of an enterohemorrhagic *Escherichia coli* O157:H7 derived from Sakai outbreak. DNA Res. *5*, 1-9.

Makino, K., Oshima, K., Kurokawa, K., Yokoyama, K., Uda, T., Tagomori, K., Iijima, Y., Najima, M., Nakano, M., Yamashita, A., *et al.* (2003). Genome sequence of *Vibrio parahaemolyticus*: a pathogenic mechanism distinct from that of *V. cholerae*. Lancet *361*, 743-749.

Mantengoli, E., and Rossolini, G.M. (2005). Tn*5393*d, a complex Tn*5393* derivative carrying the PER-1 extended-spectrum β-lactamase gene and other resistance determinants. Antimicrob. Agents Chemother. *49*, 3289-3296.

Marques, M.V., da Silva, A.M., and Gomes, S.L. (2001). Genetic organization of plasmid pXF51 from the plant pathogen *Xylella fastidiosa*. Plasmid *45*, 184-199.

Martinez, B., Tomkins, J., Wackett, L.P., Wing, R., and Sadowsky, M.J. (2001). Complete nucleotide sequence and organization of the atrazine catabolic plasmid pADP-1 from Pseudomonas sp. strain ADP. J. Bacteriol. *183*, 5684-5697.

Mazel, D. (2006). Integrons: agents of bacterial evolution. Nat. Rev. Microbiol. *4*, 608-620.

Mazel, D., Dychinco, B., Webb, V.A., and Davies, J. (1998). A distinctive class of integron in the *Vibrio cholerae* genome. Science *280*, 605-608.

Mazodier, P., and Davies, J. (1991). Gene transfer between distantly related bacteria. Annu. Rev. Genet. *25*, 147-171.

Mazodier, P., Petter, R., and Thompson, C. (1989). Intergeneric conjugation between *Escherichia coli* and *Streptomyces* species. J. Bacteriol. *171*, 3583-3585.

McGowan, J.E., Jr. (2006). Resistance in nonfermenting gram-negative bacteria: multidrug resistance to the maximum. Am. J. Med. *119*, S29-36; discussion S62-70.

McHugh, G.L., Moellering, R.C., Hopkins, C.C., and Swartz, M.N. (1975). *Salmonella typhimurium* resistant to silver nitrate, chloramphenicol, and ampicillin. Lancet *1*, 235-240.

Mergeay, M., and Gerits, J. (1978). F'-plasmid transfer from *Escherichia coli* to *Pseudomonas fluorescens*. J. Bacteriol. *135*, 18-28.

Mergeay, M., Monchy, S., Vallaeys, T., Auquier, V., Benotmane, A., Bertin, P., Taghavi, S., Dunn, J., van der Lelie, D., and Wattiez, R. (2003). *Ralstonia metallidurans*, a bacterium specifically adapted to toxic metals: towards a catalogue of metal-responsive genes. FEMS Microbiol. Rev. *27*, 385-410.

Merlin, C., Mahillon, J., Nesvera, J., and Toussaint, A. (2000). Gene recruiters and transporters: the modular structure of bacterial mobile elements. In The horizontal gene pool: bacterial plasmids and gene spread, C.M. Thomas, ed. (Amsterdam, Harwood Academic Publishers), pp. 363-409.

Minakhina, S., Kholodii, G., Mindlin, S., Yurieva, O., and Nikiforov, V. (1999). Tn*5053* family transposons are *res* site hunters sensing plasmidal *res* sites occupied by cognate resolvases. Mol. Microbiol. *33*, 1059-1068.

Mindlin, S., Minakhin, L., Petrova, M., Kholodii, G., Minakhina, S., Gorlenko, Z., and Nikiforov, V. (2005). Present-day mercury resistance transposons are common in bacteria preserved in permafrost grounds since the Upper Pleistocene. Res. Microbiol. *156*, 994-1004.

Miyakoshi, M., Shintani, M., Terabayashi, T., Kai, S., Yamane, H., and Nojiri, H. (2007). Transcriptome analysis of *Pseudomonas putida* KT2440 harboring the completely sequenced IncP-7 plasmid pCAR1. J. Bacteriol. *189*, 6849-6860.

Miyazaki, R., Sato, Y., Ito, M., Ohtsubo, Y., Nagata, Y., and Tsuda, M. (2006). Complete nucleotide sequence of an exogenously isolated plasmid, pLB1, involved in γ-hexachlorocyclohexane degradation. Appl. Environ. Microbiol. *72*, 6923-6933.

Monchy, S., Benotmane, M.A., Janssen, P., Vallaeys, T., Taghavi, S., Lelie, D.V., and Mergeay, M. (2007). Plasmids pMOL28 and pMOL30 of *Cupriavidus metallidurans* are specialised in the maximal viable response to heavy metals. J. Bacteriol. *189*, 7417-7425.

Moulin, L., Munive, A., Dreyfus, B., and Boivin-Masson, C. (2001). Nodulation of legumes by members of the β-subclass of Proteobacteria. Nature *411*, 948-950.

Mølbak, L., Licht, T.R., Kvist, T., Kroer, N., and Andersen, S.R. (2003). Plasmid transfer from *Pseudomonas putida* to the indigenous bacteria on alfalfa sprouts: characterization, direct quantification, and in situ location of transconjugant cells. Appl. Environ. Microbiol. *69*, 5536-5542.

Mølbak, L., Molin, S., and Kroer, N. (2007). Root growth and exudate production define the frequency of horizontal plasmid transfer in the rhizosphere. FEMS Microbiol. Ecol. *59*, 167-176.

Murray, K.D., Aronstein, K.A., and de Leon, J.H. (2007). Analysis of pMA67, a predicted rolling-circle replicating, mobilizable, tetracycline-resistance plasmid from the honey bee pathogen, *Paenibacillus larvae*. Plasmid *58*, 89-100.

Musovic, S., Oregaard, G., Kroer, N., and Sørensen, S.J. (2006). Cultivation-independent examination of horizontal transfer and host range of an IncP-1 plasmid among gram-positive and gram-negative bacteria indigenous to the barley rhizosphere. Appl. Environ. Microbiol. *72*, 6687-6692.

Naas, T., Fortineau, N., Snanoudj, R., Spicq, C., Durrbach, A., and Nordmann, P. (2005). First nosocomial outbreak of vancomycin-resistant *Enterococcus faecium* expressing a VanD-like phenotype associated with a *vanA* genotype. J. Clin. Microbiol. *43*, 3642-3649.

Nagata, Y., Kamakura, M., Endo, R., Miyazaki, R., Ohtsubo, Y., and Tsuda, M. (2006). Distribution of γ-hexachlorocyclohexane-degrading genes on three replicons in *Sphingobium japonicum* UT26. FEMS Microbiol. Lett. *256*, 112-118.

Nakatsu, C., Ng, J., Singh, R., Straus, N., and Wyndham, C. (1991). Chlorobenzoate catabolic transposon Tn*5271* is a composite class I element with flanking class II insertion sequences. Proc. Natl. Acad. Sci. U. S. A. *88*, 8312-8316.

Nancharaiah, Y.V., Wattiau, P., Wuertz, S., Bathe, S., Mohan, S.V., Wilderer, P.A., and Hausner, M. (2003). Dual labeling of *Pseudomonas putida* with fluorescent proteins for in situ monitoring of conjugal transfer of the TOL plasmid. Appl. Environ. Microbiol. *69*, 4846-4852.

Nash, H.A. (1996). Site-specific recombination: integration, excision, resolution, and inversion of defined DNA segments. In *Escherichia coli* and *Salmonella*, F.C. Neidhardt, ed. (Washington, D. C., ASM Press), pp. 2363-2376.

Nelson, K.E., Clayton, R.A., Gill, S.R., Gwinn, M.L., Dodson, R.J., Haft, D.H., Hickey, E.K., Peterson, J.D., Nelson, W.C., Ketchum, K.A., *et al.* (1999). Evidence for lateral gene transfer between Archaea and bacteria from genome sequence of *Thermotoga maritima*. Nature *399*, 323-329.

Newby, D.T., and Pepper, I.L. (2002). Dispersal of plasmid pJP4 in unsaturated and saturated 2,4-dichlorophenoxyacetic acid contaminated soil. FEMS Microbiol. Ecol. *39*, 157-164.

Nies, D.H. (2003). Efflux-mediated heavy metal resistance in prokaryotes. FEMS Microbiol. Rev. *27*, 313-339.

Noble, W.C., Virani, Z., and Cree, R.G. (1992). Co-transfer of vancomycin and other resistance genes from *Enterococcus faecalis* NCTC 12201 to *Staphylococcus aureus*. FEMS Microbiol. Lett. *72*, 195-198.

Nojiri, H., Shintani, M., and Omori, T. (2004). Divergence of mobile genetic elements involved in the distribution of xenobiotic-catabolic capacity. Appl. Microbiol. Biotechnol. *64*, 154-174.

Novick, R.P. (1989). Staphylococcal plasmids and their replication. Annu. Rev. Microbiol. *43*, 537-565.

Ochman, H., Lawrence, J.G., and Groisman, E.A. (2000). Lateral gene transfer and the nature of bacterial innovation. Nature *405*, 299-304.

Ogawa, N., Chakrabarty, A.M., and Zaborina, O. (2004). Degradative plasmids. In Plasmid biology, G. Phillips, and B. Funnel, eds. (Washington, D. C., ASM Press), pp. 341-376.

Ohtsubo, Y., Genka, H., Komatsu, H., Nagata, Y., and Tsuda, M. (2005). High-temperature-induced transposition of insertion elements in *Burkholderia multivorans* ATCC 17616. Appl. Environ. Microbiol. *71*, 1822-1828.

Okinaka, R.T., Cloud, K., Hampton, O., Hoffmaster, A.R., Hill, K.K., Keim, P., Koehler, T.M., Lamke, G., Kumano, S., Mahillon, J., *et al.* (1999). Sequence and organization of pXO1, the large *Bacillus anthracis* plasmid harboring the anthrax toxin genes. J. Bacteriol. *181*, 6509-6515.

Ono, A., Miyazaki, R., Sota, M., Ohtsubo, Y., Nagata, Y., and Tsuda, M. (2007). Isolation and characterization of naphthalene-catabolic genes and plasmids from oil-contaminated soil by using two cultivation-independent approaches. Appl. Microbiol. Biotechnol. *74*, 501-510.

Oppon, J.C., Sarnovsky, R.J., Craig, N.L., and Rawlings, D.E. (1998). A Tn7-like transposon is present in the *glmUS* region of the obligately chemoautolithotrophic bacterium *Thiobacillus ferrooxidans*. J. Bacteriol. *180*, 3007-3012.

Osborn, A.M., and Boltner, D. (2002). When phage, plasmids, and transposons collide: genomic islands, and conjugative- and mobilizable-transposons as a mosaic continuum. Plasmid *48*, 202-212.

Osterblad, M., Norrdahl, K., Korpimaki, E., and Huovinen, P. (2001). Antibiotic resistance. How wild are wild mammals? Nature *409*, 37-38.

Otto, H., Tezcan-Merdol, D., Girisch, R., Haag, F., Rhen, M., and Koch-Nolte, F. (2000). The *spvB* gene-product of the *Salmonella enterica* virulence plasmid is a mono(ADP-ribosyl)transferase. Mol. Microbiol. *37*, 1106-1115.

Otto, R., de Vos, W.M., and Gavrieli, J. (1982). Plasmid DNA in S*treptococcus cremoris* Wg2: Influence of pH on selection in chemostats of a variant lacking a protease plasmid. Appl. Environ. Microbiol. *43*, 1272-1277.

Park, J.E., Young, K.E., Schlegel, H.G., Rhie, H.G., and Lee, H.S. (2003). Conjugative plasmid mediated inducible nickel resistance in *Hafnia alvei* 5-5. Int. Microbiol. *6*, 57-64.

Parks, A.R., and Peters, J.E. (2007). Transposon Tn*7* is widespread in diverse bacteria and forms genomic islands. J. Bacteriol. *189*, 2170-2173.

Partridge, S.R., and Hall, R.M. (2003). The IS*1111* family members IS*4321* and IS*5075* have subterminal inverted repeats and target the terminal inverted repeats of Tn*21* family transposons. J. Bacteriol. *185*, 6371-6384.

Paterson, D.L. (2006). Resistance in gram-negative bacteria: *Enterobacteriaceae*. Am. J. Med. *119*, S20-28; discussion S62-70.

Paulsen, I.T., Banerjei, L., Myers, G.S., Nelson, K.E., Seshadri, R., Read, T.D., Fouts, D.E., Eisen, J.A., Gill, S.R., Heidelberg, J.F., *et al.* (2003). Role of mobile DNA in the evolution of vancomycin-resistant *Enterococcus faecalis*. Science *299*, 2071-2074.

Perichon, B., and Courvalin, P. (2006). Synergism between β-lactams and glycopeptides against VanA-type methicillin-resistant *Staphylococcus aureus* and heterologous expression of the *vanA* operon. Antimicrob. Agents Chemother. *50*, 3622-3630.

Perna, N.T., Plunkett, G., III, Burland, V., Mau, B., Glasner, J.D., Rose, D.J., Mayhew, G.F., Evans, P.S., Gregor, J., Kirkpatrick, H.A., *et al.* (2001). Genome sequence of enterohaemorrhagic *Escherichia coli* O157:H7. Nature *409*, 529-533.

Peters, J.E., and Craig, N.L. (2001). Tn*7*: smarter than we thought. Nat. Rev. Mol. Cell Biol. *2*, 806-814.

Pettis, G.S., and Cohen, S.N. (1994). Transfer of the pIJ101 plasmid in *Streptomyces lividans* requires a *cis*-acting function dispensable for chromosomal gene transfer. Mol. Microbiol. *13*, 955-964.

Picioreanu, C., Kreft, J.-U., and Van Loosdrecht, M.C.M. (2004). Particle-based multidimensional multispecies biofilm model. App. Environ. Microbiol. *70*, 3024-3040.

Picioreanu, C., van Loosdrecht, M.C., and Heijnen, J.J. (1998a). Mathematical modeling of biofilm structure with a hybrid differential-discrete cellular automaton approach. Biotechnol. Bioeng. *58*, 101-116.

Picioreanu, C., Van Loosdrecht, M.C.M., and Heijnen, J.J. (1998b). A new combined differential-discrete cellular automaton approach for biofilm modeling: application for growth in gel beads. Biotechnol. Bioeng. *57*, 718-731.

Pillar, C.M., and Gilmore, M.S. (2004). Virulence plasmids of nonsporulating Gram-positive pathogens. In Plasmid biology, B.E. Funnell, and G.J. Phillips, eds. (Washington, D. C., ASM Press), pp. 439-454.

Poeta, P., Costa, D., Saenz, Y., Klibi, N., Ruiz-Larrea, F., Rodrigues, J., and Torres, C. (2005). Characterization of antibiotic resistance genes and virulence factors in faecal enterococci of wild animals in Portugal. J. Vet. Med. *52*, 396-402.

Polak, J., and Novick, R.P. (1982). Closely related plasmids from *Staphylococcus aureus* and soil bacilli. Plasmid *7*, 152-162.

Prangishvili, D., Albers, S.V., Holz, I., Arnold, H.P., Stedman, K., Klein, T., Singh, H., Hiort, J., Schweier, A., Kristjansson, J.K., *et al.* (1998). Conjugation in archaea: frequent occurrence of conjugative plasmids in *Sulfolobus*. Plasmid *40*, 190-202.

Projan, S.J., Monod, M., Narayanan, C.S., and Dubnau, D. (1987). Replication properties of pIM13, a naturally occurring plasmid found in *Bacillus subtilis*, and of its close relative pE5, a plasmid native to *Staphylococcus aureus*. J. Bacteriol. *169*, 5131-5139.

Radstrom, P., Skold, O., Swedberg, G., Flensburg, J., Roy, P.H., and Sundstrom, L. (1994). Transposon Tn*5090* of plasmid R751, which carries an integron, is related to Tn*7*, Mu, and the retroelements. J. Bacteriol. *176*, 3257-3268.

Ramirez-Romero, M.A., Soberon, N., Perez-Oseguera, A., Tellez-Sosa, J., and Cevallos, M.A. (2000). Structural elements required for replication and incompatibility of the *Rhizobium etli* symbiotic plasmid. J. Bacteriol. *182*, 3117-3124.

Rasko, D.A., Altherr, M.R., Han, C.S., and Ravel, J. (2005). Genomics of the *Bacillus cereus* group of organisms. FEMS Microbiol. Rev. *29*, 303-329.

Rasko, D.A., Rosovitz, M.J., Okstad, O.A., Fouts, D.E., Jiang, L., Cer, R.Z., Kolsto, A.B., Gill, S.R., and Ravel, J. (2007). Complete sequence analysis of novel plasmids from emetic and periodontal *Bacillus cereus* isolates reveals a common evolutionary history among the *B. cereus*-group plasmids, including *Bacillus anthracis* pXO1. J. Bacteriol. *189*, 52-64.

Ravel, J., DiRuggiero, J., Robb, F.T., and Hill, R.T. (2000). Cloning and sequence analysis of the mercury resistance operon of *Streptomyces* sp. Strain CHR28 reveals a novel putative second regulatory gene. J. Bacteriol. *182*, 2345-2349.

Ravel, J., Schrempf, H., and Hill, R.T. (1998). Mercury resistance is encoded by transferable giant linear plasmids in two chesapeake bay *Streptomyces* strains. Appl. Environ. Microbiol. *64*, 3383-3388.

Rawlings, D.E., and Tietze, E. (2001). Comparative biology of IncQ and IncQ-like plasmids. Microbiol. Mol. Biol. Rev. *65*, 481-496.

Recchia, G.D., and Hall, R.M. (1995). Gene cassettes: a new class of mobile element. Microbiology *141*, 3015-3027.

Reddy, B.R., Shaw, L.E., Sayers, J.R., and Williams, P.A. (1994). Two identical copies of IS*1246*, a 1275 base pair sequence related to other bacterial insertion sequences, enclose the *xyl* genes on TOL plasmid pWW0. Microbiology *140*, 2305-2307.

Reisner, A., Haagensen, J.A., Schembri, M.A., Zechner, E.L., and Molin, S. (2003). Development and maturation of *Escherichia coli* K-12 biofilms. Mol. Microbiol. *48*, 933-946.

Reniero, D., Mozzon, E., Galli, E., and Barbieri, P. (1998). Two aberrant mercury resistance transposons in the *Pseudomonas stutzeri* plasmid pPB. Gene *208*, 37-42.

Rensing, C., Ghosh, M., and Rosen, B.P. (1999). Families of soft-metal-ion-transporting ATPases. J. Bacteriol. *181*, 5891-5897.

Reynolds, P.E. (1989). Structure, biochemistry and mechanism of action of glycopeptide antibiotics. Eur. J. Clin. Microbiol. Infect. Dis. *8*, 943-950.

Rhodes, G., Parkhill, J., Bird, C., Ambrose, K., Jones, M.C., Huys, G., Swings, J., and Pickup, R.W. (2004). Complete nucleotide sequence of the conjugative tetracycline resistance plasmid pFBAOT6, a member of a group of IncU plasmids with global ubiquity. Appl. Environ. Microbiol. *70*, 7497-7510.

Roberts, A.P., Pratten, J., Wilson, M., and Mullany, P. (1999). Transfer of a conjugative transposon, Tn*5397* in a model oral biofilm. FEMS Microbiol. Letters *177*, 63-66.

Romero, D., and Brom, S. (2004). The symbiotic plasmids of the *Rhizobiaceae*. In Plasmid biology, B.E. Funnell, and G.J. Phillips, eds. (Washington, D. C., ASM Press), pp. 271-290.

Romine, M.F., Stillwell, L.C., Wong, K.K., Thurston, S.J., Sisk, E.C., Sensen, C., Gaasterland, T., Fredrickson, J.K., and Saffer, J.D. (1999). Complete sequence of a 184-kilobase catabolic plasmid from *Sphingomonas aromaticivorans* F199. J. Bacteriol. *181*, 1585-1602.

Rood, J.I. (2004). Virulence plasmids of spore-forming bacteria. In Plasmid biology, B.E. Funnell, and G.J. Phillips, eds. (Washington, D. C., ASM Press), pp. 413-422.

Rowe-Magnus, D.A., Guerout, A.M., and Mazel, D. (2002). Bacterial resistance evolution by recruitment of super-integron gene cassettes. Mol. Microbiol. *43*, 1657-1669.

Rowe-Magnus, D.A., Guerout, A.M., Ploncard, P., Dychinco, B., Davies, J., and Mazel, D. (2001). The evolutionary history of chromosomal super-integrons provides an ancestry for multiresistant integrons. Proc. Natl. Acad. Sci. U. S. A. *98*, 652-657.

Rowe-Magnus, D.A., and Mazel, D. (2001). Integrons: natural tools for bacterial genome evolution. Curr. Opin. Microbiol. *4*, 565-569.

Rowe-Magnus, D.A., and Mazel, D. (2002). The role of integrons in antibiotic resistance gene capture. Int. J. Med. Microbiol. *292*, 115-125.

Ruhfel, R.E., Manias, D.A., and Dunny, G.M. (1993). Cloning and characterization of a region of the *Enterococcus faecalis* conjugative plasmid, pCF10, encoding a sex pheromone-binding function. J. Bacteriol. *175*, 5253-5259.

Rychlik, I., Gregorova, D., and Hradecka, H. (2006). Distribution and function of plasmids in *Salmonella enterica*. Vet. Microbiol. *112*, 1-10.

Sajjaphan, K., Shapir, N., Wackett, L.P., Palmer, M., Blackmon, B., Tomkins, J., and Sadowsky, M.J. (2004). *Arthrobacter aurescens* TC1 atrazine catabolism genes *trzN*, *atzB*, and *atzC* are linked on a 160-kilobase region and are functional in *Escherichia coli*. Appl. Environ. Microbiol. *70*, 4402-4407.

Schleper, C., Holz, I., Janekovic, D., Murphy, J., and Zillig, W. (1995). A multicopy plasmid of the extremely thermophilic archaeon *Sulfolobus* effects its transfer to recipients by mating. J. Bacteriol. *177*, 4417-4426.

Schlüter, A., Heuer, H., Szczepanowski, R., Forney, L.J., Thomas, C.M., Pühler, A., and Top, E.M. (2003). The 64 508 bp IncP-1β antibiotic multiresistance plasmid pB10 isolated from a waste-water treatment plant provides evidence for recombination between members of different branches of the IncP-1β group. Microbiology *149*, 3139-3153.

Schlüter, A., Heuer, H., Szczepanowski, R., Poler, S.M., Schneiker, S., Pühler, A., and Top, E.M. (2005). Plasmid pB8 is closely related to the prototype IncP-1β plasmid R751 but transfers poorly to *Escherichia coli* and carries a new transposon encoding a small multidrug resistance efflux protein. Plasmid *54*, 135-148.

Schlüter, A., Szczepanowski, R., Pühler, A., and Top, E.M. (2007). Genomics of IncP-1 antibiotic resistance plasmids isolated from wastewater treatment plants provides evidence for a widely accessible drug resistance gene pool. FEMS Microbiol. Rev. *31*, 449-477.

Schmidt, H., Beutin, L., and Karch, H. (1995). Molecular analysis of the plasmid-encoded hemolysin of *Escherichia coli* O157:H7 strain EDL 933. Infect. Immun. *63*, 1055-1061.

Schmidt, H., Henkel, B., and Karch, H. (1997). A gene cluster closely related to type II secretion pathway operons of gram-negative bacteria is located on the large plasmid of enterohemorrhagic *Escherichia coli* O157 strains. FEMS Microbiol. Lett. *148*, 265-272.

Schneiker, S., Keller, M., Droge, M., Lanka, E., Puhler, A., and Selbitschka, W. (2001). The genetic organization and evolution of the broad host range mercury resistance plasmid pSB102 isolated from a microbial population residing in the rhizosphere of alfalfa. Nucleic Acids Res. *29*, 5169-5181.

Schwarz, F.V., Perreten, V., and Teuber, M. (2001). Sequence of the 50-kb conjugative multiresistance plasmid pRE25 from *Enterococcus faecalis* RE25. Plasmid *46*, 170-187.

Seubert, A., Hiestand, R., de la Cruz, F., and Dehio, C. (2003). A bacterial conjugation machinery recruited for pathogenesis. Mol. Microbiol. *49*, 1253-1266.

Sevastsyanovich, Y.R., Titok, M.A., Krasowiak, R., Bingle, L.E., and Thomas, C.M. (2005). Ability of IncP-9 plasmid pM3 to replicate in *Escherichia coli* is dependent on both *rep* and *par* functions. Mol. Microbiol. *57*, 819-833.

Shapiro, J.A. (1979). Molecular model for the transposition and replication of bacteriophage Mu and other transposable elements. Proc. Natl. Acad. Sci. U. S. A. *76*, 1933-1937.

Sherratt, D. (1989). Tn*3* and related transposable elements: site-specific recombination and transposition. In Mobile DNA, D.E. Berg, and M.M. Howe, eds. (Washington DC, American Society for Microbiology), pp. 163-184.

Sherzinger, E., Haring, V., Lurz, R., and Otto, S. (1991). Plasmid RSF1010 DNA replication in vitro promoted by purified RSF1010 RepA, RepB and RepC proteins. Nucleic Acids Res. *19*, 1203-1211.

Shingler, V., and Thomas, C.M. (1989). Analysis of nonpolar insertion mutations in the *trfA* gene of IncP plasmid RK2 which affect its broad-host-range property. Biochim. Biophys. Acta *1007*, 301-308.

Shintani, M., Habe, H., Tsuda, M., Omori, T., Yamane, H., and Nojiri, H. (2005a). Recipient range of IncP-7 conjugative plasmid pCAR2 from *Pseudomonas putida* HS01 is broader than from other *Pseudomonas* strains. Biotechnol. Lett. *27*, 1847-1853.

Shintani, M., Urata, M., Inoue, K., Eto, K., Habe, H., Omori, T., Yamane, H., and Nojiri, H. (2007). The *Sphingomonas* plasmid pCAR3 is involved in complete mineralization of carbazole. J. Bacteriol. *189*, 2007-2020.

Shintani, M., Yano, H., Habe, H., Omori, T., Yamane, H., Tsuda, M., and Nojiri, H. (2006). Characterization of the replication, maintenance, and transfer features of the IncP-7 plasmid pCAR1, which carries genes involved in carbazole and dioxin degradation. Appl. Environ. Microbiol. *72*, 3206-3216.

Shintani, M., Yoshida, T., Habe, H., Omori, T., and Nojiri, H. (2005b). Large plasmid pCAR2 and class II transposon Tn*4676* are functional mobile genetic elements to distribute the carbazole/dioxin-degradative car gene cluster in different bacteria. Appl. Microbiol. Biotechnol. *67*, 370-382.

Sia, E.A., Roberts, R.C., Easter, C., Helinski, D.R., and Figurski, D.H. (1995). Different relative importances of the *par* operons and the effect of conjugal transfer on the maintenance of intact promiscuous plasmid RK2. J. Bacteriol. *177*, 2789-2797.

Signon, L., and Kleckner, N. (1995). Negative and positive regulation of Tn*10*/IS*10*-promoted recombination by IHF: two distinguishable processes inhibit transposition off of multicopy plasmid replicons and activate chromosomal events that favor evolution of new transposons. Genes Dev. *9*, 1123-1136.

Silva, R., Saadi, S., and Maas, W. (1988). A basic replicon of virulence-associated plasmids of *Shigella* spp. and enteroinvasive *Escherichia coli* is homologous with a basic replicon in plasmids of IncF groups. Infect. Immun. *56*, 836-842.

Silver, S. (1996). Bacterial resistances to toxic metal ions--a review. Gene *179*, 9-19.

Silver, S. (2003). Bacterial silver resistance: molecular biology and uses and misuses of silver compounds. FEMS Microbiol. Rev. *27*, 341-353.

Silver, S., Budd, K., Leahy, K.M., Shaw, W.V., Hammond, D., Novick, R.P., Willsky, G.R., Malamy, M.H., and Rosenberg, H. (1981). Inducible plasmid-determined resistance to arsenate, arsenite, and antimony (III) in *Escherichia coli* and *Staphylococcus aureus*. J. Bacteriol. *146*, 983-996.

Silver, S., and Phung le, T. (2005). A bacterial view of the periodic table: genes and proteins for toxic inorganic ions. J. Ind. Microbiol. Biotechnol. *32*, 587-605.

Silver, S., and Phung, L.T. (1996). Bacterial heavy metal resistance: new surprises. Annu. Rev. Microbiol. *50*, 753-789.

Simjee, S., White, D.G., McDermott, P.F., Wagner, D.D., Zervos, M.J., Donabedian, S.M., English, L.L., Hayes, J.R., and Walker, R.D. (2002). Characterization of Tn*1546* in vancomycin-resistant *Enterococcus faecium* isolated from canine urinary tract infections: evidence of gene exchange between human and animal enterococci. J. Clin. Microbiol. *40*, 4659-4665.

Simonsen, L. (1990). Dynamics of plasmid transfer on surfaces. J. Gen. Microbiol. *136*, 1001-1007.

Simonsen, L. (1991). The existence conditions for bacterial plasmids: theory and reality. Microb. Ecol. *22*, 187-205.

Simonsen, L., Gordon, D.M., Stewart, F.M., and Levin, B.R. (1990). Estimating the rate of plasmid transfer: an end point method. J. Gen. Microbiol. *136*, 2319-2325.

Singh, P.K., Schaefer, A.L., Parsek, M.R., Moninger, T.O., Welsh, M.J., and Greenberg, E.P. (2000). Quorum-sensing signals indicate that cystic fibrosis lungs are infected with bacterial biofilms. Nature *407*, 762-764.

Skurnik, D., Ruimy, R., Andremont, A., Amorin, C., Rouquet, P., Picard, B., and Denamur, E. (2006). Effect of human vicinity on antimicrobial resistance and integrons in animal faecal Escherichia coli. J. Antimicrob. Chemother. *57*, 1215-1219.

Smalla, K., Haines, A.S., Jones, K., Krogerrecklenfort, E., Heuer, H., Schloter, M., and Thomas, C.M. (2006). Increased abundance of IncP-1β plasmids and mercury resistance genes in mercury-polluted river sediments: first discovery of IncP-1β plasmids with a complex mer transposon as the sole accessory element. Appl. Environ. Microbiol. *72*, 7253-7259.

Smalla, K., and Sobecky, P.A. (2002). The prevalence and diversity of mobile genetic elements in bacterial communties of different environmental habitats: insights gained from different methodological approaches. FEMS Microbiol. Ecol. *42*, 165-175.

Smith, C.A., and Thomas, C.M. (1984). Nucleotide sequence of the *trfA* gene of broad host-range plasmid RK2. J. Mol. Biol. *175*, 251-262.

Smith, M.C., and Thorpe, H.M. (2002). Diversity in the serine recombinases. Mol. Microbiol. *44*, 299-307.

Sokurenko, E.V., Hasty, D.L., and Dykhuizen, D.E. (1999). Pathoadaptive mutations: gene loss and variation in bacterial pathogens. Trends Microbiol. *7*, 191-195.

Sota, M., Endo, M., Nitta, K., Kawasaki, H., and Tsuda, M. (2002). Characterization of a class II defective transposon carrying two haloacetate dehalogenase genes from *Delftia acidovorans* plasmid pUO1. Appl. Environ. Microbiol. *68*, 2307-2315.

Sota, M., Kawasaki, H., and Tsuda, M. (2003). Structure of haloacetate-catabolic IncP-1β plasmid pUO1 and genetic mobility of its residing haloacetate-catabolic transposon. J. Bacteriol. *185*, 6741-6745.

Sota, M., Tsuda, M., Yano, H., Suzuki, H., Forney, L.J., and Top, E.M. (2007). Region-specific insertion of transposons in combination with selection for high plasmid transferability and stability accounts for the structural similarity of IncP-1 plasmids. J. Bacteriol. *189*, 3091-3098.

Sota, M., Yano, H., Nagata, Y., Ohtsubo, Y., Genka, H., Anbutsu, H., Kawasaki, H., and Tsuda, M. (2006a). Functional analysis of unique class II insertion sequence IS*1071*. Appl. Environ. Microbiol. *72*, 291-297.

Sota, M., Yano, H., Ono, A., Miyazaki, R., Ishii, H., Genka, H., Top, E.M., and Tsuda, M. (2006b). Genomic and functional analysis of the IncP-9 naphthalene-catabolic plasmid NAH7 and its transposon Tn*4655* suggests catabolic gene spread by a tyrosine recombinase. J. Bacteriol. *188*, 4057-4067.

Sørensen, S.J., Bailey, M., Hansen, L.H., Kroer, N., and Wuertz, S. (2005). Studying plasmid horizontal transfer in situ: A critical review. Nat. Rev. Microbiol. *3*, 700-710.

Sørensen, S.J., Sørensen, A.H., Hansen, L.H., Oregaard, G., and Veal, D. (2003). Direct detection and quantification of horizontal gene transfer by using flow cytometry and *gfp* as a reporter gene. Curr. Microbiol. *47*, 129-133.

Springael, D., and Top, E.M. (2004). Horizontal gene transfer and microbial adaptation to xenobiotics: new types of mobile genetic elements and lessons from ecological studies. Trends Microbiol. *12*, 53-58.

Stanisich, V.A. (1974). Interaction between an R factor and an element conferring resistance to mercuric ions in *Pseudomonas aeruginosa*. Mol. Gen. Genet. *128*, 201-212.

Stedman, K.M., She, Q., Phan, H., Holz, I., Singh, H., Prangishvili, D., Garrett, R., and Zillig, W. (2000). pING family of conjugative plasmids from the extremely thermophilic archaeon *Sulfolobus islandicus*: insights into recombination and conjugation in Crenarchaeota. J. Bacteriol. *182*, 7014-7020.

Stewart, F.M., and Levin, B.R. (1977). The population biology of bacterial plasmids: A priori conditions for the existence of conjugationally transmitted factors. Genetics *87*, 209-228.

Stokes, H.W., and Hall, R.M. (1989). A novel family of potentially mobile DNA elements encoding site-specific gene-integration functions: integrons. Mol. Microbiol. *3*, 1669-1683.

Sullivan, J.T., and Ronson, C.W. (1998). Evolution of rhizobia by acquisition of a 500-kb symbiosis island that integrates into a phe-tRNA gene. Proc. Natl. Acad. Sci. U. S. A. *95*, 5145-5149.

Summers, A.O., Wireman, J., Vimy, M.J., Lorscheider, F.L., Marshall, B., Levy, S.B., Bennett, S., and Billard, L. (1993). Mercury released from dental "silver" fillings provokes an increase in mercury- and antibiotic-resistant bacteria in oral and intestinal floras of primates. Antimicrob. Agents Chemother. *37*, 825-834.

Sundin, G.W. (2007). Genomic insights into the contribution of phytopathogenic bacterial plasmids to the evolutionary history of their hosts. Annu. Rev. Phytopathol. *45*, 129-151.

Sundin, G.W., Mayfield, C.T., Zhao, Y., Gunasekera, T.S., Foster, G.L., and Ullrich, M.S. (2004). Complete nucleotide sequence and analysis of pPSR1 (72,601 bp), a pPT23A-family plasmid from *Pseudomonas syringae* pv. *syringae* A2. Mol. Genet. Genomics *270*, 462-476.

Sundstrom, L., Roy, P.H., and Skold, O. (1991). Site-specific insertion of three structural gene cassettes in transposon Tn*7*. J. Bacteriol. *173*, 3025-3028.

Szpirer, C., Top, E., Couturier, M., and Mergeay, M. (1999). Retrotransfer or gene capture: a feature of conjugative plasmids, with ecological and evolutionary significance. Microbiology *145*, 3321-3329.

Tan, H.M. (1999). Bacterial catabolic transposons. Appl. Microbiol. Biotechnol. *51*, 1-12.

Tanimoto, K., An, F.Y., and Clewell, D.B. (1993). Characterization of the *traC* determinant of the *Enterococcus faecalis* hemolysin-bacteriocin plasmid pAD1: binding of sex pheromone. J. Bacteriol. *175*, 5260-5264.

Tauch, A., Schlüter, A., Bischoff, N., Goesmann, A., Meyer, F., and Pühler, A. (2003). The 79,370-bp conjugative plasmid pB4 consists of an IncP-1β backbone loaded with a chromate resistance transposon, the *strA-strB* streptomycin resistance gene pair, the oxacillinase gene bla_{NPS-1}, and a tripartite antibiotic efflux system of the resistance-nodulation-division family. Mol. Genet. Genomics *268*, 570-584.

Tauch, A., Schneiker, S., Selbitschka, W., Pühler, A., van Overbeek, L.S., Smalla, K., Thomas, C.M., Bailey, M.J., Forney, L.J., Weightman, A., *et al.* (2002). The complete nucleotide sequence and environmental distribution of the cryptic, conjugative, broad-host-range plasmid pIPO2 isolated from bacteria of the wheat rhizosphere. Microbiology *148*, 1637-1653.

Taylor, D.E., Gibreel, A., Lawley, T.D., and Tracz, D.M. (2004). Antibiotic resistance plasmids. In Plasmid biology, B.E. Funnell, and G.J. Phillips, eds. (Washington, D. C., ASM Press), pp. 473-491.

Tetaz, T.J., and Luke, R.K. (1983). Plasmid-controlled resistance to copper in *Escherichia coli*. J. Bacteriol. *154*, 1263-1268.

Tett, A., Spiers, A.J., Crossman, L.C., Ager, D., Ciric, L., Dow, J.M., Fry, J.C., Harris, D., Lilley, A., Oliver, A., *et al.* (2007). Sequence-based analysis of pQBR103; a representative of a unique, transfer-proficient mega plasmid resident in the microbial community of sugar beet. ISME J. *1*, 331-340.

Tezcan-Merdol, D., Nyman, T., Lindberg, U., Haag, F., Koch-Nolte, F., and Rhen, M. (2001). Actin is ADP-ribosylated by the *Salmonella enterica* virulence-associated protein SpvB. Mol. Microbiol. *39*, 606-619.

Thomas, C.M. (2000a). Paradigms of plasmid organization. Mol. Microbiol. *37*, 485-491.

Thomas, C.M. (2000b). The horizontal gene pool (Amsterdam, Harwood Academic Publishers).

Thomas, C.M. (2004). Evolution and population genetics of bacterial plasmids. In Plasmid biology, B.E. Funnell, and G.J. Phillips, eds. (Washington, D. C., ASM Press), pp. 509-528.

Thomas, C.M. (2006). Transcription regulatory circuits in bacterial plasmids. Biochem. Soc. Trans. *34*, 1072-1074.

Thomas, C.M., and Haines, A.S. (2004). Plasmids of the genus *Pseudomonas*. In *Pseudomonas*, J.-L. Ramos, ed. (New York, U. S. A., Plenum Publishing Corporation), pp. 197-231.

Thomas, C.M., and Nielsen, K.M. (2005). Mechanisms of, and barriers to, horizontal gene transfer between bacteria. Nat. Rev. Microbiol. *3*, 711-721.

Thomas, C.M., and Smith, C.A. (1987). Incompatibility group P plasmids: genetics, evolution, and use in genetic manipulation. Annu. Rev. Microbiol. *41*, 77-101.

Thompson, J.K., and Collins, M.A. (2003). Completed sequence of plasmid pIP501 and origin of spontaneous deletion derivatives. Plasmid *50*, 28-35.

Thorsted, P.B., Macartney, D.P., Akhtar, P., Haines, A.S., Ali, N., Davidson, P., Stafford, T., Pocklington, M.J., Pansegrau, W., Wilkins, B.M., *et al.* (1998). Complete sequence of the IncPβ plasmid R751: implications for evolution and organisation of the IncP backbone. J. Mol. Biol. *282*, 969-990.

Tilman, D., and Kareiva, P., eds. (1997). Spatial Ecology (Princeton, NY, Princeton Univ. Press).

Tolker-Nielsen, T., and Molin, S. (2004). Biofilm development by pseudomonas. Paper presented at: European Symposium on Environmental Biotechnology - ESEB 2004 (Oostende, Belgium, A.A. Balkema Publishers).

Tomita, H., and Ike, Y. (2005). Genetic analysis of transfer-related regions of the vancomycin resistance *Enterococcus* conjugative plasmid pHTβ: identification of *oriT* and a putative relaxase gene. J. Bacteriol. *187*, 7727-7737.

Tomita, H., Tanimoto, K., Hayakawa, S., Morinaga, K., Ezaki, K., Oshima, H., and Ike, Y. (2003). Highly conjugative pMG1-like plasmids carrying Tn*1546*-like transposons that encode vancomycin resistance in *Enterococcus faecium*. J. Bacteriol. *185*, 7024-7028.

Top, E., Mergeay, M., Springael, D., and Verstraete, W. (1990). Gene escape model: transfer of heavy metal resistance genes from *Escherichia coli* to *Alcaligenes eutrophus* on agar plates and in soil samples. Appl. Environ. Microbiol. *56*, 2471-2479.

Top, E.M., Moenne-Loccoz, Y., Pembroke, T., and Thomas, C.M. (2000). Phenotypic traits conferred by plasmids. In The Horizontal Gene Pool, C.M. Thomas, ed. (Amsterdam, Harwood Academic Publishers), pp. 249-286.

Top, E.M., and Springael, D. (2003). The role of mobile genetic elements in bacterial adaptation to xenobiotic organic compounds. Curr. Opin. Biotechnol. *14*, 262-269.

Top, E.M., Springael, D., and Boon, N. (2002). Catabolic mobile genetic elements and their possible use in bioaugmentation of polluted soils and waters. FEMS Microbiol. Ecol. *42*, 199-208.

Trefault, N., De la Iglesia, R., Molina, A.M., Manzano, M., Ledger, T., Perez-Pantoja, D., Sanchez, M.A., Stuardo, M., and Gonzalez, B. (2004). Genetic organization of the catabolic plasmid pJP4 from *Ralstonia eutropha* JMP134 (pJP4) reveals mechanisms of adaptation to chloroaromatic pollutants and evolution of specialized chloroaromatic degradation pathways. Environ. Microbiol. *6*, 655-668.

Tsuda, M., and Iino, T. (1988). Identification and characterization of Tn*4653*, a transposon covering the toluene transposon Tn*4651* on TOL plasmid pWW0. Mol. Gen. Genet. *213*, 72-77.

Tsuda, M., and Iino, T. (1990). Naphthalene degrading genes on plasmid NAH7 are on a defective transposon. Mol. Gen. Genet. *223*, 33-39.

Tsuda, M., Minegishi, K., and Iino, T. (1989). Toluene transposons Tn*4651* and Tn*4653* are class II transposons. J. Bacteriol. *171*, 1386-1393.

Tsuda, M., Tan, H.M., Nishi, A., and Furukawa, K. (1999). Mobile catabolic genes in bacteria. J. Biosci. Bioeng. *87*, 401-410.

Tuffin, I.M., de Groot, P., Deane, S.M., and Rawlings, D.E. (2005). An unusual Tn*21*-like transposon containing an ars operon is present in highly arsenic-resistant strains of the biomining bacterium *Acidithiobacillus caldus*. Microbiology *151*, 3027-3039.

Udo, E.E., Jacob, L.E., and Mathew, B. (2000). A cadmium resistance plasmid, pXU5, in *Staphylococcus aureus*, strain ATCC25923. FEMS Microbiol. Lett. *189*, 79-80.

Vaisvila, R., Morgan, R.D., Posfai, J., and Raleigh, E.A. (2001). Discovery and distribution of super-integrons among pseudomonads. Mol. Microbiol. *42*, 587-601.

Van der Auwera, G., and Mahillon, J. (2005). Tn*XO1*, a germination-associated class II transposon from *Bacillus anthracis*. Plasmid *53*, 251-257.

van der Meer, J.R., Zehnder, A.J., and de Vos, W.M. (1991). Identification of a novel composite transposable element, Tn*5280*, carrying chlorobenzene dioxygenase genes of *Pseudomonas* sp. strain P51. J. Bacteriol. *173*, 7077-7083.

van Elsas, J.D., and Bailey, M.J. (2002). The ecology of transfer of mobile genetic elements. FEMS Microbiol. Ecol. *42*, 187-197.

Vavrova, S., Valkova, D., Drahovska, H., Kokavec, J., Mravec, J., and Turna, J. (2006). Analysis of the tellurite resistance determinant on the pNT3B derivative of the pTE53 plasmid from uropathogenic *Escherichia coli*. Biometals *19*, 453-460.

Venkatesan, M.M., and Burland, V. (2004). Genome-scale analysis of virulence plasmids: the contribution of plasmid-borne virulence genes to enterobacterial pathogenesis. In Plasmid biology, B.E. Funnell, and G.J. Phillips, eds. (Washington, D. C., ASM Press), pp. 395-411.

Venkatesan, M.M., Goldberg, M.B., Rose, D.J., Grotbeck, E.J., Burland, V., and Blattner, F.R. (2001). Complete DNA sequence and analysis of the large virulence plasmid of *Shigella flexneri*. Infect. Immun. *69*, 3271-3285.

Waddell, C.S., and Craig, N.L. (1988). Tn*7* transposition: two transposition pathways directed by five Tn*7*-encoded genes. Genes Dev. *2*, 137-149.

Wang, L., Chen, S., Xiao, X., Huang, X., You, D., Zhou, X., and Deng, Z. (2006). *arsRBOCT* arsenic resistance system encoded by linear plasmid pHZ227 in *Streptomyces* sp. strain FR-008. Appl. Environ. Microbiol. *72*, 3738-3742.

Ward, J.M., and Grinsted, J. (1978). Mapping of functions in the R-plasmid R388 by examination of deletion mutants generated in vitro. Gene *3*, 87-95.

Waters, V.L. (2001). Conjugation between bacterial and mammalian cells. Nat. Genet. *29*, 375-376.

Wei, W., and Krone, S.M. (2005). Spatial invasion by a mutant pathogen. J. Theor. Biol. *236*, 335-348.

Weigel, L.M., Clewell, D.B., Gill, S.R., Clark, N.C., McDougal, L.K., Flannagan, S.E., Kolonay, J.F., Shetty, J., Killgore, G.E., and Tenover, F.C. (2003). Genetic analysis of a high-level vancomycin-resistant isolate of *Staphylococcus aureus*. Science *302*, 1569-1571.

Weinreich, M.D., and Reznikoff, W.S. (1992). Fis plays a role in Tn*5* and IS*50* transposition. J. Bacteriol. *174*, 4530-4537.

Weiss, A.A., Murphy, S.D., and Silver, S. (1977). Mercury and organomercurial resistances determined by plasmids in *Staphylococcus aureus*. J. Bacteriol. *132*, 197-208.

Welch, R.A., Burland, V., Plunkett, G., 3rd, Redford, P., Roesch, P., Rasko, D., Buckles, E.L., Liou, S.R., Boutin, A., Hackett, J., *et al.* (2002). Extensive mosaic structure revealed by the complete genome sequence of uropathogenic *Escherichia coli*. Proc. Natl. Acad. Sci. U. S. A. *99*, 17020-17024.

Wiater, L.A., and Grindley, N.D. (1991). γδ transposase. Purification and analysis of its interaction with a transposon end. J. Biol. Chem. *266*, 1841-1849.

Will, W.R., and Frost, L.S. (2006). Hfq is a regulator of F-plasmid TraJ and TraM synthesis in *Escherichia coli*. J. Bacteriol. *188*, 124-131.

Willetts, N. (1977). The transcriptional control of fertility in F-like plasmids. J. Mol. Biol. *112*, 141-148.

Williams, P.A., Jones, R.M., and Zylstra, G. (2004). Genomics of catabolic plasmids. In *Pseudomonas*, J.-L. Ramos, ed. (New York, U. S. A., Plenum Publishing Corporation), pp. 165-195.

Williams, P.A., and Sayers, J.R. (1994). The evolution of pathways for aromatic hydrocarbon oxidation in *Pseudomonas*. Biodegradation *5*, 195-217.

Wimpenny, J.W.T., and Colasanti, R. (1997). The structure of microbial biofilms: a cellular automaton study. FEMS Microbiol. Ecol. *22*, 1-16.

Wu, J., and Rosen, B.P. (1991). The ArsR protein is a trans-acting regulatory protein. Mol. Microbiol. *5*, 1331-1336.

Wuertz, S., Hendrickx, L., Kuehn, M., Rodenacker, K., and Hausner, M. (2001). In situ quantification of gene transfer in biofilms. Methods Enzymol. *336*, 129–143.

Wyndham, R.C., Cashore, A.E., Nakatsu, C.H., and Peel, M.C. (1994). Catabolic transposons. Biodegradation *5*, 323-342.

Yang, F., Yang, J., Zhang, X., Chen, L., Jiang, Y., Yan, Y., Tang, X., Wang, J., Xiong, Z., Dong, J., *et al.* (2005). Genome dynamics and diversity of *Shigella* species, the etiologic agents of bacillary dysentery. Nucleic Acids Res *33*, 6445-6458.

Yang, J.C., Lessard, P.A., and Sinskey, A.J. (2007). Characterization of the mobilization determinants of pAN12, a small replicon from *Rhodococcus erythropolis* AN12. Plasmid *57*, 71-81.

Yano, H., Garruto, C.E., Sota, M., Ohtsubo, Y., Nagata, Y., Zylstra, G.J., Williams, P.A., and Tsuda, M. (2007). Complete sequence determination combined with analysis of transposition/site-specific recombination events to explain genetic organization of IncP-7 TOL plasmid pWW53 and related mobile genetic elements. J. Mol. Biol. *369*, 11-26.

Yen, K.M., and Serdar, C.M. (1988). Genetics of naphthalene catabolism in pseudomonads. Crit. Rev. Microbiol. *15*, 247-268.

Yoshida, K., Takegami, T., Katoh, A., Nishikawa, M., and Nishida, T. (1997). Construction of a novel conjugative plasmid harboring a GFP reporter gene and its introduction into animal cells by transfection and trans-kingdom conjugation. Nucleic Acids Symp. Ser., 157-158.

Zatyka, M., Jagura-Burdzy, G., and Thomas, C.M. (1994). Regulation of transfer genes of promiscuous IncPα plasmid RK2: repression of Tra1 region transcription both by relaxosome proteins and by the Tra2 regulator TrbA. Microbiology *140*, 2981-2990.

Zechner, E.L., de la Cruz, F., Eisenbrandt, R., Grahn, A.M., Koraimann, G., Lanka, E., Muth, G., Pansegrau, W., Thomas, C.M., Wilkins, B.M., *et al.* (2000). Conjugative-DNA transfer process. In The Horizontal Gene Pool, C.M. Thomas, ed. (Amsterdam, Harwood Academic Publishers), pp. 87-174.

Zhong, Z., Helinski, D., and Toukdarian, A. (2005). Plasmid host-range: restrictions to F replication in *Pseudomonas*. Plasmid *54*, 48-56.

Chapter 6

Molecular Machinery for DNA Translocation in Bacterial Conjugation

Silvia Russi, Roeland Boer and Miquel Coll

ABSTRACT

Whatever the route used, horizontal transfer of DNA requires elaborated multi-protein machinery to enable the long and charged nucleic acid polymer to cross the cell envelope barriers. The best-studied system for cell-to-cell DNA translocation is bacterial conjugation. This system can be divided in two discrete specialized modules: the relaxosome, which triggers and takes part in plasmid DNA processing and replication, and a type IV secretion system (T4SS), which impels protein and single-stranded DNA through the membranes. In addition, a coupling protein (CP), linking both modules, and a number of ancillary proteins are needed. Over the last decades research efforts in the field have resulted in the clarification of many aspects of this system and its machinery assembly. In particular, structural biology has provided details of the molecular architecture of several of the pieces involved in this intricate scenario.

INTRODUCTION

Single-stranded DNA (ssDNA) transfer in conjugation, DNA uptake in transformation, and double-stranded DNA (dsDNA) transfer -via bacteriophages- in transduction, are all different routes by which a DNA molecule can be translocated across the bacterial cell envelope (Boucher et al., 2003; Lederberg, 2000; Ochman et al., 2000). During the translocation process the DNA molecule has to cross several barriers of different physicochemical properties. For example, in Gram-negative bacteria, the hydrophobicity of the inner membrane, the presence of nucleases in the periplasm and the negative charge of the outer membrane are some of the hurdles associated to this transport (Dreiseikelmann, 1994). To overcome these barriers, sophisticated DNA translocation molecular machinery is employed.

Two of the DNA transfer routes, conjugation (including T-DNA transfer from bacteria to plants) and transformation, have some functional and structural elements in common (Chen et al., 2005). The third route, transduction, is not related to the others and requires the dsDNA packaging and injection machinery of a phage (Johnson and Chiu, 2007). In conjugation and transformation the transferred DNA molecule is single-stranded. Still, there are marked differences in the two systems (Chen and Dubnau, 2004).

Among the different DNA translocation systems, conjugative DNA transfer is the best studied (Lanka and Wilkins, 1995; Silverman, 1997). The ability of bacteria to conjugate, had already been discovered in *E. coli* cultures in 1946 (Lederberg and Tatum, 1946). Gram-positive bacteria have also been found to harbour conjugative plasmids (Grohmann et al., 2003; Scott and Churchward, 1995; Ton-That and Schneewind, 2003). This has attracted much interest because of the clinical problems associated with it (Alekshun and Levy, 2007). The spread of antibiotic resistance genes in just a few decades since antibiotics started to be widely used, is largely due to this fast and powerful horizontal gene transfer mechanism.

The transfer of conjugative DNA is mediated by a multi-component molecular machinery, whose protein components are usually encoded on the (plasmid) DNA involved in the conjugation process (Llosa et al., 2002). Bacterial conjugation transmembrane transport channels are a subfamily of the type IV secretion system (T4SS) involved in the transfer of macromolecules, toxins and DNA-protein complexes across the cell envelope (Backert and Meyer, 2006; Burns, 2003; Cascales and Christie, 2003; Chen et al., 2005; Christie, 2004; Llosa and O'Callaghan, 2004; Sexton and Vogel, 2002). T4SS are therefore not restricted to the translocation of DNA. Example of non-DNA transporting T4SS are those from *Bordetella pertussis* (Ptl), responsible for whooping cough (Weiss et al., 1993); *Helicobacter pylori* (Cag), related to peptic ulcers, MALT-lymphoma and adenocarcinoma of the stomach (Graham, 2000; Wotherspoon, 1998); and *Bartonella henselae* (VirB/VirD4) causing the cat scratch disease (Cascales and Christie, 2003). Bacterial conjugation of DNA may be, in fact, an adaptation of ancestral translocation machinery that transfers proteins (Christie et al., 2005). These transport systems are functionally diverse, not only in terms of the transported substrate, but also with respect to the recipient cell, which can be bacteria of the same or different species, fungi, plants or mammalian cells (Christie, 2001; Christie and Cascales, 2005; Christie and Vogel, 2000; Ding et al., 2003; Seubert et al., 2003; Waters, 2001). In general, the substrate delivery to the recipient cell occurs via direct cell-to-cell contact, although there are examples of contact-independent protein export and DNA release (Dillard and Seifert, 2001; Hofreuter et al., 2001).

In recent years, X-ray crystallography and NMR spectroscopy studies have provided valuable structural information on various proteins involved in the conjugation process (Gomis-Ruth and Coll, 2006; Remaut and Waksman, 2004; Yeo and Waksman, 2004). This information, revealing the molecular architecture of some of the central pieces of the mechanism, has answered some of the fundamental questions about T4SS-mediated DNA transport machinery. Here we will review relaxosome and T4SS structures determined by X-ray crystallography and NMR spectroscopy.

COMPONENTS OF THE TYPE IV SECRETION SYSTEM

The T4SS components can be grouped into two modules according to their function: i) the relaxosome (Lanka and Wilkins, 1995; Wilkins and Lanka, 1993), ii) the trans-membranal conduit or channel. In addition, a coupling protein (CP) links both modules (Gomis-Ruth et al., 2004; Llosa and de la Cruz, 2005). The relaxosome is a nucleoprotein complex that prepares the T-DNA for transfer by specifically nicking the DNA at the origin of transfer (*oriT*) (Ziegelin et al., 1989) after the donor and recipient cells have come into contact (Durrenberger et al., 1991). It also is responsible for the delivery of the DNA to the T4SS secretion system (Fekete and Frost, 2000) through the action of the coupling protein (T4CP). Its protein subunits are encoded on a gene region called the DNA transfer replication region (*Dtr*). It consists of a relaxase, which also works as the pilot protein, and two or more accessory proteins.

The T4SS trans-membranal conduit is a membrane-spanning multi-protein complex that comprises copies of around 12 different proteins, which are encoded on the mating pair formation (Mtf) region (Backert and Meyer, 2006; Christie et al., 2005; Silverman, 1997). Its extracellular part, called the pilus, appears as an elongated tubular appendage of the donor cell in micrographs. The T4S system mediates the association between the host and recipient cells through the pilus, which binds to the surface of the recipient cell and brings both cell envelopes into close contact. What happens to the pilus after establishing the mating junction is not clear. In F plasmids, which encode for a group of proteins with little homology to most other conjugation machines, the pilus retracts after establishing cell-to-cell contact (Lawley et al., 2003). However, pilus retraction has not been demonstrated in other systems (Christie et al. 2005). Some systems lack the subunits dedicated to pilus retraction, which include plasmids RP4, R388 and pKM101 and the Gram-negative bacterium *Agrobacterium tumefaciens* (Harris and Silverman, 2004). Whether DNA passes through the extracellular part of the pilus during conjugation is not yet clear (Kalkum et al., 2004).

T4S systems can be classified into types IV-A, IV-B and into an as of yet unclassified group, according to their homology among the functional subunits. The IV-A systems are homologous to that of the *Agrobacterium tumefaciens* VirB/D system, which is the archetype of type IV-A secretion systems and has been studied extensively at biochemical and structural levels. This Gram-negative bacterium introduces an oncogenic DNA fragment, the T-DNA fragment, into plant cells (Christie, 2004). The T4SS components identified and mapped on the genome are found in the VirB and VirD genes, which correspond to the transmembrane conduit and relaxosome respectively. The *virB* gene products comprise eleven proteins (VirB1 to VirB11) encoded by the *virB* operon. So far, two of the *virD* gene products, VirD2 and VirD4, have been shown to be involved in T4SS DNA translocation. VirD2 is responsible for correct DNA processing and is transported through the T4SS as a covalent complex with the T-DNA (Howard et al., 1992; Vergunst et al., 2000). Its function is comparable to that of TrwC of the relaxosome of plasmid R388 (Draper et al., 2005). VirD4 is the coupling protein in this system, analogous to TrwB of plasmid R388 (Christie, 1997).

Further along this paper the structural organization and the reaction mechanisms of the relaxosome components and the T4SS will be discussed in more detail, based on the available structural information. The section on the relaxosome will focus on the systems studied by us and others in the Gram-negative *E. coli* plasmids R388 and F and will include a discussion on the structure of the coupling protein, determined in our laboratory. In a subsequent section, the description of the structural information available for the secretion apparatus will be based on proteins of the DNA translocation membrane conduit of *Agrobacterium tumefaciens*, encoded on the bacterial non-translational chromosome. Many of the structures of the T4SS constituents of *A. tumefaciens* or closely related organisms have been determined in the laboratory of G. Waksman.

THE RELAXOSOME: OVERVIEW

The relaxosome is a nucleoprotein structure involved in the initiation of DNA-processing reactions that occur during bacterial conjugation. The proteins of the relaxosome bind specifically to the origin of transfer (*oriT*), a short DNA sequence where the process starts and ends. It comprises a relaxase/helicase protein, responsible for nicking the DNA and, in many cases, for unwinding the supercoiled plasmid DNA for further processing. This protein is also responsible for the reconstitution of a single-stranded circular DNA by transesterification of the 3' and 5' ends of the translocated DNA strand, presumably, in the recipient cell. In addition to the helicase, host factors such as the integration host factor (IHF) may form part of the complex and a transcriptional auto-regulatory protein is often included. Furthermore, an ancillary protein (TraM) is required in several plasmids, like F and R1 (Fig. 1a).

The architecture of the relaxosome of the *E. coli* plasmid R388 (Fig. 1b), one of the representative plasmids of the plasmid incompatibility group IncW (Moncalian et al., 1997; Moncalian et al., 1999b), has been studied in detail. In this plasmid, two of the encoded proteins (TrwA and TrwC) form the relaxosome, and a third protein (TrwB) is responsible for coupling the relaxosome to the base of the secretion channel. TrwC is the relaxase/helicase protein of this system. TrwA and IHF are the transcription regulator and the integration host factor, respectively.

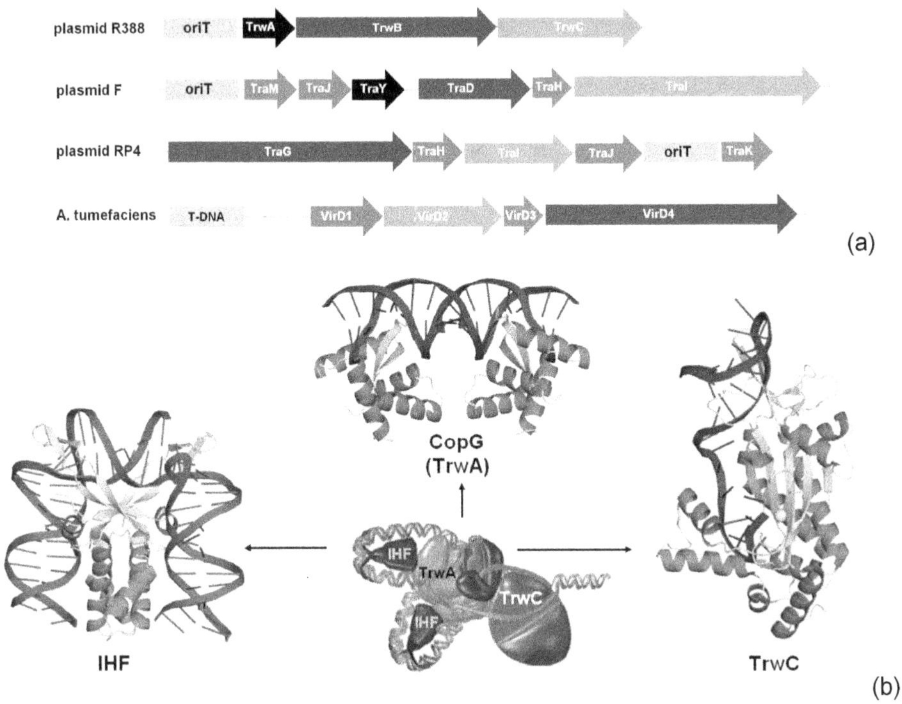

Fig. 1. a) Gene organization and nomenclature of the relaxosome proteins of conjugative plasmids R388, F, RP4 and Ti. b) Structural models of the relaxosome components in the *E. coli* plasmid R388 system: the TrwA-related transcriptional regulator CopG dimer in complex with its cognate 19-bp dsDNA (PDB code 1b01, upper central panel), *E. coli* host-encoded IHF in complex with a 35-bp dsDNA (PDB code 1ihf, left), relaxase/helicase TrwC relaxase domain in complex with a 27-base DNA hairpin (PDB code 2cdm, right). The central panel shows a modeled scheme of the assembly of the relaxosome complex.

The R388 system lacks a TraM ortholog and the relaxosome is assembled at the origin of transfer (*oriT*) DNA. The helicase activity of TrwC on the plasmid DNA separates the nicked parental strand from the complementary strand and allows the regeneration of duplex plasmid DNA through replicative synthesis of the latter by host polymerases (Kingsman and Willetts, 1978). The mechanism of this process is a variant of the rolling-circle replication mechanism (RCR) (Furuya and Komano, 2000; Kramer et al., 1997).

DNA BINDING FACTORS

TrwA is a 53 kDa tetrameric DNA-binding protein that specifically recognizes the oriT sequence of plasmid R388. The two binding sites (*sbaA* and *sbaB*) are not close to the *nic* site. Site *sbaA* and *sbaB* are located 125 to 150 bp and 67 to 96 bp downstream from *nic*, respectively. The relative location of these sites is similar to the recognition sites *sbyA* and *sbyB* of the homologous protein TraY of plasmid F (Moncalian et al., 1997). TrwA has two biochemically independent roles in R388 conjugation. First, the binding of TrwA to a specific recognition sequence in *oriT*, different from that of TrwC, results in a transcriptional repression of the *trwABC* operon that encodes TrwA, TrwB, and TrwC. Secondly, TrwA has a direct role in *oriT* processing, enhancing the relaxation reaction catalyzed by TrwC on supercoiled DNA R388 *oriT* (Moncalian and de la Cruz, 2004). Analogous behaviour has been observed in the TrwC homolog TraY with the enhancement of the relaxase/helicase activity of TraI and the regulation of the expression of the *tra* genes (Lum et al., 2002). Since TrwA expression is autoregulated, it may be expected to play an important role on the regulation of the frequency of plasmid conjugation.

The N-terminal domain of TrwA and TraY is a DNA-binding domain that recognises the two palindromic sites around *oriT*. The C-terminal is a dimerization domain. The secondary structure prediction of the DNA-binding domain suggests a ribbon-helix-helix motif, as found in the Arc/MetJ/CopG superfamily of transcriptional repressors (Suzuki, 1995). The crystal structure of a representative member of this family of transcriptional regulators, CopG of plasmid pMV158, in complex with its cognate double-stranded DNA has been solved (Gomis-Ruth et al., 1998). It was found to recognize the DNA as a dimer of dimers (Fig. 1b). An antiparallel β-ribbon, formed by the ribbon-helix-helix motif of each of the protomers of the dimer, enters the DNA major groove. The cognate DNA is curved over 120° per protein tetramer. The homology between the N-terminal domain of CopG and that of TrwA and TraY, suggests that TrwA and TraY have a similar mode of action and help the DNA acquire the topology necessary for the optimal cleavage by way of this relaxase (del Solar et al., 2002).

The integration host factor (IHF) is a heterodimeric protein that binds specifically to the R388 oriT region at two different sites (*ihfA* and *ihfB*). Site *ihfA* lies between *nic* and *sbaB*, whereas site *ihfB* is located between *sbaA* and *sbaB* (Moncalian et al., 1999b). This protein is known to assist in many prokaryotic processes, such as replication, transcriptional regulation, and several site-specific recombination events, in which special DNA axis distortions are required. In the plasmid F system, IHF facilitates the cleavage reaction of the relaxase by affecting the topology of the *nic* DNA site (Howard et al., 1995), whereas it was reported to inhibit the TrwC *nic* cleavage reaction (Moncalian et al., 1999b). The crystal structure of an IHF-DNA complex (Rice et al., 1996) shows that the two 10 kDa component subunits of IHF, which share about 30% sequence identity, are structurally equivalent (Fig. 1b). An N-terminal segment, formed by two consecutive α-helices and three β-strands featuring an antiparallel sheet, is responsible for the dimerization of the protein. The β–ribbon shape of the DNA-binding domain allows it to penetrate the DNA minor groove and curl around the DNA producing a sharp bend into the double helix which results in the formation of a ~160° U-turn (Rice et al., 1996). Two large kinks in the phosphate backbone, mediated by prolines at the tip of each β-ribbon, are responsible for most of the bending.

INITIATION AND PREPARATION

TrwC is a large protein (966 residues long) of 108 kDa harbouring two functionally different domains (Llosa et al., 1996). The recognition of the cleavage point at *oriT* and endonuclease, relaxase and transesterification activities take place at the N-terminal domain of TrwC, whereas the C-terminal domain displays a 5' to 3' DNA helicase activity. TrwC contains two active-site tyrosyl residues (instead of the single one in P-family relaxases) that catalyze the initiation and termination steps in conjugative DNA replication and a highly conserved histidine triad, a site for the binding of a metal ion, also required for the relaxase activity of the protein (Grandoso et al., 2000). Biochemical studies demonstrated that *in vitro*, the N-terminal relaxase domain, like the full-length TrwC, cleaves single-stranded oligonucleotides containing the *nic* site and performs complete strand transfer reactions in a way that resembles type I topoisomerases (Champoux, 2001; Wang, 1996). Furthermore, these studies stressed the importance of metal ions for driving catalysis.

TrwC recognizes the cleavage point at the *nic* site because the distinctive feature of the *nic* sequence: a six-base pair inverted repeat that forms a cruciform structure (Guasch et al., 2003). The supercoiled DNA plasmid is locally deformed and stabilized by TrwA, the *trw* gene promoter repressor, and the IHF protein. After the initial cleavage of the T-strand, its 3' end to *nic* remains covalently bound to a catalytic TrwC tyrosine by a 5'-phosphotyrosyl linkage during transfer to the recipient cell, whereas the 3' end is released (Llosa et al., 2002; Wilkins and Lanka, 1993). TrwC, acting as a helicase, now migrates processively in the 5' to 3' direction (away from the *trw* genes) on the displaced strand, unwinding the DNA. The accompanying host DNA polymerase III binds the uncleaved strand for complementary strand synthesis. TrwC mediated helicase activity has been shown to depend on an external ssDNA-dependent ATPase activity (Grandoso et al., 1994). There is evidence that TrwC stays bound to the T-DNA strand until it reaches the reconstituted *oriT* hairpin, upon which a second cleavage occurs, followed by a termination reaction resulting in the formation of the single stranded circular T-DNA. In this model, the termination reaction would occur in the host cell and the circular ssDNA would subsequently be transported through the T4SS. However, it was demonstrated that the relaxase protein is also transported to the recipient cell by the T4SS, where it is anticipated to track the incoming DNA until the passage of the *oriT* hairpin and then to carry out the termination reaction (Draper et al., 2005; Garcillan-Barcia et al., 2007). A model was proposed in which the initial nicking is performed by a TrwC dimer, which dissociates afterwards, as shown in Fig. 2. The

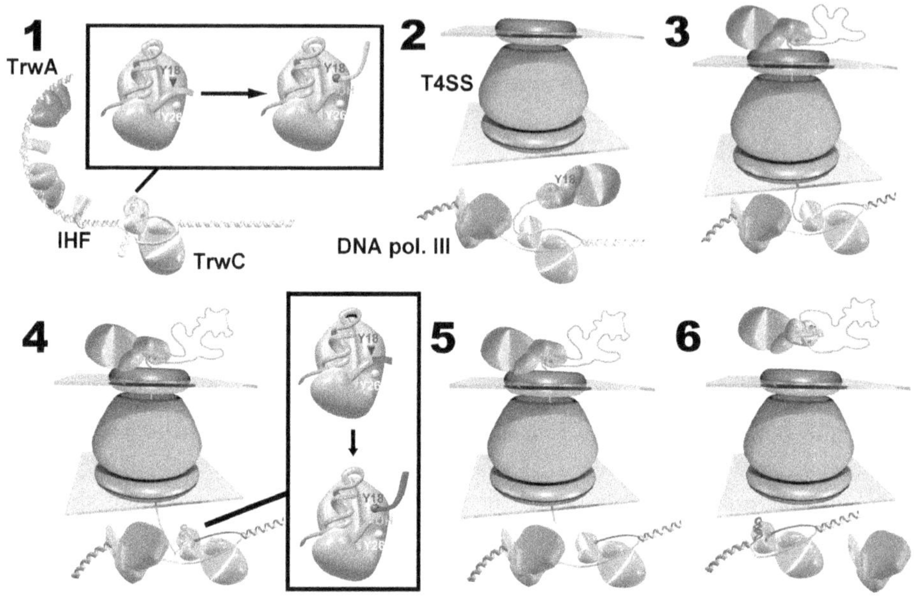

Fig. 2. A schematic representation of the proposed sequence of events during bacterial conjugation of plasmid R388, encoding the *trw* genes. A discontinuous process is assumed which allows conjugative translocation of a single copy of the parental strand. Initiation requires TrwA, IHF and dimeric TrwC to bind to the *oriT* region of the plasmid. Subsequent nicking of the parental strand releases the 3' end (1) and one of the TrwC monomers, covalently linked to the 5' end, passes through the T4SS (2). Simultaneously, the free 3' end is elongated by a host polymerase (new DNA shown in dark grey). The parental strand of the plasmid DNA passes to the recipient cell where it is scanned by the translocated TrwC monomer (3). After processing the full plasmid, the donor-cell TrwC again reaches the *oriT* hairpin and a second cut occurs (4), which frees the original parental strand for passage to the host cell (5). The host polymerase duplicates the template strand up to the *nic* site and dissociates from the complex (6). At both the donor and recipient side, the TrwC-DNA complexes recognise the reconstituted hairpins and finish their tasks by closure of the two parental DNA strands.

monomer, that is covalently bound to the T-DNA, is passed on to the recipient cell, whereas the other monomer stays in the host to separate the plasmid strands.

MECHANISM OF TrwC AND HOMOLOGS

The crystal structures of the relaxase N-terminal domain of TrwC (Guasch et al., 2003) and its homolog TraI from plasmid F (Datta et al., 2003) show a two-layer α/β plate or open sandwich core (Fig. 3a). The core structure bears resemblance to DNA polymerases and looks like a right hand, hence the nickname 'palm' (Fig. 3b). A recent structure of a 186 residue N-terminal domain of MobA of plasmid R1162 by X-ray crystallography shows a similar fold (Monzingo et al., 2007). The main secondary structure elements of the 'palm' are a central β-sheet and two parallel α-helices ($\alpha 5$ and $\alpha 7$), diagonally crossing the sheet on one of its sides. Two other

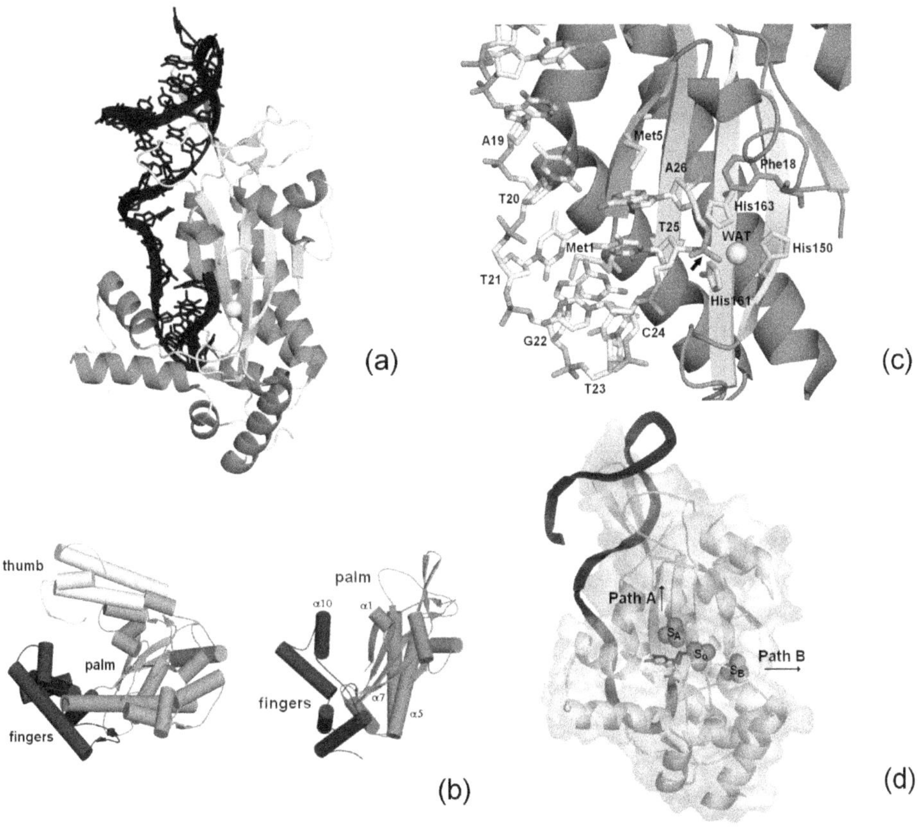

Fig. 3. a) *E. coli* R388 TrwC relaxase domain Y18F mutant in complex with a 27-base ssDNA that mimics the recognised *oriT* region (PDB code 2cdm). The hairpin is recognized by the top part of the protein. The active site is located at the lower end of the β-sheet, after the point where the DNA makes a sharp turn upwards. The direction of the DNA follows one of the possible exit paths of the DNA (see panel d and text). b) Overall topological similarity between DNA polymerase I Klenow fragment (left) and TrwC (right), showing the domain characteristics of polymerases, namely the thumb, the palm and the fingers, but only palm and fingers in the relaxase c) Ribbon representation of the TrwCY18F-DNA27 complex showing the U-turn of the DNA near the active site. The interactions of the G22:T25 mismatch pairing and bases G22 and T23 are shown. The interactions of Met1 at the hydrophobic cage formed by T21, G22 and T23 are shown, as well as the T25/A26/Met5 stack. The black arrow indicates the scissile bond. The position of the metal ion is occupied, in this structure, by a water molecule labelled WAT. d) Surface representation of the TrwCY18F-DNA27 complex. The 27-mer DNA oligonucleotide is depicted as a ribbon. Three sulphate ions, labelled SA, SO and SB, that are present in the active site crevice of the TrwC-DNA25 complex are also shown. Two putative exit paths for the DNA (A and B) are indicated with black arrows.

helices (α1 and α10) are at the opposite side of the β-sheet. They are arranged in a parallel fashion that gives rise to a narrow and long crevice that sits on the central β-sheet. The DNA runs in a direction parallel to the major longitudinal axis of the 'palm' until the 'fingers' engulf it. In TrwC, the active site is located at the crevice and is formed by the catalytic tyrosine (Tyr18) in charge of the T-strand DNA cleavage and the triad of histidines (His 150, 161 and 163) coordinating a metal ion. The interactions between the protein and the DNA are essentially electrostatic in nature, where most of the TrwC surface in contact with the DNA is electropositive. Specific interactions –including many hydrogen bonds- also occur between the DNA and the protein. A β-ribbon, which is an extension of a loop connecting two strands of the central β-sheet, enters the major groove of the dsDNA stem of the hairpin, while another loop segment, flanked by two turns, enters the opposite minor groove. After the hairpin, the DNA strand follows until it forms a peculiar U-turn just before entering the active site pocket. The U-turn serves as an anchoring point to TrwC via Met1, while the DNA continues antiparallel to an extended (G17–T21) stretch previous to the U-turn (Fig. 3c). Bases G22 and T25, conserved in different *nic* sequences of the F-family of conjugative plasmids, were found to form a mismatch pair. It is likely that this interaction serves to orient the DNA to the active site entrance and to place the scissile phosphate within proximity of the metal binding site, important for the cleavage reaction. More recent complex studies of TrwC and TraI confirmed that the scissile phosphate is directly coordinated to the metal ion (Boer et al., 2006; Larkin et al., 2005), which suggests that it may polarize the scissile phosphate facilitating the nucleophilic attack of the hydroxyl group of the catalytic tyrosine, and stabilize the subsequent penta-coordinate intermediate.

In the first structure of TrwC in complex with a 25 nucleotide *oriT* mimicking a ssDNA fragment, three sulphate ions were found near the cleavage site. Since they can mirror phosphate groups, their positions could indicate the directions of two possible functional paths, bifurcating at the metal binding site (Guasch et al., 2003). One path is nearly along the major longitudinal axis of TrwC, whereas the other runs perpendicular to it (Fig. 3d). A second structure of a T18F mutated TrwC in complex with an elongated DNA fragment is in agreement with the initial proposal for the DNA exit pathways (Boer et al., 2006), where the DNA could follow either path according to the timeline of the transfer process. For the initiation reaction, the DNA follows the longitudinal path. This configuration might be the preferred one in the constrained binding to supercoiled DNA, where only Tyr18 can cleave the DNA. After the first cleavage, Tyr18 forms a covalent bond with the DNA 5' to the cleavage site (see Fig. 2). The DNA hairpin containing the cleaved 3'-OH leaves the active site after which the RCR mechanism follows its course as described above and the TrwC monomer with the 5' end of the parental bound DNA is transported to the recipient cell. Passage of the DNA through the perpendicular pathway of TrwC as it enters the recipient cell allows the protein to scan the incoming DNA for the *oriT* sequence in order to complete the termination reaction. Upon passage of the *oriT* hairpin, the DNA is locked into place within TrwC (Garcillan-Barcia et al., 2007), forming a structure similar to that described previously with the additional hairpin bound. The 3' end of the hairpin is subsequently covalently linked to the 5' end of the bound DNA, which reconstitutes the parental strand in the recipient cell.

The overall fold of the relaxase domain of TrwC/TraI is similar to that observed in viral Rep proteins (Campos-Olivas et al., 2002; Hickman et al., 2002), the DNA-binding domain of the replication initiation protein E1 from papillomavirus (Enemark et al., 2002) and the origin-of-replication DNA-binding domain of SV40 T-antigen (Meinke et al., 2006). All these proteins are grouped into the same topological family (Murzin et al., 1995), and exhibit the typical 'origin-of-replication-binding domain' (OBD) fold, which comprises a central antiparallel five-folded β-sheet, which acts as support to two diagonally crossing helices. TrwC also bears some resemblance to DNA polymerases. Although the folds are similar, there are considerable differences between the proteins of this family, both in the domain constitution and size, as well as in function (Fig. 4). Rep proteins and TrwC/TraI are transesterification enzymes with a catalytic tyrosine and a HXH metal-binding motif, while the DNA-binding domains of E1 and the SV40 T-antigen have only DNA-binding function and lack these catalytic residues and an active-site pocket. A comparison of relaxases and viral Rep proteins shows that a circular permutation has occurred between them, resulting in the positioning of the catalytic tyrosine close to the N-terminus in the relaxases, but close to the C-terminus in the Reps (Fig. 4). More distantly, TrwC/TraI folding resembles the DNA polymerases folding, although in the former only the 'palm' and 'finger' subdomains are recognisable (Fig. 3b). In spite of the fact that the function of these proteins is different, the catalytic residues of the relaxase at the 'palm' spatially coincide with the corresponding residues of the Klenow fragment of the DNA polymerase I (Ollis et al., 1985).

ACTIVE SITE AND REACTION MECHANISM

The nicking and transesterification reactions are catalyzed through the action of a metal ion that is coordinated by the histidine triad, composed of His150,161 and 163. It has been shown that TrwC-bound Mn^{2+}, Ni^{2+}, Zn^{2+} and Cu^{2+} are able to cleave DNA at low concentration, and that Mg^{2+} and Ca^{2+} infer cleavage activity at high concentrations. X-ray structures of complexes of TrwC with Zn^{2+}, Ni^{2+} and Cu^{2+} have been resolved and show that these bind to the active site pocket. However, the nature of the endogenous metal ligand is as of yet unknown. The interaction of the metal ion at the active site is stabilized through a network of hydrogen bonds with origins at the nitrogen atoms of the histidines distal to the metal ion. Nitrogen Nδ of His161 is coordinated to the amide carboxyl oxygen of Gln159 and Nε of His150 to the carboxyl oxygen atom of Tyr18. Nδ of His163 is hydrogen bonded to a oxygen atom of the carboxyl group of Asp85, which in turn interacts with the hydroxyl group of Tyr18. It is the oxygen atom of the Tyr18 hydroxyl group that effectuates the nucleophilic attack on the DNA scissile phosphate. The fact that mutation of Asp85 impairs cleavage and that it occupies a pivotal location in the

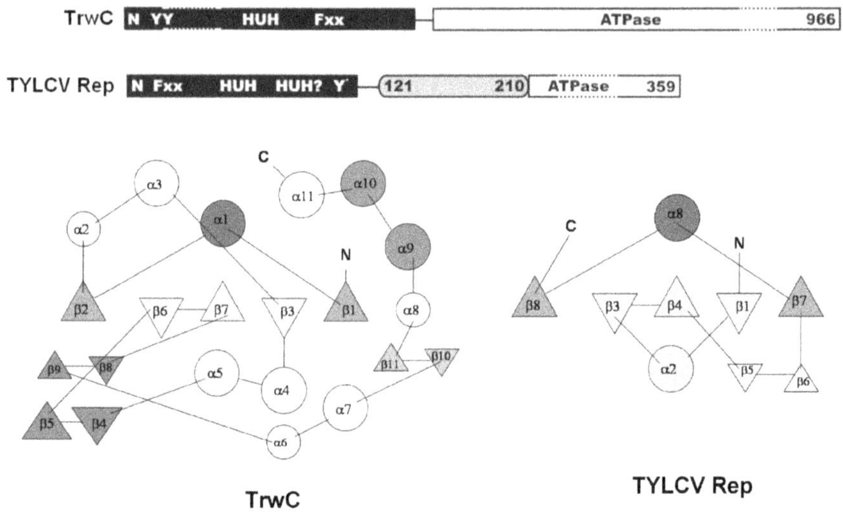

Fig. 4. Topological organization of the functional domains and secondary structural elements of two representative proteins of the origin-binding domain (OBD) family, *i.e.* relaxase/helicase TrwC of plasmid R388 and the replication initiator of tomato yellow leaf virus (TYLCV Rep). The top panel shows the domains present in the sequence of each protein and the relative location of the metal-binding (HUH) motifs and active site tyrosines (Y' or YY) in the OBD (solid black bars). The lower panels shows topology diagrams of the OBD of both proteins. The similarity in fold is evidenced by the homology between the central five β–sheets and the two contacting α-helices on both sides. The catalytic tyrosines are located in α1 and α8 in the relaxase and Rep protein, respectively. The order of the TrwC structural elements HUH and Y can be reconstituted in TYLCV Rep by applying a circular permutation, that is by breaking the connection between β6 and β7 and connecting the C to N terminal ends.

reaction site, suggests that it plays a role in the activation of Tyr18 for cleavage. A plausible function is that it acts as a general base through the abstraction of the Tyr18 hydroxyl proton in order to facilitate the nucleophilic attack of its hydroxyl oxygen on the phosphate. The phosphor atom of the scissile phosphate, in turn, is polarized by the coordination to the metal ion, which makes it more susceptible for cleavage. The presence of the DNA in the active site is stabilized by positively charged residues in the vicinity of the active site, *i.e.* Arg14, Arg154 and Arg266.

A continuing issue that has recently been resolved is the role and significance of a second tyrosine in the relaxases involved in the conjugational process. Originally, it had been proposed that two tyrosines are required for the nicking and transesterification reactions through a flip-flop mechanism as mutation of either tyrosine in TrwC led to impairment of conjugation (Grandoso et al., 2000). In this model, the DNA stays covalently attached to a first tyrosine after nicking, followed by transesterification through a second tyrosine. In the structures available, however, these tyrosines are spatially separated. Thus, the second tyrosine can reach the active site in the monomer only by extensive movement of the loop in which it is integrated. Evidence exists that TrwC forms a dimer in solution, which would allow the Tyr26 of a second molecule to come into close proximity of the active site. This model, however, seems to contradict the recent finding that TrwC passes to the recipient cell (see above). A recent paper confirmed that Tyr26 is responsible for the transesterifcation reaction in a N-terminal truncated form of R388 TrwC (Gonzalez-Perez et al., 2007). In addition, the authors show that this reaction occurs intramolecularly and that Y26 supplied 'in trans' to a Y26F mutant is not capable of transesterification. Interestingly, the initial cleavage reaction, but not the Tyr26 transesterification activity, could be detected for the dimeric full-length protein. This is consistent with the notion that transesterification takes place in the recipient cell by a monomer of TrwC (see above).

RELAXASE-BASED ANTIBIOTIC DESIGN

An exciting development in the field of structural biology of plasmid relaxases is exemplified by a recent report that describes the identification of biphosphonate inhibitors of F plasmid TraI (Lujan et al., 2007). The authors present a structure of a phosphate-bound Mg^{2+} ion intermediate that was obtained by soaking imidobisphosphate into crystals of an Y16F mutant of TraI. The resulting structure has been refined with the phosphate group attached to the Mg^{2+} ion. The covalent bond between the Mg^{2+} ion and the phosphate group sheds new light on the role of the metal ion in the reaction, hitherto assumed to be auxiliary through the polarization of the phoshodiester of the DNA. Efficacy assays of the compounds on the inhibition of conjugation demonstrated that some biphosphonates are highly effective *in vitro*. The presented structure of the impaired complex marks the starting point of the design of a new class of antibiotic (helper) agents that attack at the very heart of the bacterium's defence system against antibiotics.

INTERACTION WITH THE T4SS

To function as a DNA translocation apparatus, the type IV secrction system requires the assistance of a component called the type IV coupling protein or T4CP (Cabezon et al., 1997; Firth et al., 1996; Gomis-Ruth et al., 2004; Moncalian et al., 1999a; Santini and Stanisich, 1998). T4S systems dedicated exclusively to protein export generally lack T4CPs, as in the *Bordetella pertussis* Ptl system (Weiss et al., 1993). However, examples exist where coupling proteins mediate protein substrate translocation, as in *Agrobacterium tumefaciens* and *Helicobacter pylori* (Covacci et al., 1999; Christie, 2001; Vergunst et al., 2000). T4CPs are integral inner membrane proteins, with an N-terminal transmembrane moiety and a C-terminal cytoplasmic domain. They possess the typical Walker A and B motifs present in NTP-binding proteins (Walker et al., 1982) and their NTPase activity probably contributes to energize the secretion machinery (Cabezon and de la Cruz, 2006). The T4CPs family comprises members like TrwB from plasmid R388, TraD and TraG from various gram-negative plasmids, VirD4 from *Agrobacterium tumefaciens*, and other

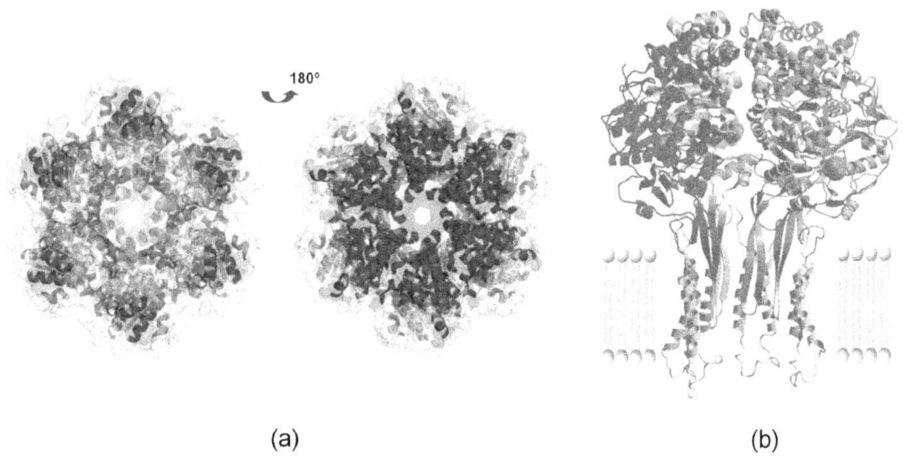

Fig. 5. a) The crystal structure of the conjugative T4CP TrwBΔN70 (PDB code 1e9r) viewed along the central channel on both sides of the hexamer. b) Ribbon representation of the full-length TrwB, including a model for the transmembrane part, based on two consecutive transmembrane helices of the photosynthetic reaction centre structure (PDB code 1mps). Only four of the six protomers are shown for clarity.

related proteins. The structure of TrwB has been determined and will be discussed in detail below (Gomis-Ruth et al., 2001).

A REPRESENTATIVE CRYSTAL STRUCTURE OF THE T4CP

TrwB is a 507-residue protein that is anchored to the inner membrane by a 70-residue N-terminal transmembrane domain. It binds ssDNA and dsDNA non-specifically and independently of NTP-binding. The crystal structure of the soluble part of the TrwB protein (TrwBΔN70) (Gomis-Ruth et al., 2002; Gomis-Ruth et al., 2001) consists of six equivalent monomers associated in an almost spherical structure, flattened at both poles, with a central channel (Fig. 5). Each monomer has two domains: a cytosol-oriented all-α helical C-terminal domain, AAD, comprised of seven α-helices; and an N-terminal membrane-proximal nucleotide-binding domain, NBD, a central twisted β-sheet flanked by helices on both sides. The central channel that connects the cytoplasm with the periplasm for the full-length protein, is 20Å wide along its entire length, except at the cytosol pole. A ring of asparagine residues plug the cytoplasmic channel entrance and restrict its diameter to ~8Å (Fig. 5a). The nucleotide-binding site is located at the C-terminal edge of the β-sheet. The architecture of this site, where the Walker A and B motif are located, is determined by two segments comprising the β2-αC and β6-αN loops. The six nucleotide-binding sites are 32Å apart from each other, located in superficial cavities at the interfaces among vicinal protomers.

TrwB has structural homology with the dsDNA translocase FtsK protein, a hexameric helicase involved in the resolution of chromosomal dimers during bacterial cell division, which supposedly functions as a molecular motor in the transport of DNA across the membrane. The X-ray structure of the soluble FtsK (Massey et al., 2006) shows a hexameric arrangement, similar to TrwB, as shown in Fig. 6b. Each protein monomer has three domains (α, β and γ). Due to structural disorder around the γ domain region, this domain could not be observed in the structure. There is no counterpart for the FtsK α-domain in the sequence of TrwB, but the β domain, which

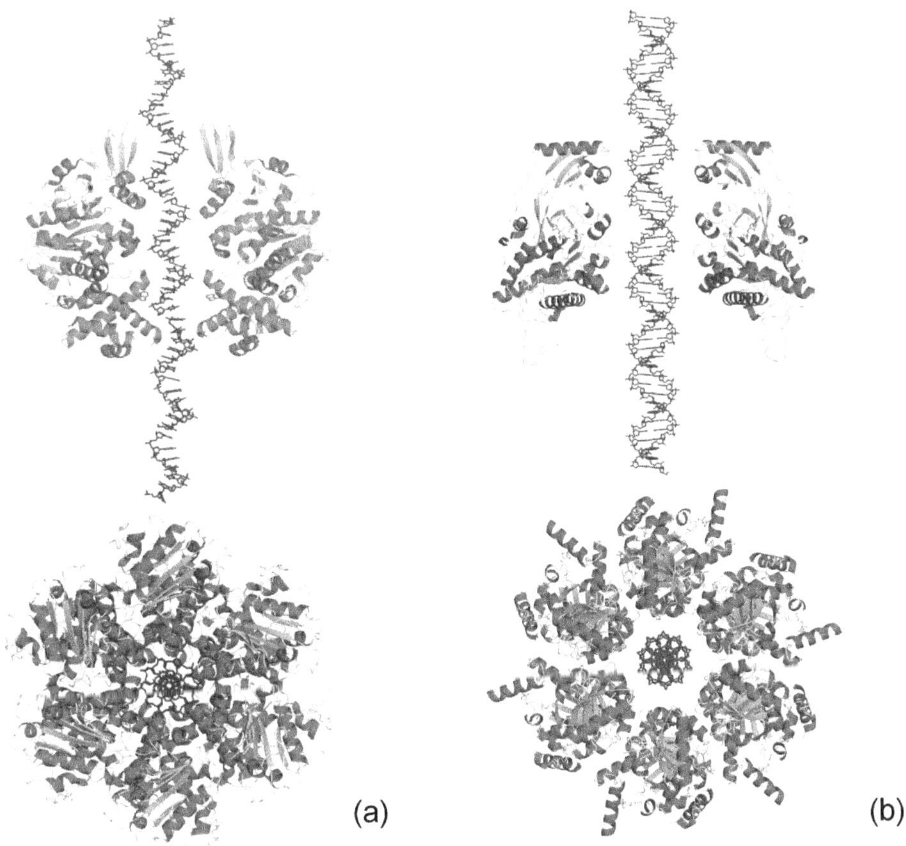

Fig. 6. a) The crystal structure of the conjugative T4CP TrwBΔN70 (PDB code 1e9r) with a ssDNA modelled along the central channel based on the 13-base ssDNA described in the papillomavirus E1 hexameric helicase ssDNA complex (PDB code 2gxa). b) FtsK DNA translocase (PDB code 2iuu) with a standard dsDNA modelled along the central channel. Only two opposite monomers flanking the DNA are shown in the upper panels. ATPase domains are depicted in equivalent orientations after their superimposition. Both DNA translocator proteins share a common central ATPase domain (DALI Z-score: TrwB/FtsK=14.2, DALI URL: http://www.ebi.ac.uk/dali), but the adjacent domains are different. The corresponding hexamers, viewed along the channel, are depicted in the lower panels.

is the NTP-binding domain, is structurally similar to the TrwB NBD. The central channel of the FtsK hexamer is ~30Å wide and, as shown by electron micrographs, is large enough to accommodate a dsDNA helix (Massey et al., 2006). In addition, there is a strong structural similarity between TrwB and the α- and β-subunits of the heterohexameric F_1-ATPase protein, a RecA-like ATPase protein related to the mitochondrial proton-driven molecular machine for ATP synthesis (Abrahams et al., 1994). The structural analogy is much stronger with the β-subunits in the occupied conformation (β_{DP} and β_{TP}) than with the empty one (β_E). The superposition of the NTP-binding domains shows the same relative orientation with respect to the membrane, although TrwB has six equally potential catalytic sites (like hexameric helicases), instead of the three catalytic subunits of F_1-ATPase.

DNA TRANSLOCATION BY T4CP

Structural and biochemical data suggest that a single-strand DNA might pass through the TrwB central channel (Fig. 6a), as in the distantly related E1 ATPase from bovine papillomavirus I (Enemark and Joshua-Tor, 2006). However, the asparagine plug at the cytosol pole narrows the entrance of the channel and could impede the passage of DNA. This apparent contradiction can be explained by assuming that the TrwBDN70 structure represents a closed conformation of the protein and the cytoplasmic gate opens on activation when the protein couples with the relaxosome. The DNA-dependent ATPase activity of TrwB (Tato et al., 2005) and the structural changes of the central channel observed in TrwB structures due to the binding of phosphate or nucleotides at the active site (Gomis-Ruth et al., 2002), reinforce the idea that the coupling protein is a molecular motor. Based on the structure, a model was proposed in which active DNA transport is achieved by cyclic conformational changes of the ring subunits that would push the ssDNA across the inner membrane during conjugal transfer. The conformational changes associated with pumping occur at the nucleotide binding site and are conferred to the inner lining of the central core of the hexamer. The region involved in the transfer has homology to a conserved sequence of AAA-ATPases and is suggested to aid in DNA displacement.

The ATPase driven pumping model of ssDNA through the TrwB hexamer can only occur after the relaxase, covalently bound to the T-strand, has entered the pilus. Therefore, the TrwB hexamer somehow needs to be assembled on the T-DNA, downstream of the TrwC-DNA complex. In a process proposed by Cabezon and de la Cruz (2006), the nucleoprotein complex is guided towards the inner membrane by a monomeric form of TrwB. After establishing the connection between TrwC and the secretion apparatus, the relaxase is transported to the recipient cell, piloting the trailing DNA. The energy needed for this first step could be provided by VirB11 and/or VirB4 ATPases from the T4SS. Once the proteic part of the relaxase-DNA complex has crossed the inner membrane, TrwB wraps around the DNA and oligomerizes as a hexamer. The hexameric assembly of TrwB in the inner membrane allows the pumping of DNA across the inner membrane through its ATPase activity and cyclic conformational changes. TrwB can be considered an adapter to the T4S system that changes the substrate specificity from proteins to DNA, as it establishes the connection between the *oriT*-specific relaxase TrwC and the T4S system, responsible for TrwC translocation.

TraM, AN ANCILLARY PROTEIN

TraM is a small 14 kDa cytoplasmic DNA-binding protein coded in the *tra* region of a number of plasmids, like F and R1, but not R388 (see Fig. 1a). It has been associated with various ancillary functions required for T-strand processing and complementing relaxase/helicase, IHF, and the transcriptional regulator. TraM binds to three sites within *oriT*, repressing its own gene promoter. It could be a constituent of the relaxosome and/or could signal the cell that mating pair formation is completed; however, it is not essential for relaxosome-dependent nicking. Other investigations showed that TraM enhances relaxase activity and that it specifically interacts with the coupling protein of plasmid R1, TraD (Beranek et al., 2004). Thus, TraM may be a helper protein in the coupling of the relaxosome with the DNA-transfer apparatus (Disque-Kochem and Dreiseikelmann, 1997; Kupelwieser et al., 1998; Stockner et al., 2001). Consistent with this idea, the C-terminal domain of coupling protein TraD that interacts with TraM, is missing in the TrwB, although the cytoplasmic domains of TraD and TrwB bear strong sequence similarity (Beranek et al., 2004; Llosa et al., 1994).

A N-terminal stretch of 54 residues of the 127-residue long TraM from R1 was studied by NMR- spectroscopy (Stockner et al., 2001). This portion of TraM fully retains the specific recognizing and binding capacity to DNA. The NMR study revealed that the domain is all α-helical and consists of three helical regions. The first two helical regions of TraM are connected by a loop while the last two are linked through a glycine residue. Overall, the N-terminal domain of TraM displays topological similarity with a region within the C-terminal all-α domain of TrwB (Gomis-Ruth et al., 2002). It can be speculated that, similar to the N-terminal domain of TraM, the C-terminal DNA-binding domain TrwB comes into contact with the relaxosome.

A structural analysis by X-ray crystallography of the C-terminal fragment of TraM from plasmid F containing the TraD-binding segment revealed that four protomers interact in a compact eight-helical bundle (Lu et al., 2006). Here, the N-terminal helices from each protomer interact to form a central, parallel four-stranded coiled-coil, whereas each C-terminal helix packs in an antiparallel arrangement around the periphery of the structure. This oligomerization creates a central tunnel lined by the inner N-terminal helices, where four protonated glutamate residues provided by each of the protomers create a central solvent-rich ring. Deprotonation of this acidic residue relaxes the TraM structure and this affects interactions with TraD.

T4SS COMPONENTS: STRUCTURE AND ASSEMBLY

As mentioned before, the *Agrobacterium tumefaciens* VirB/VirD4 secretion system has been extensively characterized, both biochemically and structurally and it is considered the archetype of T4SS (Fig. 7). Its components can be classified into three groups, according to function (mostly putative) and cellular location: i) the cytoplasm- or inner membrane associated ATPases: VirB4, VirB11 and VirD4; ii) the core complex: VirB6, VirB7, VirB8, VirB9 and VirB10; and iii) the pilus components: VirB2 and VirB5 (Baron et al., 2002; Christie, 2001). The VirB2 channel has its origin in the periplasmic space, then spans the outer membrane, ending in the extracellular milieu. The translocation pathway for DNA through this secretory apparatus (Cascales and Christie, 2004b) has been shown to roughly

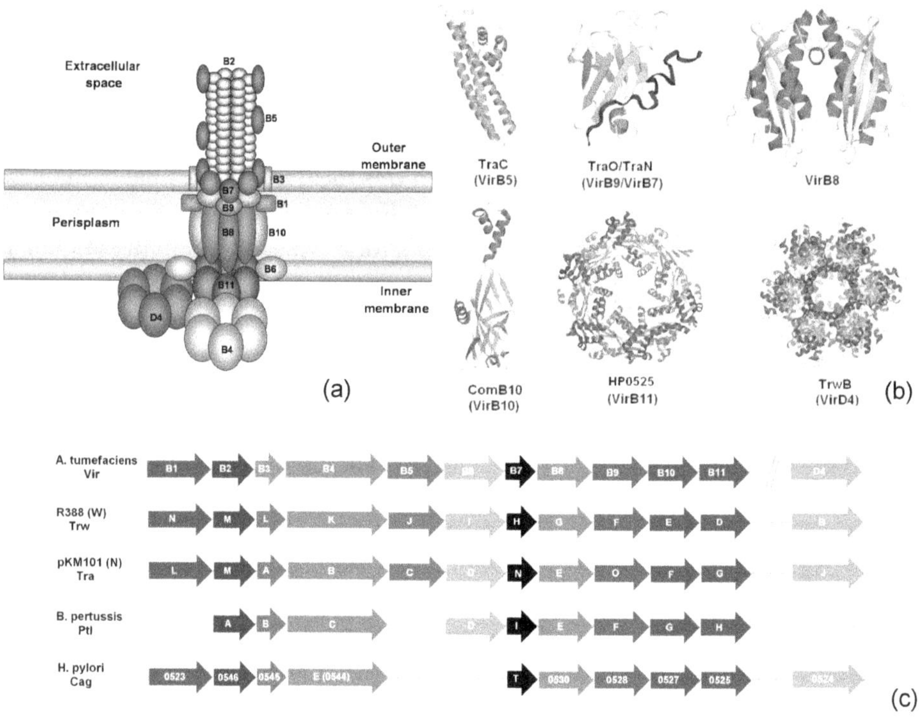

Fig. 7. a) Tentative assembly model of the type IV secretion system of *A. tumefaciens* showing the cytoplasm- or inner membrane associated ATPases: VirB4, VirB11 and VirD4; the core complex: VirB6, VirB7, VirB8, VirB9 and VirB10; and the pilus components: VirB2 and VirB5. b) Resolved structures of components of the T4SS: X-ray structures of the VirB5 homologue, TraC (pdb code 1r8i), VirB8 (PDB code 2cc3), the VirB10 homologue, ComB10 (PDB code 2bhv), the VirB11 homologue, HP0525 (PDB code 1nlz), and the VirD4 homologue, TrwB (PDB code 1e9r) and the NMR-spectroscopy structure of the interacting domains of the complex of VirB7/VirB9 homologous TraN/TraO (PDB code 2ofq). c) Gene organization and nomenclature of *A. tumefaciens* VirB/VirD4, *E. coli* R388 Trw, *E. coli* pKM101 Tra, *B. pertussis* Ptl and *H. pylori* Cag, secretion systems.

follow the order of the subcomponents mentioned here. All the components of the VirB operon and the VirD2 and VirD4 gene products are essential for the secretion of DNA and/or protein substrates, with the exception of VirB1, which is involved in the localized lysis of the peptidoglycan layer (Berger and Christie, 1994).

The structures of several of these proteins, or their homologous, are nowadays known (Fig. 7b). These include X-ray structures of a VirB5 homologue, TraC (Yeo et al., 2003), VirB8 from *Brucella suis* (Terradot et al., 2005) and *Agrobacterium tumefaciens* (Bailey et al., 2006), a VirB10 homologue, ComB10 (Terradot et al., 2005), a VirB11 homologue, HP0525 (Yeo et al., 2000), and a VirD4 homologue, TrwB (Gomis-Ruth et al., 2001). In addition, NMR-spectroscopy structures of the interacting domains of the complex of VirB7/VirB9 homologous TraN/TraO are available (Bayliss et al., 2007).

INNER MEMBRANE ATPases

VirB4, VirB11 and the coupling protein VirD4 are all associated inner membrane ATPases that have been shown to mutually interact (Atmakuri et al., 2004). They are believed to provide the energy to the T4SS, based on the observation that they all contain the conserved Walker A and B sequence motifs. The structure of a VirD4 analogue, TrwB from the R388 plasmid, was discussed in the section on relaxosomes. Crystal structures of the VirB11 homologue protein HP0525 from *Helicobacter pylori* have been determined in an ADP-bound form (Yeo et al., 2000) as well as in an ATPγS-bound and a free form (Savvides et al., 2003). In addition, a structure of a homologue from *Brucella suis* has been determined by X-ray crystallography, showing an interesting topological domain swap (Hare et al., 2006). HP0525 consists of a hexameric ring formed by monomers consisting of a C-terminal RecA-like α/β domain and a grapple-shaped N-terminal domain. Conformational differences between the ADP-bound form and the free form occur at the N-terminal domain, which rotates away from the central edge. These motions are different for each of the monomers and result in an asymmetric ring. Based on these nucleotide-induced conformational changes a mechanism of sequential binding and hydrolysis of ATP by HP0525 has been proposed (Savvides et al., 2003), in which the protein alternates between symmetrical and asymmetrical hexamer forms upon substrate binding, hydrolysis and product release.

CORE COMPLEX

The components that span the periplasmic space consist of components VirB7 to VirB10, of which VirB8 and VirB10 appear to be the major constituents of the periplasmic channel. VirB7 and VirB9 are generally considered to be outer-membrane associated proteins. Atomic structures are available for (analogues of) both VirB8 and VirB10.

PERIPLASMIC SPACE-SPANNING COMPONENTS

VirB8 is an inner membrane protein with a short N-terminal domain, exposed to the cytoplasm, and a large C-terminal domain, exposed to the periplasm, linked by a transmembrane helix (Buhrdorf et al., 2003; Das and Xie, 1998; Thorstenson and Zambryski, 1994). It has been proposed to function as a periplasmic nucleation centre recruiting VirB9 and VirB10 to the sites of the complex assembly (Kumar et al., 2000) In addition to VirB9 and VirB10, VirB8 is able to interact with itself (Das and Xie, 2000), which was confirmed by structural studies using X-ray crystallography of the *Agrobacterium tumefaciens* VirB8 (Bailey et al., 2006) and its *Brucella suis* homologue (Terradot et al., 2005). The structures of the periplasmic domain of VirB8 represent dimers in which each monomer has five α-helices interacting with one side of a four-strand antiparallel β-sheet which extends over the lengths of the molecule and curves around the first helix (Fig. 7b). A highly conserved surface region extends over the surface of the central region of the β-sheet at a groove opposite to the dimer interface. It was suggested that this region corresponds to a protein-protein interaction site for other T4SS proteins.

VirB10 seems to alternate between two states depending on the cellular ATP level of the system. These states are controlled through interactions with VirB11 and VirD4, but not VirB4 (Cascales and Christie, 2004a). Only the 'energised' conformation of VirB10 is able to form complexes with the outer membrane component VirB9 (Terradot et al., 2005). In the reported crystal structure of the C-terminal, periplasmic domain of the VirB10-like protein ComB of *Helicobacter pylori*, the protein crystallizes as a dimer in which each monomer folds as a C-terminal distorted β-barrel with a large connected N-terminal α-helical antenna (Fig. 7b). The dimer forms through mutual interactions between the C-terminal and the N-terminal domains of the crystallized fragment. The authors propose that the dimer in the crystal structure represents the unenergised conformation of the protein, which dissociates in presence of ATP, releasing the protruding helices for interaction with VirB9 (Cascales and Christie, 2004a). The N-terminal domains of the VirB10 dimer found in the crystal structure are positioned at opposite ends, which is consistent with the finding that VirB10 localizes to both membranes (Fernandez et al., 1996a; Kumar et al., 2000). The thickness of the periplasmic space (20-30 nm) of *H. pylori* (Wen et al., 1997) and the distance between the two N-terminal membrane spanning regions (~11 nm) of only the C-terminal domain of VirB10 that was crystallized, do not contradict this possibility.

THE OUTER MEMBRANE COMPONENTS

The outer membrane components consist of VirB7 and VirB9. VirB7 is a small lipoprotein localized in the periplasmic side of the outer membrane and at the extracellular milieu, which is involved in stabilising the integrity of the pilus (Beaupre et al., 1997; Fernandez et al., 1996a; Sagulenko et al., 2001). VirB7 is able to associate with itself and with VirB9 through the formation of intra- and intermolecular disulfide bridges (Anderson et al., 1996; Baron et al., 1997; Das et al., 1997; Farizo et al., 1996; Fernandez et al., 1996b; Harris et al., 2001; Spudich et al., 1996). The NMR-spectroscopy structure of the C-terminal domain of the plasmid pKM101 encoded TraO homologue of VirB9 in complex with the corresponding TraN homologue of VirB7 has recently been reported (Bayliss et al., 2007). TraO adopts an immunoglobulin-like β-sandwich folding, comprising nine β-strands and a short stretch of 3_{10}-helix located between the two first strands (Fig. 7b). TraN wraps around TraO perpendicularly to the direction of the β-sandwich and introduces one of its residues between two β-sheets. The part of the protein, that has been demonstrated to span the outer membrane, protrudes from the structure supporting the general model that VirB9-like proteins are a component of an outer membrane channel complex.

PILUS COMPONENTS

Pili are formed by the association of small subunits called pilins (VirB2-like proteins), which assemble into a filamentous pilus structure at the cell surface (Eisenbrandt et al., 1999). Two additional proteins are also localized to the pilus, VirB5 and VirB7 (Sagulenko et al., 2001). The minor component VirB5-like proteins (Schmidt-Eisenlohr et al., 1999a; Schmidt-Eisenlohr et al., 1999b), are ~220 residue proteins containing an N-terminal signal sequence that targets the protein to the periplasmic space. The crystal structure of the VirB5-like protein TraC of IncN plasmid pKM101 revealed an α-helical elongated structure, as shown in Fig. 7b (Yeo et al., 2003). Although the protein dimerizes in the crystal, structured-based mutagenesis studies of activity and characterization of the oligomeric state of TraC by sedimentation equilibrium methods, demonstrated that this dimeric state was an artefact. The extended protein surface region involved in these crystalline contacts is probably the region of protein-protein interactions with other component of the T4SS system. Since the connection between the pilus of the host cell and the outer membrane of the recipient cell is achieved through interactions between VirB5 of T4SS and adhesins on the recipient membrane (Soto and Hultgren, 1999), it has been proposed that VirB5 proteins have an adhesin-like function (Yeo et al., 2003).

T4SS ASSEMBLY AND MECHANISM

A four-stage model of the assembly pathway for the VirB/D4 secretion system has been proposed by Christie and collaborators based on the available literature on the subject (Christie et al., 2005). It comprises (i) the formation of the core complex, (ii) the recruitment of pilus associated proteins, (iii) the recruitment of the VirB11 ATPase and (iv) the formation of either the T pilus or a secretion channel. In the first stage VirB4, VirB7, VirB8, VirB9 and VirB10 assemble across the cell envelope forming the core complex. Once the core complex is assembled, the recruitment of the pilus proteins VirB2, VirB3 and VirB5 starts through VirB4, which functions as a scaffold. In the next stage, VirB11 ATPase is recruited to assist in the completion of the inner-membrane basal structure, the formation of a transenvelope structure comprising a VirB2 polymer and the outer-membrane pore structure. Finally, the last stage of the assembly pathway results in the formation of either a secretion channel or a pilus.

In order to describe the DNA pathway across the inner membrane, different models have been proposed, which differ in the role attributed to the ATPases VirB4, VirB11 and the T4CP. In theory, both the T DNA-relaxase and the T-DNA substrates could be translocated by any of the ATPases. However, VirB4 is generally not considered important as supplier of translocation energy, a role which is reserved for VirB11. In the models that have been proposed, VirB11 unfolds the relaxase and then it either 1) passes the unfolded peptide chain and DNA substrate through the inner membrane (channel model), or 2) it passes the unfolded peptide chain to the T4CP for passage through the inner membrane (ping-pong model, Atmakuri et al., 2004), or 3) it passes the peptide chain through the inner membrane and the DNA is then pumped through by T4CP ('shoot-and-pump', Llosa et al., 2002). The shoot-and-pump model is consistent with the origin of T4S systems as protein translocation machines and the function of the coupling protein in DNA substrate recognition and delivery. In this model, the coupling protein acts as an DNA adaptor to the protein secretion machine. The assembly of VirD4 monomers around the DNA-relaxase has been proposed to account for the structural reorganization required for this mode of action (Cabezon and de la Cruz, 2006). The ping-pong model also agrees well with the available biochemical and functional data and relies on recent findings that, in some systems, both protein and DNA translocation requires the T4CP. Future experiments will determine how the T4SS functions on a molecular level.

CONCLUSION

A large body of experimental data and literature has accumulated concerning DNA translocation in bacterial conjugation. As a result, the location of the genes important for these processes and the assembly topology of the functional machinery have been accurately mapped. Furthermore, a discrete number of specialised sub-systems, such as the relaxosome and the type IV secretion system (T4SS), have been identified. The recent increase in the number of sequenced genomes has helped identify these sub-systems in different bacteria and has allowed the determination of similarities and differences between them. In addition, the particular function of many of the protein components has been ascertained, although the details of many remain unknown.

Several protein components have been analyzed at the structural level, which has laid the foundation for the analysis of the function and molecular mechanisms of the molecular machines of the bacterial conjugation mechanism at a (near-)atomic level. *E.g.*, many relaxosome components have been structurally characterised. These include ternary complexes of the relaxase domains of TrwC from the R388 plasmid and TraI from F plasmid with DNA and the catalytic metal ions (Datta et al., 2003; Guasch et al., 2003). This revealed important details of the initial steps of bacterial conjugation, especially about the recognition of the *nic* site and the possible routes by which DNA is processed by the relaxase at different stages of conjugation. Other components of the relaxosome that have been studied at the structural level are the CopG homologue of the *trw* gene repressor TrwA, the integration host factor (IHF) of *Escherichia coli* (Rice et al., 1996) and DNA binding domain of the ancillary transfer protein (TraM) of plasmid R1 (Stockner et al., 2001). Furthermore, the atomic structure of a coupling protein, *i.e.* R388 plasmid TrwB, has been determined, along with several related helicases that are somehow involved in DNA translocation processes. These structures allowed the postulation of a step-wise mechanism of DNA translocation mediated through this family of ATPases.

Structures of many representative components for the type IV secretion system (T4SS) have also been determined. Full-length inner membrane associated ATPase VirB11 has been crystallized both in the apo- and ATP-bound forms (Savvides et al., 2003; Yeo et al., 2000). These crystal structures provided further insight into the mode of action of the unit, which, via nucleotide hydrolysis, powers the transport of substrates from the cytoplasmic/inner-membrane end of the secretion channel. Of the proteins that form the core, the NMR spectroscopy solution structure of full-length TraN (a homologue of VirB7 from plasmid pKM101) in complex with the C-terminal domain of TraO (a VirB9 homologue from that plasmid) shed some light on how two pieces of the core form a stable association (Bayliss et al., 2007). The elucidation of this structure, together with the structures of VirB8, the periplasmic nucleation centre for complex assembly, and VirB10, the sensor of the energetic state of VirB11, allowed the spatial mapping of several protein-protein interactions on the protein surface. The structure of a homologue of VirB5, TraC from plasmid pKM101, served to speculate about its role in attaching the pilus to the membrane (Yeo et al., 2003).

FUTURE TRENDS

Overall, many of the full-length or partial individual components of the relaxosome and T4SS have now been structurally characterized. From these puzzle pieces, a crude picture of the fully assembled pilus apparatus can be put together. However, the full understanding of the T4SS-mediated bacterial conjugation process is still far away. First of all, several individual components have not, or only partially, been structurally characterized to date. For instance, the structure of VirB4, one of the engines of the power plant of the complex, remains unresolved. The same is true for VirB3, the outer-membrane component of the complex, and for VirB1, the disruptor of the peptidoglycan layer for facilitating complex assembly.

Although the structural analysis of the proteins involved in the conjugal transfer of DNA does provide valuable information, they nonetheless represent individual components of a integrated multicomponent assembly. With the exception of the complex of the VirB7 and VirB9 analogues determined by NMR spectroscopy, little is known about the way the individual components interact with one another from a structural point of view. A structural analysis of multiprotein macromolecular assemblies needs to be performed. As some of the protein-protein interactions can be expected to be transient and often weak, difficulties arising from technical problems, such as the stabilization of interactions among a variety of protein components sufficient for structural characterization and the crystallization of components with trans-membrane segments, will have to be overcome. Electron microscopy and SAXS are likely to contribute importantly to the characterization of these large assemblies. If electron microscopy 3D reconstructions of the complexes were available, the individual proteins or domains determined by X-ray or NMR at high resolution could be fitted in the electron microscopy envelopes yielding quasi-atomic models of the whole assemblies, as has been demonstrated in other systems.

Another important aspect of the study of large and dynamic machinery of proteins conglomerates lies in the inherent flexibility and rearrangements of components that are associated with function. The individual proteins and multi-component complexes can be expected to change conformation, oligomerisation state, interaction partners and reaction mechanisms depending on the processing stage on the conjugational timeline. As a consequence, interpretation of the available structural information is complicated by the fact that they represent still images of very dynamic and constantly varying structures. For a full understanding of the conjugational process, structural biologists will need to focus on the resolution of multicomponent complexes with well-defined functions at various points of the conjugational process.

Some of the fundamental questions appear elusive to be answered. What is the *in vivo* factor that triggers *oriT* cleavage and DNA transfer? How does the pilus sense and attach to the recipient cell? How is the recipient cell membrane barrier penetrated? What is the structure of the helicase domain of the relaxase like and how does it move processively along the DNA? What are the precise mutual interactions of the components of the relaxosome with those of the coupling protein? What is the mechanism by which the T4SS translocates the substrate? In the case of the VirD/VirB system of *Agrobacterium tumefaciens*, what are the specific functions of each of the three ATPases involved in the process (VirD4, VirB4 and VirB11)?

The answer to these questions will require a time-resolved structural analysis of complicated arrangements of proteins and protein-DNA complexes. The achievement of this formidable task will likely require the integrated use of all techniques available to the structural biologist, including SAXS, NMR, EM and X-ray crystallography. Elucidation of the interactions that the components form could yield a satisfactory explanation of the observed biochemical data, while at the same time may trigger further questions at the academic as well as the applied scientific areas.

ACKNOWLEDGEMENTS

Funding provided by research grants BFU2005-06758/BMC and GEN2003-20642 from the Ministerio de Educación y Ciencia of Spain, as well as 2005SGR-00280 from the Generalitat de Catalunya are acknowledged. We are most grateful to F. de la Cruz in whose laboratory the R388 Trw proteins were prepared and for many appealing ideas on the functioning of the conjugation machinery described in this chapter. We thank A.G. Blanco for making part of Fig. 1 and 2.

REFERENCES

Abrahams, J.P., Leslie, A.G., Lutter, R., and Walker, J.E. (1994). Structure at 2.8 A resolution of F1-ATPase from bovine heart mitochondria. Nature *370*, 621-628.

Alekshun, M.N., and Levy, S.B. (2007). Molecular mechanisms of antibacterial multidrug resistance. Cell *128*, 1037-1050.

Anderson, L.B., Hertzel, A.V., and Das, A. (1996). Agrobacterium tumefaciens VirB7 and VirB9 form a disulfide-linked protein complex. Proc. Natl. Acad. Sci. U. S. A. *93*, 8889-8894.

Atmakuri, K., Cascales, E., and Christie, P.J. (2004). Energetic components VirD4, VirB11 and VirB4 mediate early DNA transfer reactions required for bacterial type IV secretion. Mol. Microbiol. *54*, 1199-1211.

Backert, S., and Meyer, T.F. (2006). Type IV secretion systems and their effectors in bacterial pathogenesis. Curr. Opin. Microbiol. *9*, 207-217.

Bailey, S., Ward, D., Middleton, R., Grossmann, J.G., and Zambryski, P.C. (2006). Agrobacterium tumefaciens VirB8 structure reveals potential protein-protein interaction sites. Proc. Natl. Acad. Sci. U. S. A. *103*, 2582-2587.

Baron, C., D, O.C., and Lanka, E. (2002). Bacterial secrets of secretion: EuroConference on the biology of type IV secretion processes. Mol. Microbiol. *43*, 1359-1365.

Baron, C., Thorstenson, Y.R., and Zambryski, P.C. (1997). The lipoprotein VirB7 interacts with VirB9 in the membranes of Agrobacterium tumefaciens. J. Bacteriol. *179*, 1211-1218.

Bayliss, R., Harris, R., Coutte, L., Monier, A., Fronzes, R., Christie, P.J., Driscoll, P.C., and Waksman, G. (2007). NMR structure of a complex between the VirB9/VirB7 interaction domains of the pKM101 type IV secretion system. Proc. Natl. Acad. Sci. U. S. A. *104*, 1673-1678.

Beaupre, C.E., Bohne, J., Dale, E.M., and Binns, A.N. (1997). Interactions between VirB9 and VirB10 membrane proteins involved in movement of DNA from Agrobacterium tumefaciens into plant cells. J. Bacteriol. *179*, 78-89.

Beranek, A., Zettl, M., Lorenzoni, K., Schauer, A., Manhart, M., and Koraimann, G. (2004). Thirty-eight C-terminal amino acids of the coupling protein TraD of the F-like conjugative resistance plasmid R1 are required and sufficient to confer binding to the substrate selector protein TraM. J. Bacteriol. *186*, 6999-7006.

Berger, B.R., and Christie, P.J. (1994). Genetic complementation analysis of the Agrobacterium tumefaciens virB operon: virB2 through virB11 are essential virulence genes. J. Bacteriol. *176*, 3646-3660.

Boer, R., Russi, S., Guasch, A., Lucas, M., Blanco, A.G., Perez-Luque, R., Coll, M., and de la Cruz, F. (2006). Unveiling the molecular mechanism of a conjugative relaxase: The structure of TrwC complexed with a 27-mer DNA comprising the recognition hairpin and the cleavage site. J Mol Biol *358*, 857-869.

Boucher, Y., Douady, C.J., Papke, R.T., Walsh, D.A., Boudreau, M.E., Nesbo, C.L., Case, R.J., and Doolittle, W.F. (2003). Lateral gene transfer and the origins of prokaryotic groups. Annu. Rev. Genet. *37*, 283-328.

Buhrdorf, R., Forster, C., Haas, R., and Fischer, W. (2003). Topological analysis of a putative virB8 homologue essential for the cag type IV secretion system in Helicobacter pylori. Int. J. Med. Microbiol. *293*, 213-217.

Burns, D.L. (2003). Type IV transporters of pathogenic bacteria. Curr. Opin. Microbiol. *6*, 29-34.

Cabezon, E., and de la Cruz, F. (2006). TrwB: an F(1)-ATPase-like molecular motor involved in DNA transport during bacterial conjugation. Res. Microbiol. *157*, 299-305.

Cabezon, E., Sastre, J.I., and de la Cruz, F. (1997). Genetic evidence of a coupling role for the TraG protein family in bacterial conjugation. Mol. Gen. Genet. *254*, 400-406.

Campos-Olivas, R., Louis, J.M., Clerot, D., Gronenborn, B., and Gronenborn, A.M. (2002). The structure of a replication initiator unites diverse aspects of nucleic acid metabolism. Proc. Natl. Acad. Sci. U. S. A. *99*, 10310-10315.

Cascales, E., and Christie, P.J. (2003). The versatile bacterial type IV secretion systems. Nat. Rev. Microbiol. *1*, 137-149.

Cascales, E., and Christie, P.J. (2004a). Agrobacterium VirB10, an ATP energy sensor required for type IV secretion. Proc. Natl. Acad. Sci. U. S. A. *101*, 17228-17233.

Cascales, E., and Christie, P.J. (2004b). Definition of a bacterial type IV secretion pathway for a DNA substrate. Science *304*, 1170-1173.

Covacci, A., Telford, J.L., Del Giudice, G., Parsonnet, J., and Rappuoli, R. (1999). Helicobacter pylori virulence and genetic geography. Science *284*, 1328-1333.

Champoux, J.J. (2001). DNA topoisomerases: structure, function, and mechanism. Annu. Rev. Biochem. *70*, 369-413.

Chen, I., Christie, P.J., and Dubnau, D. (2005). The ins and outs of DNA transfer in bacteria. Science *310*, 1456-1460.

Chen, I., and Dubnau, D. (2004). DNA uptake during bacterial transformation. Nat. Rev. Microbiol. *2*, 241-249.

Christie, P.J. (1997). The Agrobacterium T-complex transport apparatus: a paradigm for a new family of multifunctional transporters in eubacteria. J. Bacteriol. *179*, 3085-3304.

Christie, P.J. (2001). Type IV secretion: intercellular transfer of macromolecules by systems ancestrally related to conjugation machines. Mol. Microbiol. *40*, 294-305.

Christie, P.J. (2004). Type IV secretion: the Agrobacterium VirB/D4 and related conjugation systems. Biochim. Biophys. Acta *1694*, 219-234.

Christie, P.J., Atmakuri, K., Krishnamoorthy, V., Jakubowski, S., and Cascales, E. (2005). Biogenesis, architecture, and function of bacterial type IV secretion systems. Annu. Rev. Microbiol. *59*, 451-485.

Christie, P.J., and Cascales, E. (2005). Structural and dynamic properties of bacterial type IV secretion systems (review). Mol. Membr. Biol. *22*, 51-61.

Christie, P.J., and Vogel, J.P. (2000). Bacterial type IV secretion: conjugation systems adapted to deliver effector molecules to host cells. Trends Microbiol. *8*, 354-360.

Das, A., Anderson, L.B., and Xie, Y.H. (1997). Delineation of the interaction domains of Agrobacterium tumefaciens VirB7 and VirB9 by use of the yeast two-hybrid assay. J. Bacteriol. *179*, 3404-3409.

Das, A., and Xie, Y.H. (1998). Construction of transposon Tn3phoA: its application in defining the membrane topology of the Agrobacterium tumefaciens DNA transfer proteins. Mol. Microbiol. *27*, 405-414.

Das, A., and Xie, Y.H. (2000). The Agrobacterium T-DNA transport pore proteins VirB8, VirB9, and VirB10 interact with one another. J. Bacteriol. *182*, 758-763.

Datta, S., Larkin, C., and Schildbach, J.F. (2003). Structural insights into single-stranded DNA binding and cleavage by F factor TraI. Structure *11*, 1369-1379.

del Solar, G., Hernandez-Arriaga, A.M., Gomis-Ruth, F.X., Coll, M., and Espinosa, M. (2002). A genetically economical family of plasmid-encoded transcriptional repressors involved in control of plasmid copy number. J. Bacteriol. *184*, 4943-4951.

Dillard, J.P., and Seifert, H.S. (2001). A variable genetic island specific for Neisseria gonorrhoeae is involved in providing DNA for natural transforation and is found more often in disseminated infection isolates. Mol. Microbiol. *41*, 263-277.

Ding, Z., Atmakuri, K., and Christie, P.J. (2003). The outs and ins of bacterial type IV secretion substrates. Trends Microbiol. *11*, 527-535.

Disque-Kochem, C., and Dreiseikelmann, B. (1997). The cytoplasmic DNA-binding protein TraM binds to the inner membrane protein TraD *in vitro*. J. Bacteriol. *179*, 6133-6137.

Draper, O., Cesar, C.E., Machon, C., de la Cruz, F., and Llosa, M. (2005). Site-specific recombinase and integrase activities of a conjugative relaxase in recipient cells. Proc. Natl. Acad. Sci. U. S. A. *102*, 16385-16390.

Dreiseikelmann, B. (1994). Translocation of DNA across bacterial membranes. Microbiol. Rev. *58*, 293-316.

Durrenberger, M.B., Villiger, W., and Bachi, T. (1991). Conjugational junctions: morphology of specific contacts in conjugating Escherichia coli bacteria. J. Struct. Biol. *107*, 146-156.

Eisenbrandt, R., Kalkum, M., Lai, E.M., Lurz, R., Kado, C.I., and Lanka, E. (1999). Conjugative pili of IncP plasmids, and the Ti plasmid T pilus are composed of cyclic subunits. J. Biol. Chem. *274*, 22548-22555.

Enemark, E.J., and Joshua-Tor, L. (2006). Mechanism of DNA translocation in a replicative hexameric helicase. Nature *442*, 270-275.

Enemark, E.J., Stenlund, A., and Joshua-Tor, L. (2002). Crystal structures of two intermediates in the assembly of the papillomavirus replication initiation complex. EMBO J. *21*, 1487-1496.

Farizo, K.M., Cafarella, T.G., and Burns, D.L. (1996). Evidence for a ninth gene, ptlI, in the locus encoding the pertussis toxin secretion system of Bordetella pertussis and formation of a PtlI-PtlF complex. J. Biol. Chem. *271*, 31643-31649.

Fekete, R.A., and Frost, L.S. (2000). Mobilization of chimeric oriT plasmids by F and R100-1: role of relaxosome formation in defining plasmid specificity. J. Bacteriol. *182*, 4022-4027.

Fernandez, D., Dang, T.A., Spudich, G.M., Zhou, X.R., Berger, B.R., and Christie, P.J. (1996a). The Agrobacterium tumefaciens virB7 gene product, a proposed component of the T-complex transport apparatus, is a membrane-associated lipoprotein exposed at the periplasmic surface. J. Bacteriol. *178*, 3156-3167.

Fernandez, D., Spudich, G.M., Zhou, X.R., and Christie, P.J. (1996b). The Agrobacterium tumefaciens VirB7 lipoprotein is required for stabilization of VirB proteins during assembly of the T-complex transport apparatus. J. Bacteriol. *178*, 3168-3176.

Firth, N., Ippen-Ihler, K., and Skurray, R.A. (1996). Structure and function of the F factor and mechanism of conjugation, In Escherichia Coli and Salmonella: Cellular and Molecular Biology, N. FC, ed. (American Sociaty of Microbiology), pp. 2377-2401.

Furuya, N., and Komano, T. (2000). Initiation and termination of DNA transfer during conjugation of IncI1 plasmid R64: roles of two sets of inverted repeat sequences within oriT in termination of R64 transfer. J. Bacteriol. *182*, 3191-3196.

Garcillan-Barcia, M.P., Jurado, P., Gonzalez-Perez, B., Moncalian, G., Fernandez, L.A., and de la Cruz, F. (2007). Conjugative transfer can be inhibited by blocking relaxase activity within recipient cells with intrabodies. Mol. Microbiol. *63*, 404-416.

Gomis-Ruth, F.X., and Coll, M. (2006). Cut and move: protein machinery for DNA processing in bacterial conjugation. Curr. Opin. Struct. Biol. *16*, 744-752.

Gomis-Ruth, F.X., Moncalian, G., de la Cruz, F., and Coll, M. (2002). Conjugative plasmid protein TrwB, an integral membrane type IV secretion system coupling protein. Detailed structural features and mapping of the active site cleft. J. Biol. Chem. *277*, 7556-7566.

Gomis-Ruth, F.X., Moncalian, G., Perez-Luque, R., Gonzalez, A., Cabezon, E., de la Cruz, F., and Coll, M. (2001). The bacterial conjugation protein TrwB resembles ring helicases and F1-ATPase. Nature *409*, 637-641.

Gomis-Ruth, F.X., Sola, M., Acebo, P., Parraga, A., Guasch, A., Eritja, R., Gonzalez, A., Espinosa, M., del Solar, G., and Coll, M. (1998). The structure of plasmid-encoded transcriptional repressor CopG unliganded and bound to its operator. EMBO J. *17*, 7404-7415.

Gomis-Ruth, F.X., Sola, M., de la Cruz, F., and Coll, M. (2004). Coupling factors in macromolecular type-IV secretion machineries. Curr. Pharm. Des. *10*, 1551-1565.

Gonzalez-Perez, B., Lucas, M., Cooke, L.A., Vyle, J.S., de la Cruz, F., and Moncalian, G. (2007). Analysis of DNA processing reactions in bacterial conjugation by using suicide oligonucleotides. Embo J. *26*, 3847-3857.

Graham, D.Y. (2000). Helicobacter pylori infection is the primary cause of gastric cancer. J. Gastroenterol. *35 Suppl 12*, 90-97.

Grandoso, G., Avila, P., Cayon, A., Hernando, M.A., Llosa, M., and de la Cruz, F. (2000). Two active-site tyrosyl residues of protein TrwC act sequentially at the origin of transfer during plasmid R388 conjugation. J. Mol. Biol. *295*, 1163-1172.

Grandoso, G., Llosa, M., Zabala, J.C., and de la Cruz, F. (1994). Purification and biochemical characterization of TrwC, the helicase involved in plasmid R388 conjugal DNA transfer. Eur. J. Biochem. *226*, 403-412.

Grohmann, E., Muth, G., and Espinosa, M. (2003). Conjugative plasmid transfer in gram-positive bacteria. Microbiol. Mol. Biol. Rev. *67*, 277-301, table of contents.

Guasch, A., Lucas, M., Moncalian, G., Cabezas, M., Perez-Luque, R., Gomis-Ruth, F.X., de la Cruz, F., and Coll, M. (2003). Recognition and processing of the origin of transfer DNA by conjugative relaxase TrwC. Nat. Struct. Biol. *10*, 1002-1010.

Hare, S., Bayliss, R., Baron, C., and Waksman, G. (2006). A large domain swap in the VirB11 ATPase of Brucella suis leaves the hexameric assembly intact. J. Mol. Biol. *360*, 56-66.

Harris, R.L., Hombs, V., and Silverman, P.M. (2001). Evidence that F-plasmid proteins TraV, TraK and TraB assemble into an envelope-spanning structure in Escherichia coli. Mol. Microbiol. *42*, 757-766.

Harris, R.L., and Silverman, P.M. (2004). Tra proteins characteristic of F-like type IV secretion systems constitute an interaction group by yeast two-hybrid analysis. J. Bacteriol. *186*, 5480-5485.

Hickman, A.B., Ronning, D.R., Kotin, R.M., and Dyda, F. (2002). Structural unity among viral origin binding proteins: crystal structure of the nuclease domain of adeno-associated virus Rep. Mol. Cell *10*, 327-337.

Hofreuter, D., Odenbreit, S., and Haas, R. (2001). Natural transformation competence in Helicobacter pylori is mediated by the basic components of a type IV secretion system. Mol. Microbiol. *41*, 379-391.

Howard, E.A., Zupan, J.R., Citovsky, V., and Zambryski, P.C. (1992). The VirD2 protein of A. tumefaciens contains a C-terminal bipartite nuclear localization signal: implications for nuclear uptake of DNA in plant cells. Cell *68*, 109-118.

Howard, M.T., Nelson, W.C., and Matson, S.W. (1995). Stepwise assembly of a relaxosome at the F plasmid origin of transfer. J. Biol. Chem. *270*, 28381-28386.

Johnson, J.E., and Chiu, W. (2007). DNA packaging and delivery machines in tailed bacteriophages. Curr. Opin. Struct. Biol. *17*, 237-243.

Kalkum, M., Eisenbrandt, R., and Lanka, E. (2004). Protein circlets as sex pilus subunits. Curr. Protein Pept. Sci. *5*, 417-424.

Kingsman, A., and Willetts, N. (1978). The requirements for conjugal DNA synthesis in the donor strain during flac transfer. J. Mol. Biol. *122*, 287-300.

Kramer, M.G., Khan, S.A., and Espinosa, M. (1997). Plasmid rolling circle replication: identification of the RNA polymerase-directed primer RNA and requirement for DNA polymerase I for lagging strand synthesis. EMBO J. *16*, 5784-5795.

Kumar, R.B., Xie, Y.H., and Das, A. (2000). Subcellular localization of the Agrobacterium tumefaciens T-DNA transport pore proteins: VirB8 is essential for the assembly of the transport pore. Mol. Microbiol. *36*, 608-617.
Kupelwieser, G., Schwab, M., Hogenauer, G., Koraimann, G., and Zechner, E.L. (1998). Transfer protein TraM stimulates TraI-catalyzed cleavage of the transfer origin of plasmid R1 *in vivo*. J. Mol. Biol. *275*, 81-94.
Lanka, E., and Wilkins, B.M. (1995). DNA processing reactions in bacterial conjugation. Annu. Rev. Biochem. *64*, 141-169.
Larkin, C., Datta, S., Harley, M.J., Anderson, B.J., Ebie, A., Hargreaves, V., and Schildbach, J.F. (2005). Inter- and intramolecular determinants of the specificity of single-stranded DNA binding and cleavage by the F factor relaxase. Structure *13*, 1533-1544.
Lawley, T.D., Klimke, W.A., Gubbins, M.J., and Frost, L.S. (2003). F factor conjugation is a true type IV secretion system. FEMS Microbiol. Lett. *224*, 1-15.
Lederberg, J. (2000). Infectious history. Science *288*, 287-293.
Lederberg, J., and Tatum, E. (1946). Gene recombination in E. coli. Nature *158*, 558.
Llosa, M., Bolland, S., and de la Cruz, F. (1994). Genetic organization of the conjugal DNA processing region of the IncW plasmid R388. J. Mol. Biol. *235*, 448-464.
Llosa, M., and de la Cruz, F. (2005). Bacterial conjugation: a potential tool for genomic engineering. Res. Microbiol. *156*, 1-6.
Llosa, M., Gomis-Ruth, F.X., Coll, M., and de la Cruz Fd, F. (2002). Bacterial conjugation: a two-step mechanism for DNA transport. Mol. Microbiol. *45*, 1-8.
Llosa, M., Grandoso, G., Hernando, M.A., and de la Cruz, F. (1996). Functional domains in protein TrwC of plasmid R388: dissected DNA strand transferase and DNA helicase activities reconstitute protein function. J. Mol. Biol. *264*, 56-67.
Llosa, M., and O'Callaghan, D. (2004). Euroconference on the Biology of Type IV Secretion Processes: bacterial gates into the outer world. Mol. Microbiol. *53*, 1-8.
Lu, J., Edwards, R.A., Wong, J.J., Manchak, J., Scott, P.G., Frost, L.S., and Glover, J.N. (2006). Protonation-mediated structural flexibility in the F conjugation regulatory protein, TraM. EMBO J. *25*, 2930-2939.
Lujan, S.A., Guogas, L.M., Ragonese, H., Matson, S.W., and Redinbo, M.R. (2007). Disrupting antibiotic resistance propagation by inhibiting the conjugative DNA relaxase. Proc. Natl. Acad. Sci. U. S. A. *104*, 12282-12287.
Lum, P.L., Rodgers, M.E., and Schildbach, J.F. (2002). TraY DNA recognition of its two F factor binding sites. J. Mol. Biol. *321*, 563-578.
Massey, T.H., Mercogliano, C.P., Yates, J., Sherratt, D.J., and Lowe, J. (2006). Double-stranded DNA translocation: structure and mechanism of hexameric FtsK. Mol. Cell *23*, 457-469.
Meinke, G., Bullock, P.A., and Bohm, A. (2006). Crystal structure of the simian virus 40 large T-antigen origin-binding domain. J. Virol. *80*, 4304-4312.
Moncalian, G., Cabezon, E., Alkorta, I., Valle, M., Moro, F., Valpuesta, J.M., Goni, F.M., and de La Cruz, F. (1999a). Characterization of ATP and DNA binding activities of TrwB, the coupling protein essential in plasmid R388 conjugation. J. Biol. Chem. *274*, 36117-36124.
Moncalian, G., and de la Cruz, F. (2004). DNA binding properties of protein TrwA, a possible structural variant of the Arc repressor superfamily. Biochim. Biophys. Acta *1701*, 15-23.
Moncalian, G., Grandoso, G., Llosa, M., and de la Cruz, F. (1997). oriT-processing and regulatory roles of TrwA protein in plasmid R388 conjugation. J. Mol. Biol. *270*, 188-200.
Moncalian, G., Valle, M., Valpuesta, J.M., and de la Cruz, F. (1999b). IHF protein inhibits cleavage but not assembly of plasmid R388 relaxosomes. Mol. Microbiol. *31*, 1643-1652.
Monzingo, A.F., Ozburn, A., Xia, S., Meyer, R.J., and Robertus, J.D. (2007). The structure of the minimal relaxase domain of MobA at 2.1 A resolution. J. Mol. Biol. *366*, 165-178.
Murzin, A.G., Brenner, S.E., Hubbard, T., and Chothia, C. (1995). SCOP: a structural classification of proteins database for the investigation of sequences and structures. J. Mol. Biol. *247*, 536-540.

Ochman, H., Lawrence, J.G., and Groisman, E.A. (2000). Lateral gene transfer and the nature of bacterial innovation. Nature *405*, 299-304.

Ollis, D.L., Brick, P., Hamlin, R., Xuong, N.G., and Steitz, T.A. (1985). Structure of large fragment of Escherichia coli DNA polymerase I complexed with dTMP. Nature *313*, 762-766.

Remaut, H., and Waksman, G. (2004). Structural biology of bacterial pathogenesis. Curr. Opin. Struct. Biol. *14*, 161-170.

Rice, P.A., Yang, S., Mizuuchi, K., and Nash, H.A. (1996). Crystal structure of an IHF-DNA complex: a protein-induced DNA U-turn. Cell *87*, 1295-1306.

Sagulenko, V., Sagulenko, E., Jakubowski, S., Spudich, E., and Christie, P.J. (2001). VirB7 lipoprotein is exocellular and associates with the Agrobacterium tumefaciens T pilus. J. Bacteriol. *183*, 3642-3651.

Santini, J.M., and Stanisich, V.A. (1998). Both the fipA gene of pKM101 and the pifC gene of F inhibit conjugal transfer of RP1 by an effect on traG. J. Bacteriol. *180*, 4093-4101.

Savvides, S.N., Yeo, H.J., Beck, M.R., Blaesing, F., Lurz, R., Lanka, E., Buhrdorf, R., Fischer, W., Haas, R., and Waksman, G. (2003). VirB11 ATPases are dynamic hexameric assemblies: new insights into bacterial type IV secretion. EMBO J. *22*, 1969-1980.

Scott, J.R., and Churchward, G.G. (1995). Conjugative transposition. Annu. Rev. Microbiol. *49*, 367-397.

Schmidt-Eisenlohr, H., Domke, N., Angerer, C., Wanner, G., Zambryski, P.C., and Baron, C. (1999a). Vir proteins stabilize VirB5 and mediate its association with the T pilus of Agrobacterium tumefaciens. J. Bacteriol. *181*, 7485-7492.

Schmidt-Eisenlohr, H., Domke, N., and Baron, C. (1999b). TraC of IncN plasmid pKM101 associates with membranes and extracellular high-molecular-weight structures in Escherichia coli. J. Bacteriol. *181*, 5563-5571.

Seubert, A., Hiestand, R., de la Cruz, F., and Dehio, C. (2003). A bacterial conjugation machinery recruited for pathogenesis. Mol. Microbiol. *49*, 1253-1266.

Sexton, J.A., and Vogel, J.P. (2002). Type IVB secretion by intracellular pathogens. Traffic *3*, 178-185.

Silverman, P.M. (1997). Towards a structural biology of bacterial conjugation. Mol. Microbiol. *23*, 423-429.

Soto, G.E., and Hultgren, S.J. (1999). Bacterial adhesins: common themes and variations in architecture and assembly. J. Bacteriol. *181*, 1059-1071.

Spudich, G.M., Fernandez, D., Zhou, X.R., and Christie, P.J. (1996). Intermolecular disulfide bonds stabilize VirB7 homodimers and VirB7/VirB9 heterodimers during biogenesis of the Agrobacterium tumefaciens T-complex transport apparatus. Proc. Natl. Acad. Sci. U. S. A. *93*, 7512-7517.

Stockner, T., Plugariu, C., Koraimann, G., Hogenauer, G., Bermel, W., Prytulla, S., and Sterk, H. (2001). Solution structure of the DNA-binding domain of TraM. Biochemistry *40*, 3370-3377.

Suzuki, M. (1995). DNA recognition by a beta-sheet. Protein Eng. *8*, 1-4.

Tato, I., Zunzunegui, S., de la Cruz, F., and Cabezon, E. (2005). TrwB, the coupling protein involved in DNA transport during bacterial conjugation, is a DNA-dependent ATPase. Proc. Natl. Acad. Sci. U. S. A. *102*, 8156-8161.

Terradot, L., Bayliss, R., Oomen, C., Leonard, G.A., Baron, C., and Waksman, G. (2005). Structures of two core subunits of the bacterial type IV secretion system, VirB8 from Brucella suis and ComB10 from Helicobacter pylori. Proc. Natl. Acad. Sci. U. S. A. *102*, 4596-4601.

Thorstenson, Y.R., and Zambryski, P.C. (1994). The essential virulence protein VirB8 localizes to the inner membrane of Agrobacterium tumefaciens. J. Bacteriol. *176*, 1711-1717.

Ton-That, H., and Schneewind, O. (2003). Assembly of pili on the surface of Corynebacterium diphtheriae. Mol. Microbiol. *50*, 1429-1438.

Vergunst, A.C., Schrammeijer, B., den Dulk-Ras, A., de Vlaam, C.M., Regensburg-Tuink, T.J., and Hooykaas, P.J. (2000). VirB/D4-dependent protein translocation from Agrobacterium into plant cells. Science *290*, 979-982.

Walker, J.E., Saraste, M., Runswick, M.J., and Gay, N.J. (1982). Distantly related sequences in the alpha- and beta-subunits of ATP synthase, myosin, kinases and other ATP-requiring enzymes and a common nucleotide binding fold. EMBO J. *1*, 945-951.

Wang, J.C. (1996). DNA topoisomerases. Annu. Rev. Biochem. *65*, 635-692.

Waters, V.L. (2001). Conjugation between bacterial and mammalian cells. Nat. Genet. *29*, 375-376.

Weiss, A.A., Johnson, F.D., and Burns, D.L. (1993). Molecular characterization of an operon required for pertussis toxin secretion. Proc. Natl. Acad. Sci. U. S. A. *90*, 2970-2974.

Wen, M., Yamada, N., Zhang, Y., and Matsuhisa, T. (1997). Morphological changes of Helicobacter pylori after antibacterial therapy: an electron microscope study. Med. Electron. Microsc. *30*, 131-137.

Wilkins, B., and Lanka, E. (1993). DNA processing and replication during plasmid transfer between Gram-negative bacteria (New York: Plenum Press).

Wotherspoon, A.C. (1998). Gastric lymphoma of mucosa-associated lymphoid tissue and Helicobacter pylori. Annu. Rev. Med. *49*, 289-299.

Yeo, H.J., Savvides, S.N., Herr, A.B., Lanka, E., and Waksman, G. (2000). Crystal structure of the hexameric traffic ATPase of the Helicobacter pylori type IV secretion system. Mol. Cell *6*, 1461-1472.

Yeo, H.J., and Waksman, G. (2004). Unveiling molecular scaffolds of the type IV secretion system. J. Bacteriol. *186*, 1919-1926.

Yeo, H.J., Yuan, Q., Beck, M.R., Baron, C., and Waksman, G. (2003). Structural and functional characterization of the VirB5 protein from the type IV secretion system encoded by the conjugative plasmid pKM101. Proc. Natl. Acad. Sci. U. S. A. *100*, 15947-15952.

Ziegelin, G., Furste, J.P., and Lanka, E. (1989). TraJ protein of plasmid RP4 binds to a 19-base pair invert sequence repetition within the transfer origin. J. Biol. Chem. *264*, 11989-11994.

Chapter 7

Plasmids for Gene Therapy

Karthikeyan Kandavelou and Srinivasan Chandrasegaran

ABSTRACT

The success of gene therapy depends on the efficient insertion of therapeutic genes at the appropriate chromosomal target sites within the human genome, without causing cell injury, oncogenic mutations or an immune response. Although viral vectors offer excellent vehicles for highly efficient transduction of human cells, the associated safety concerns make non-viral delivery of therapeutic genes by using plasmid DNA into cells more attractive. The construction of plasmid vectors is simple and straightforward. Custom-designed zinc finger nucleases (ZFNs) that combine the non-specific cleavage domain (N) of FokI endonuclease with zinc finger proteins (ZFPs) offer a general way to deliver a site-specific double strand break (DSB) to the genome, and stimulate local homologous recombination (HR) by several orders of magnitude. This makes targeted gene correction or genome editing a viable option in human cells. Since ZFN-encoded plasmids could be used to transiently express ZFNs to target a DSB to a specific gene locus in human cells, they offer an excellent way for targeted delivery of the therapeutic genes to a pre-selected chromosomal site. The ZFN-encoded plasmid-based approach has the potential to circumvent all the problems associated with the viral delivery of therapeutic genes.

INTRODUCTION

There are about $3x10^9$ base pairs (bp) in the haploid genome of a human cell. Now that the nucleotide sequence of the human genome is available, biologists estimate that the genome codes for about 20,000 to 30,000 different genes. Scientists have shown that many human monogenic diseases are caused by changes in single genes, that is, they result from mutations within the coding or regulatory regions of critical genes in the genome giving rise to disease phenotypes (Porteus et al., 2006). Several thousands of human diseases appear to fall into the category of monogenic diseases. The "Holy Grail" of genetic medicine and biomedical researchers is to treat human genetic diseases by correcting the causative genetic defects within the human genome by targeted engineering of human cells (Chandrasegaran and Smith, 1999). The concept of gene therapy, envisions that treating certain human diseases, particularly the monogenic diseases, at the gene level by correcting the causative gene mutations responsible for the disease to wild type sequence, will result in the reversion of the disease phenotypes into normal phenotypes. This is obviously a new paradigm to treat human disease at its origins.

The complete decoding of the nucleotide sequence of the human genome and the recent advances in recombinant DNA technology have made it possible to identify the genetic origin of human diseases and invent ways for targeted manipulation of the DNA at the gene level in cells. Over the past decade, attempts have been made in several clinical trials to use gene therapy to treat various human diseases throughout the world (Porteus et al., 2006; Thomas et al., 2003). While some of these gene therapy trials have shown promise, they also have revealed numerous hidden pitfalls in the various methodologies used for gene therapy highlighting the need for further fine tuning and refinement of the approach. Gene therapy still requires many modifications in the vector design and methodology before it can become a common therapeutic regimen in clinical practice (Li and Ma, 2001). The success of gene therapy depends largely on efficient insertion of the therapeutic genes at the appropriate targeted chromosomal locus within the human genome, without causing cell injury or oncogenic mutations or inflammation (Porteus et al., 2006; Wu et al., 2007).

ADVERSE EVENTS FROM GENE THERAPY CLINICAL TRIALS

The beginning of gene therapy was perceived as a revolutionary technology with the dream to cure all diseases, once we understood the genetic basis of those diseases. But later this enthusiasm waned rather quickly as the adverse effects of the gene therapy started to show up in various clinical trials. A few tragic deaths that have occurred in the past in some of these clinical trials have focused on the pitfalls of using viral vectors in gene therapy. These incidents revealed the poor understanding of the vector biology and the host interaction with the viral vectors. The main challenges to gene therapy include immunogenicity to viral vectors and mutagenesis resulting from the random integration of the viral vectors into the genome. On the other hand, such adverse results have also motivated biomedical researchers more than ever to find solutions to the technical problems associated with gene therapy.

IMMUNOGENICITY

One of the major hurdles of gene therapy is that many of the body's defense mechanisms that tackle the infections are triggered to act against the viral vectors, which are recognized as foreign. Adenoviral vectors in particular are notorious for triggering the immune response. In 1999, an 18 year old patient was treated with adenovirus vectors containing a therapeutic gene in a clinical trial with the intention of curing the deficiency of ornithine transcarbamylase, a liver enzyme required in clearing the excessive nitrogen from the body. Within four hours after treatment the patient developed severe fever and later developed liver injury and died of multi-organ failure within four days (2002; Marshall, 1999). Subsequent studies in monkeys showed that viral capsid proteins can elicit a massive inflammatory reaction (Schnell et al., 2001).

INSERTIONAL MUTAGENESIS

The major success of retroviral vectors in achieving stable gene transfer into dividing cells like hematopoietic cells was undermined by their random integration. In the French SCID clinical trial, a murine leukemia based viral vector was used to treat the $CD34^+$ cells ex vivo to transfer the human *ILR2*γ gene, the deficiency of which causes the severe combined immunodeficiency disease (SCID). Three of eleven patients later developed a leukemia-like disorder, due to activation of a proto-oncogene *LMO2*, caused by the transgene insertion in or near that gene (Cavazzana-Calvo et al., 2005; Hacein-Bey-Abina et al., 2003). In this case, it was clear that there was selection of the transduced cells to proliferate which might have favored the development of the leukemia like malignancy. A later study showed that retroviral vector integration is not as random as was originally thought. Analysis of the HIV integration sites showed that HIV prefers integration into transcriptionally active sites at the non-coding regions within the genome (Schroder et al., 2002). Though the underlying mechanism in the SCID trial is not yet clear, this event could be the result of an interaction between the expression of the inserted gene, the site of insertion and the dividing cell type. The important problem to address is: How does one overcome the adverse effects that result from the random integration of the viral vectors?

PLASMID-MEDIATED GENE TRANSFER

The success of gene therapy depends on developing an efficient vector system or a vehicle that will transfer the therapeutic genes into the diseased cells at the appropriate chromosomal locus within their nucleus. Viral vectors are highly efficient in transducing cells, but randomly integrate into the genome. The various safety concerns, which include the possibility of introducing oncogenic mutations or the development of an immune response to viral proteins by the host, make other non-viral gene delivery methods even more attractive. The simplest approach to non-viral delivery is direct gene transfer by using plasmid DNA (Li and Ma, 2001). Various techniques have been developed in the past years for direct transfer of genes into cells based on polymers and lipids (Li and Ma, 2001). The primary limitation of current direct DNA mediated gene transfer is the short duration of gene expression that is seen compared to the viral vector approach. This is due to the low rate of stable integration of naked DNA into the host genome. Even those few events of stable integrations are also random recombination, similar to recombination from viral-based vectors. This illegitimate recombination, although gives a long term expression, but also could possibly give rise to insertional mutagenesis of the neighboring genes, resulting in detrimental results.

The construction of plasmid vectors is simple and straightforward. Most plasmids carry a strong ubiquitous mammalian promoter or a tissue specific promoter, a multiple cloning site region for inserting the therapeutic gene of interest and an appropriate transcriptional terminator to stabilize the transcripts (Li and Huang, 2000). A tissue specific promoter has the advantage of restricting the expression of the therapeutic gene of interest to the targeted tissue. Most of the plasmid based gene therapy protocols yield transient expression of the therapeutic transgene; but when selection methods using antibiotic resistance or other approaches are employed, they have been shown to result in stably transduced cells that persistently express the transferred therapeutic gene.

ENHANCING GENE EXPRESSION FROM PLASMIDS

While many researchers are trying to enhance efficiency of plasmid mediated gene transfer by improving the delivery vehicle, the overall efficiency can be enhanced by improving the gene expression from the transfected plasmids. Several studies have shown that modifying various regulatory elements in the plasmids such as the promoters, enhancers, introns, polyadenylation and transcriptional terminators can increase the gene expression (see also chapter Plasmid-based Expression Systems for Mammalian Cells).

One of the key limitations of plasmid-mediated gene therapy is the short duration of gene expression. The major reason for this appears to be gene dilution, which is very prominent in replicating cells. In the absence of stable integration, with each cycle of replication the extra chromosomal plasmid decreases exponentially leading to decreased gene expression. This problem can be solved by modifying the plasmid to permit replication in the cells. This is achieved by incorporating viral replicons from viruses like EBV, BK and SV40(Cooper et al., 1997; Van Craenenbroeck et al., 2000). These replicons typically include a viral replication origin and an early gene product that acts as a replication transactivator. One of the concerns using such replicons is that some of these viral transactivators could be oncogenic. This could be overcome by using mutated transactivators.

Another major limitation is the gene inactivation by histone deacetylation and chromatin condensation. Treatment of cells with the specific histone deacetylase inhibitor trichostatin A resulted in rapid reactivation of the genes (Chen and Townes, 2000). But the use of such inhibitors may not be useful due to their nonspecific toxicity. The mechanism of such inactivation is not clearly understood. Some studies suggest that the recruitment of histone deacetylase complex (HDAC) to the target site could be involved in the process of inactivation. The interaction of HDAC with the transgene could be mediated through specific sequences on the transgene. If a clear understanding of this interaction is achieved, plasmid lacking such sequences can be constructed to overcome such interactions.

TARGETED PLASMID-MEDIATED GENE DELIVERY

The stability of the introduced DNA could be achieved by inserting the therapeutic gene by replacing the diseased gene via homologous recombination (HR). This approach will also avoid the insertional mutagenesis that could result from the random integration of the plasmid at critical gene locations of the human genome. HR is not a very efficient process in human cells. In fact, only about one in a million treated cells undergo homologous recombination with the investigator-provided donor DNA (Kandavelou et al., 2005; Vasquez et al., 2001; Wilson, 2003). However, when a chromosomal DSB is introduced at the target site, HR is induced at that gene locus, in a large fraction of cells in a population (Durai et al., 2005; Porteus, 2006). This being the case, there was then an obvious need to develop molecular scissors to induce a site specific and targeted chromosomal DSB within the human genome to stimulate local HR by several orders of magnitude. Custom zinc finger nucleases (ZFNs) offer a general way to deliver a site-specific DSB to the genome to stimulate local HR with the exogenous investigator-provided donor DNA.

ZINC FINGER NUCLEASES (ZFNS): DESIGNER MOLECULAR SCISSORS FOR PRECISION GENOME SURGERY

Zinc finger nucleases (ZFNs) are proteins designed to cut at specific DNA sequences (Fig. 1). They combine the non-specific cleavage domain (N) of Fok I restriction endonuclease with zinc finger proteins (ZFPs) (Chandrasegaran and Smith, 1999; Durai et al., 2005; Kandavelou et al., 2005). The recognition specificities of the ZFPs can easily be manipulated experimentally, to create custom ZFPs that will bind to a pre-determined DNA target site. Zinc finger nucleases (ZFNs) were originally developed in our lab to create designer restriction enzymes with novel sequence specificities, also known as chimeric nucleases (Kim et al., 1996). We had previously shown that the bacterial type IIs (where 's' stands for shifted cleavage from the recognition site) restriction enzyme, FokI endonuclease, comprised of two distinct protein domains: an N-terminal DNA binding domain and a C-terminal cleavage domain with non-specific cleavage activity (Li et al., 1992). The modular nature of Fok I restriction endonuclease suggested that it might be feasible to engineer chimeric nucleases by fusing other DNA-binding proteins to the Fok I cleavage domain (Li et al., 1992). Several novel fusion nucleases were successfully created by linking the non-specific Fok I cleavage domain to other DNA-binding proteins (Kim et al., 1996; Kim et al., 1998). Chimeric nucleases based on the zinc finger proteins (ZFPs) have proven to be the most versatile of all DNA binding motifs to create designer enzymes with novel sequence specificities, because individual zinc finger (ZF) motifs within ZFPs can be designed to bind a large range of DNA sites. Cys_2His_2 ZFPs bind DNA by inserting an α-helix into the major groove of the double helix (Elrod-Erickson and Pabo, 1999; Pavletich and Pabo, 1991; Shi and Berg, 1995). Each ZF motif primarily binds to a triplet within the DNA substrate. Key amino acids, at positions –1, 2, 3, and 6 relative to the start of the

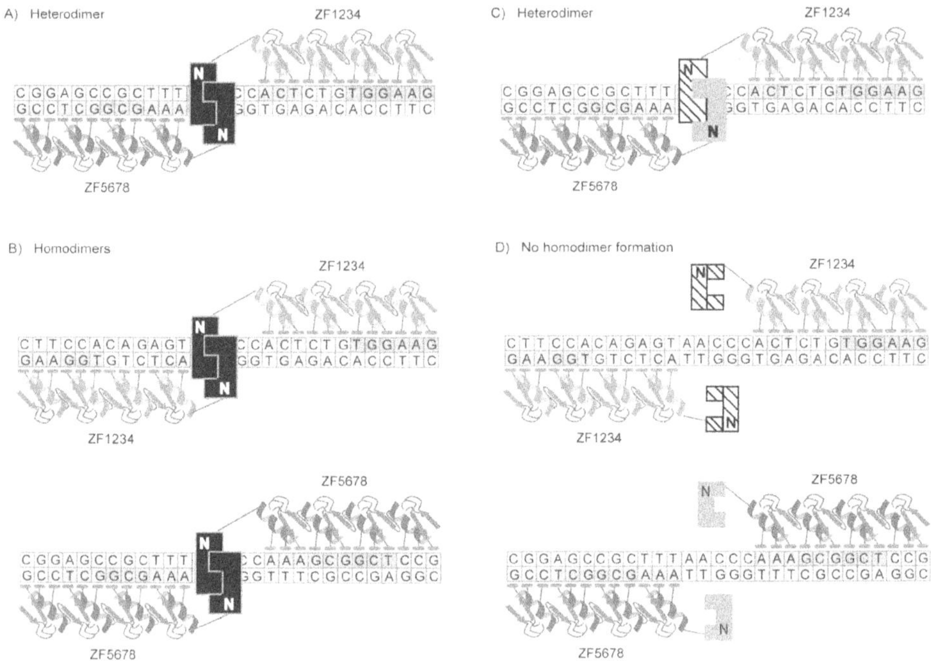

Fig. 1. Designer molecular scissors for precision genome surgery. **A**) Schematic representation of a pair of 4-finger ZFNs bound to their cognate heterodimer sites in the IL2Rγ gene. Since a pair of ZFNs requires two copies of the 12-bp recognition sites in a tail-to-tail orientation, they effectively have a 24-bp recognition site, which is long enough to specify a unique genomic address in human cells. Since the ZFN binding sites in the IL2Rγ gene are not identical, both ZFNs that bind these sites need to be introduced into the cell to induce a targeted genomic DSB. The ZFN sites in the IL2Rγ gene are separated by 5 bp (Urnov et al., 2005). **B**) Potential off-target cleavage resulting from the homodimer formation between the wild type FokI cleavage domains (N). Two different homodimers could result from the two individual ZFNs with different sequence specificities, and more often than not, they do occur. The homodimers, although are not relevant for gene modification but potentially could affect the safety and efficacy of ZFN-mediated gene targeting. **C and D**) The architecture of the ZFN dimer interface could be redesigned to select for obligate heterodimer formation (C) and inhibit homodimer formation (D) between a pair of ZFNs to greatly reduce the off-target cleavage, and hence ZFNs cytotoxicity (Miller et al., 2007; Szczepek et al., 2007). This increases the efficiency and efficacy of ZFN-mediated gene targeting in cells.

α-helix, contribute most of the specific interactions by the ZF motifs. These amino acids can be changed, while maintaining the remaining amino acids as a consensus backbone, to generate ZFPs with different sequence specificities (Desjarlais and Berg, 1993; Wolfe et al., 1999). Furthermore, the ZFP strategy has the additional advantage that greater specificity can be achieved by adding more ZF motifs to the ZFPs (Beerli et al., 1998; Kim and Pabo, 1998; Liu et al., 1997). Because of their modular structure and function, ZFPs offered an attractive framework for designing ZFNs with custom sequence-specificities (Chandrasegaran and Smith,

1999; Liu et al., 1997). Several 3-finger ZFPs, each recognizing a 9 base pair (bp) sequence, were fused to the non-specific endonuclease domain of Fok I to form ZFNs (Kim and Chandrasegaran, 1994; Mani et al., 2005). The binding specificity of a ZFP correlated directly with the cleavage specificity of the corresponding ZFN (Smith et al., 1999). ZFNs, like FokI restriction enzyme, require dimerization of the nuclease domain in order to cut DNA (Mani et al., 2005; Smith et al., 2000). Dimerization of ZFNs, and hence double-strand cleavage, is facilitated by two adjacent, oppositely oriented binding sites (Mani et al., 2005; Smith et al., 2000). The binding sites need not be identical, provided the ZFN that bind both sites are present in order to bind and induce cleavage of the substrate (Fig. 1). We have shown that two 3-finger ZFNs with different sequence-specificities collaborate to produce a DSB when their binding sites are appropriately placed and oriented with respect to each other (Smith et al., 2000). The creation of highly specific ZFNs that recognize and cleave any target sequence depends on the reliable design and/or selection of ZFPs that can recognize and bind that chosen target with high affinity and specificity. The ZF motif designs that specifically recognize each and every one of the possible 64 DNA triplets are not yet available. A wealth of literature data based on design and phage display methodologies are available for the ZF modules that specifically recognize 5'-GNN-3' and 5'-ANN-3' triplets, but ZF motif preferences for 5'-CNN-3' and 5'-TNN-3' triplets are known to a lesser extent (Dreier et al., 2001; Dreier et al., 2005; Dreier et al., 2000; Durai et al., 2006; Liu et al., 2002). Currently, two web-based ZF design softwares are available, which could be accessed at URLs http://www.zincfingertools.org (Mandell and Barbas, 2006) and http://www.zincfingers.org/software-tools.htm. Since a DSB delivered a few hundred bp away from the chosen site induces HR, almost all the genes encoded in the genome of a cell are amenable to ZFN-mediated gene targeting. Design strategies do not always give rise to the ZFNs with exquisite sequence specificity and affinity. Further refinement of the ZFNs could be done by substituting the individual ZF motifs of the initially designed ZFPs with other modules that are available for the recognizing that particular triplet sequence. Although tedious and labor-intensive, selection approaches by using phage display appear to yield ZFPs with high affinity and specificity. Several cell-based selection strategies including bacterial one-hybrid and two-hybrid systems, for producing highly specific ZFPs have also been reported in literature (Durai et al., 2006; Hurt et al., 2003; Joung et al., 2000). The ability to design and/or select zinc fingers with desired sequence specificity has made it possible for ZFPs to be made to order. As a consequence, it is now possible to design ZFNs that will cut DNA at a pre-determined or a specific site within a genome by making fusions of ZFPs to the cleavage domain of FoKI endonuclease.

ZFN-MEDIATED GENE TARGETING: A NOVEL AND EFFECTIVE METHOD FOR SITE-SPECIFIC AND PERMANENT MODIFICATION OF PLANT AND MAMMALIAN GENOMES, INCLUDING THE HUMAN GENOME

Cells use the universal process of homologous recombination (HR) to maintain their genomic integrity, particularly in the repair of a double-strand break (DSB), which otherwise would be lethal. Normally, DSB repair of a damaged chromosome by HR is achieved using the homologous DNA segment from the undamaged chromosomal partner as a template. In gene targeting, the extra-chromosomal homologous donor DNA is provided in excess by the investigator to drive the sequence exchange in cells. Gene targeting is a very rare event in mammalian cells; only about one in a million cells provided with excess donor sequence undergo the desired gene modification. Gene targeting, however, could be stimulated several orders of magnitude in a large fraction of cells in a population by inducing a targeted DSB. Designer ZFNs offer a general way to deliver a site-specific DSB to the genome, and stimulate local HR (Fig. 2). A relevant issue to ZFN-mediated gene targeting is the ability of ZFNs to distinguish the chosen target for cleavage from all other sequences in the human genome. Any sequence shorter that 16-bp would be present in multiple copies in the human genome. For instance, any 9-bp sequence is expected to occur 10^4 times within the human genome. As discussed above, since binding of two 3-finger ZFNs (each recognizing a 9-bp inverted site) is required to facilitate a DSB, a pair of ZFNs effectively has an 18-bp recognition site, which is long enough to specify a unique genomic address in human cells.

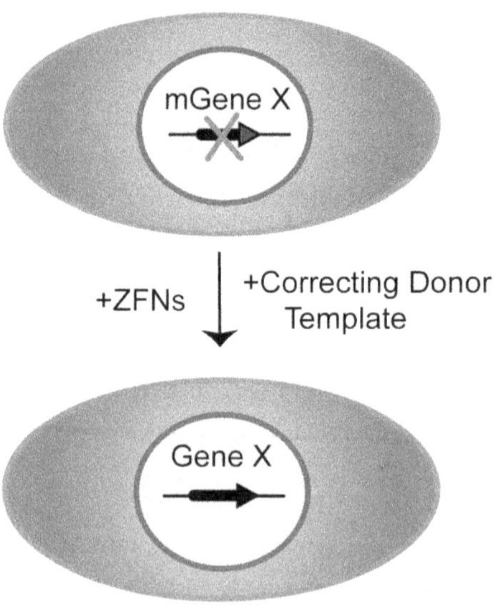

Fig. 2. ZFN-mediated gene targeting. To achieve targeted gene correction (or gene editing), cells with the mutant Gene X (shown as mGene X) are transfected with both designer ZFNs and the homologous correcting donor DNA fragment of Gene X to induce recombinogenic repair and generate normal cells. The ZFNs induce a targeted genomic DSB in cells, which stimulates local HR in the presence of homologous donor DNA via the recombinogenic repair pathway. Alternatively, site specific modification of the Gene X by ZFN-evoked HR occurs when the normal cells are co-transfected with both custom ZFNs and the homologous donor DNA containing a pre-selected Gene X modification to generate a specific desired mutant phenotype of the targeted gene.

Several labs have shown that 3-finger ZFNs find and cleave their chromosomal target in cells; and as expected, they induce local HR to repair the DSB. In the absence of homology-directed repair via HR (for example, if both alleles of a gene are damaged), cells repair the DSB by simple ligation via non-homologous end joining (NHEJ); repair by NHEJ is mutagenic. Thus, ZFNs can be used with and without homologous donor DNA sequence to induce "directed" mutations, by HR and NHEJ, respectively. Thus, ZFNs have become powerful tools for enhancing gene targeting by several orders of magnitude. Moreover, Urnov et al. have shown that the sequence specificity ZFNs could be further increased by using a pair of 4-finger ZFNs that recognize a 24 bp sequence, at the same time greatly reducing cytotoxicity when introduced into human cells (Urnov et al., 2005).

The steps involved in ZFN-mediated gene targeting are as follows:

- Identify ZFN binding sites at or near the targeted chromosomal locus within the genome of the desired cells (Mani et al., 2005; Carroll et al., 2006; Pabo et al., 2001; Wright et al., 2006).

- Design and/or select ZFPs that recognize the chosen ZFN target sites (Mani et al., 2005; Carroll et al., 2006; Pabo et al., 2001; Wright et al., 2006).

- Convert the designed and/or selected ZFPs into ZFNs (Mani et al.; Carroll et al., 2006; 2005; Pabo et al., 2001; Wright et al., 2006).

- Deliver ZFNs and the mutant target gene donor fragment into normal cells to induce a targeted DSB and stimulate HR to generate a specific mutant of the target, *i.e.* "knock-outs" or "knock-ins" of the desired gene as the case may be (Fig. 2). Alternatively, to achieve gene editing (or gene correction), we need to deliver ZFNs and the correcting wild type donor gene fragment into the mutant cells (Fig. 2) (Urnov et al., 2005).

- Monitor for gene correction (or mutagenesis) at the targeted gene loci as per the investigator- provided donor gene template. It is also critical to show that the donor DNA fragment has not integrated elsewhere in the genome of the cell by using Southern blots (Urnov et al., 2005).

ZFN-MEDIATED GENE TARGETING AS A FORM OF GENE THERAPY

A gene targeting model GFP reporter system has been developed to study the efficiency and efficacy of gene targeting in human cells by using custom ZFNs (Portcus and Baltimore, 2003). In the gene targeting model GFP reporter system, a non-functional eGFP gene was integrated into the genome of the human HEK 293 cells. Only 1-2 cells of the 500,000 treated cells were corrected for GFP, when the wild type eGFP donor fragment alone was introduced into HEK 293 cells. However, when expression construct encoding ZFNs targeting the mutant GFP locus was introduced along with the donor fragment, 2.2% of asynchronously growing cells and 10.2% of cells arrested for 30 hours at the G2/M cell cycle boundary, respectively, had their eGFP function restored (Porteus and Baltimore, 2003).

The most relevant work on ZFN-mediated gene targeting of primary human cells is the study by Urnov et al. 2005. After establishing the potential of ZFNs for gene correction in human cells using the GFP reporter system, Urnov et al. (2005) engineered 4-finger ZFNs were designed to target the human *IL2Rγ* gene, mutations in which cause X-linked SCID. They used optimized 4-finger ZFNs to modify the sequence at the IL2Rγ gene locus in the K562 cell line and in primary $CD4^{+}$ T lymphocytes. The two custom 4-finger ZFNs were cloned separately into plasmids and then introduced into the target cell substrates along with the donor DNA. The exogenous donor template plasmid used in their study contained a fragment of the IL2Rγ gene with a silent point mutation that creates a novel restriction enzyme recognition site in the exon 5 sequence. ZFN-evoked HR introduced this restriction site into the cognate chromosomal location. Four days after nucleofection of K562 cells with donor template plasmid plus ZFN plasmids, the HR frequency was 18%. The donor template specified restriction site was confirmed by sequencing the IL2Rγ gene locus in the gene-modified cells. Southern blotting failed to detect ZFN-induced donor plasmid misintegration or any aberrant rearrangements of the IL2Rγ locus in cell samples harvested after 4 days or 1 month of cell expansion. Notably, HR frequency of cells harvested at 4 days after transfection was the same as that of cells harvested one month after transfection, in contrast to published studies using 3-finger ZFNs, which showed a decline (Porteus and Baltimore, 2003). This is promising because it suggested that the 4-finger ZFNs may be less toxic to cells than the 3-finger ZFNs. Furthermore, by doing limiting dilution, single clones were isolated and genotyped. The results indicated that 13.2% of the clones had a single allele mutated, and 6.6% achieved biallelic gene modification (Urnov et al., 2005). Using the same protocol for IL2Rγ-targeting by ZFNs in human $CD4^{+}$ T lymphocytes, they observed a HR frequency of 5%. This was comparable to the HR frequency in K562 cells when adjusted for the lower transfection efficiency in T lymphocytes (30%). Thus, the Urnov et al. (Urnov et al., 2005) study demonstrates that designed ZFNs can evoke HR to generate permanent and precise genome-modification at the IL2Rγ gene locus in ~20% of treated cells, without selection.

More recently, Moehle et al. (Moehle et al., 2007) have shown efficient incorporation of novel DNA sequences at a specific location within the genome of living human cells using designed ZFNs. A precisely placed DSB induced by designed 4-finger ZFNs stimulated integration of long DNA stretches at a predetermined genomic locus, resulting in site specific high efficiency gene addition. The authors showed targeted integration frequencies of 15%, 6% and 5% respectively for the extra-chromosomal donor DNA carrying a 12 bp tag, a 900 bp ORF or a 1.5 kb promoter–transcription unit flanked by locus specific homology arms after 72 hours of treatment, without selection for desired recombinants. They also showed that ZFNs can drive the addition of an 8 kb sequence comprising of three distinct promoter-transcription units into an endogenous locus at a frequency of 6%, without selection.

Several labs have now firmly established ZFN-mediated gene targeting as a powerful research tool for site-specific and precise modification of not only plant (Lloyd et al., 2005; Wright et al., 2005) and mammalian genomes including the human genome (Moehle et al., 2007; Porteus and Baltimore, 2003; Urnov et al., 2005), but also for the targeted modification of the genomes of many other organisms (Morton et al., 2006). Furthermore, the high efficiency of ZFN-mediated gene targeting in human cells are especially exciting and encouraging because of its promise and great potential to the ultimate goal of applying ZFN-based strategies for human therapeutics that is to use ZFN-mediated gene targeting as a form of gene therapy to treat monogenic human diseases.

One issue associated with ZFN-mediated gene targeting that hinders it from widespread use in biomedical applications is the cytotoxicity of the designer ZFNs, particularly if the ZFPs used to create the ZFNs do not show exquisite sequence specificity and high affinity for the chosen DNA targets. Although custom 3-finger ZFN designs based on the original wild type FokI nuclease domains have been successfully used to target and cleave specific DNA sites in cells, sustained ectopic expression of the custom 3-finger ZFNs appear to be toxic to cells (Porteus and Baltimore, 2003). This is likely due to off-target cleavage as the concentration of 3-finger ZFNs increase in cells and particularly if the designed ZFNs are not highly specific. The higher the specificity of the engineered ZFNs, the lower should be their cellular toxicity. There are several ways to lower the cytotoxicity of the engineered ZFNs: (1) One approach is to design highly specific 4-finger ZFNs, thereby, increase the sequence specific recognition from 18 bp (for a pair of 3-finger designs) to 24-bp (for a pair of 4-finger designs) as was done by Urnov et al. (2005). (2) The cytotoxicity of custom ZFNs to cells could be further reduced by redesigning the protein-protein interactions involved at the dimer interface of the wild type FokI nuclease domain to select for heterodimer formation as discussed below. (3) The third approach is to control and regulate the expression of the ZFNs within cells. This could be done by cloning the ZFNs downstream of an inducible or a regulated promoter. Alternatively, the cleavage activity of ZFNs could be regulated by engineering conditional protein stability to ZFNs by using biologically silent small molecules. A recent report describes an experimental system in which the stability of a small protein domain, for example, rapamycin-binding protein (FKBP12) expressed in mammalian cells depends on the presence of a high affinity ligand. This ligand-dependent stability is conferred to any fused partner protein (Banaszynski et al., 2006).

IMPROVED ZFNS FOR TARGETED GENOME EDITING (OR GENE CORRECTION)

ZFNs are rapidly emerging as versatile reagents for gene correction or genome editing in a variety of cells and cell types. Targeted genome engineering using ZFNs yield high gene modification efficiencies (>10%) by delivering a recombinogenic DSB to the targeted chromosomal locus of the genome. The original ZFN architecture requires dimerization of the cleavage domain to induce a DSB, which necessitates the design and engineering of two different ZFNs for binding to adjacent half sites (Fig. 1A). Since the protein-protein interaction for the cleavage domain dimer formation is weak, the ZFNs, like FokI, are likely to remain monomeric at concentrations of at least 15 μM; and dimerize only when bound to their specific targets. Although this requirement for dimerization of the cleavage domain restricts cleavage by a pair of ZFNs to long sequences, it could also introduce a potential problem, since this dimer interaction does not select for the heterodimer species, homodimer species which form could lead to unwanted off-target cleavage elsewhere in the genome of cells. Two different homodimers could result from the two individual ZFNs with different sequence specificities, and more often than not, they do occur (Fig. 1B). The homodimers, although are not relevant for gene modification, but potentially could and often do, affect the safety and efficacy of ZFN-mediated gene targeting. Two recent articles (Miller et al., 2007; Szczepek et al., 2007) have addressed this issue of ZFN cytotoxicity by redesigning the ZFN dimer interface to inhibit homodimerization, thereby, greatly reducing off-target cleavage by ZFNs, and hence their cytotoxicity (Fig. 1C and 1D).

WORK IN PROGRESS

Several labs including ours are currently working towards efficient ZFN-evoked targeted disruption of the CCR5 gene locus in human cells. HIV-1 entry into cells involves specific interactions between the viral envelope glycoprotein and two target cellular proteins, namely CD4 and either of the two chemokine receptors, CCR5 or CXCR4, respectively. Macrophage (M)-tropic viruses require the chemokine receptor CCR5 for entry into cells while lymphocyte (T)-tropic viruses require the chemokine receptor CXCR4 for infection. Several studies suggest that CCR5-expressing cells are the critical first targets for HIV-1 infection and that the CCR5 expression levels in patients correlate strongly with disease progression. Importantly, individuals with a homozygous 32 bp deletion (Δ32) in their CCR5 gene lack functional CCR5 expression; these individuals, who are otherwise healthy, are highly protected against HIV-1 infection. Individuals who are heterozygous for CCR5Δ32 have reduced levels of CCR5, and their disease progression to AIDS is delayed by 1-2 years (Huang et al., 1996). Thus, the CCR5Δ32 mutation appears to function as a harmless polymorphism, which is protective against HIV infection and progression. The translational goal is to induce directed mutagenesis at the endogenous chromosomal sites of the CCR5 gene in human cells that are valuable for laboratory investigations and also potentially transplantable - primary human $CD34^+$ haematopoietic stem-progenitor cells (HSPCs) and human embryonic stem cells (HESCs). The major focus of this research is to use custom ZFNs to modify genes in human stem cells that can be used for functional studies in the laboratory, *e.g*. to make possible in primary human $CD34^+$ HSPCs and HESCs the kinds of studies that are already done using mouse embryonic stem cells (MESCs). The ultimate goal of this work is to apply ZFN-mediated gene targeting of CCR5 clinically to individuals who are already HIV infected or at very high risk for HIV infection. Furthermore, the chromosomal CCR5 locus could possibly be an ideal "safe harbor" site within the human genome for targeted insertion of other therapeutic transgenes for ectopic expression in human cells, in future studies. Other labs are also pursuing research along similar lines to achieve ZFN-mediated functional deletion of the CCR5 gene and to use the CCR5 chromosomal locus for targeted insertion of other therapeutic genes in human cells. As an example the structure of plasmids which can be used for gene targeting are depicted in Fig. 3. The shuttle plasmid, pNTKV, has the following elements: origin of replication, MC1 promoter which drives the expression of thymidine kinase gene (HSV-tk), PGK promoter which drives the expression of the GFP gene and ampicillin resistance gene for selection in *E. coli*.

The structure of the plasmids, pNTKV: CCR5-GFP exogenous donor DNA and pIRES: ZFNs, for use in the targeted disruption of the human CCR5 locus are shown in Fig. 3. The pNTKV: CCR5/GFP encoding the gene for GFP allows for sorting of the recombinant clones by FACS. The GFP model system is used mainly for analytical purposes of testing the efficiency and efficacy of the designed ZFNs in cell cultures. A pair of ZFNs that bind to a chosen CCR5 target sites, could be cloned into a single pIRES plasmid at the two distinct multiple cloning sites (MCS) downstream of CMV promoter. Alternatively, both ZFNs could be cloned individually into the pIRES plasmid and then both the plasmids co-transfected simultaneously into the desired target cell substrates along with donor DNA fragment during the ZFN-mediated gene targeting experiments.

Several other biomedical applications using plasmid-based therapy incorporating custom ZFNs are being contemplated and being developed by academic and commercial scientists.

LIMITATIONS OF PLASMID-BASED GENE THERAPY

The major limitation of plasmid based gene therapy has been low transfection efficiency when compared to viral vectors. This can be easily overcome with the recent advances in various methods of enhancing gene transfer, which include electroporation, lipid and polymer mediated gene transfer, gene gun, nucleofection, etc. The other drawback of the plasmid-mediated gene transfer is the immuno-

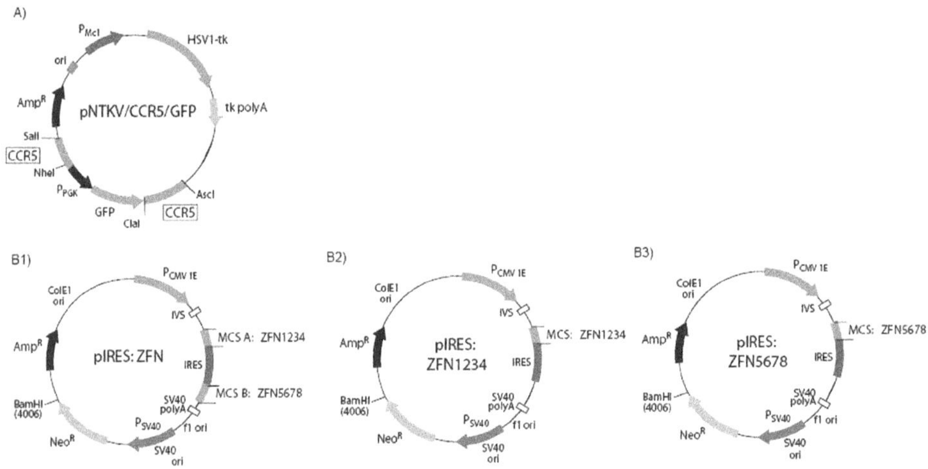

Fig. 3. Structure of pNTKV: CCR5-GFP exogenous donor DNA and pIRES: ZFNs. The donor plasmid, pNTKV: CCR5/GFP, encodes the gene for GFP, which allows for sorting of the recombinant clones by FACS. The model GFP system is used for analytical purposes only (A). The pair of ZFNs that bind to a chosen target site, could be cloned into a single pIRES plasmid at two different multiple cloning sites (MCS) downstream of CMV promoter as shown in B1. Alternatively, both ZFNs could be cloned individually into the pIRES plasmid as shown in B2 and B3. In the latter case, both the ZFN plasmids are co-transfected into the target cells along with the donor DNA fragment during ZFN-mediated gene targeting experiments.

stimulatory effect of the unmethylated CpG motifs present in the plasmid DNA from bacteria. This could be ameliorated by decreasing the number of CpGs that are present in the plasmid DNA (Yew et al., 1999; Yew et al., 1997). Plasmid-mediated co-delivery of two designed ZFNs that target the desired chromosomal locus and the donor DNA into primary human cells may pose a major hurdle particularly for the use of ZFN-directed genetic manipulation of human stem cells. In a recent publication, scientists have circumvented this problem by using integrase-defective lentiviral vector (IDLV) for delivery of ZFNs and the donor DNA into primary human cells and showed efficient gene editing and targeted transgene addition in most cell types, including human primary hematopoietic cells and ES cells (Lombardo et al., 2007).

FUTURE DIRECTION

ZFN-mediated gene targeting has become a powerful and versatile tool for targeted genome engineering of the genomes of many different organisms and cells, including the human cells. ZFN-based strategies as a form of gene therapy, in particular by modifying human stem cells *ex-vivo*, may provide a new paradigm for treating monogenic human diseases by correcting the causative genetic defect at its origins. Many of the difficulties associated with gene therapy are likely to be overcome if one could insert the corrected version of the mutation at the precise location of the genetic defect within the human genome. The current gene therapy vectors lack the requisite sequence specificity necessary for targeted correction of the defective site within the human genome. ZFN-based strategies for gene correction of human stem cells may provide a viable option to treat monogenic human diseases in the future (Porteus et al., 2006). Recent developments including ZFN–mediated gene targeting could vastly increase and improve the efficiency of plasmid-based gene therapy by providing for transient expression of ZFNs in cells in conjunction with the therapeutic gene fragment without the need for stable integration; thereby, circumventing many of the problems associated with using viral vectors as vehicles for gene delivery into cells. The ZFN technology is still at its infancy and many technical difficulties need to be addressed and overcome before it can become a routine clinical approach for treating monogenic human diseases. The ZFN technology has the potential to fulfill its promise for human therapeutics, in particular of curing monogenic diseases over the next decade.

ACKNOWLEDGEMENTS

We thank Dr. Sundar Durai for drawing the figures. The research on ZFNs in our lab has been supported by various grants from National Institutes of Health, USA, over the past 13 years; and it is currently being supported by the research grant GM077291 from NIGMS. Our work on ZFN-mediated gene targeting in human stem cells is partially supported by a grant from Maryland Stem Cell Research Fund.

REFERENCES

National Institutes of Health Recombinant DNA Advisory Committee (2002). Assessment of adenoviral vector safety and toxicity: report of the National Institutes of Health Recombinant DNA Advisory Committee. Hum. Gene. Ther. *13*, 3-13.

Banaszynski, L.A., Chen, L.C., Maynard-Smith, L.A., Ooi, A.G., and Wandless, T.J. (2006). A rapid, reversible, and tunable method to regulate protein function in living cells using synthetic small molecules. Cell *126*, 995-1004.

Beerli, R.R., Segal, D.J., Dreier, B., and Barbas, C.F., 3rd (1998). Toward controlling gene expression at will: specific regulation of the erbB-2/HER-2 promoter by using polydactyl zinc finger proteins constructed from modular building blocks. Proc. Natl. Acad. Sci. U S A *95*, 14628-14633.

Carroll, D., Morton, J.J., Beumer, K.J., and Segal, D.J. (2006). Design, construction and *in vitro* testing of zinc finger nucleases. Nat. Protoc. *1*, 1329-1341.

Cavazzana-Calvo, M., Lagresle, C., Hacein-Bey-Abina, S., and Fischer, A. (2005). Gene therapy for severe combined immunodeficiency. Annu. Rev. Med. *56*, 585-602.

Chandrasegaran, S., and Smith, J. (1999). Chimeric restriction enzymes: what is next? Biol. Chem. *380*, 841-848.

Chen, W.Y., and Townes, T.M. (2000). Molecular mechanism for silencing virally transduced genes involves histone deacetylation and chromatin condensation. Proc. Natl. Acad. Sci. U S A *97*, 377-382.

Cooper, M.J., Lippa, M., Payne, J.M., Hatzivassiliou, G., Reifenberg, E., Fayazi, B., Perales, J.C., Morrison, L.J., Templeton, D., Piekarz, R.L., and Tan, J. (1997). Safety-modified episomal vectors for human gene therapy. Proc. Natl. Acad. Sci. U S A *94*, 6450-6455.

Desjarlais, J.R., and Berg, J.M. (1993). Use of a zinc-finger consensus sequence framework and specificity rules to design specific DNA binding proteins. Proc. Natl. Acad. Sci. U S A *90*, 2256-2260.

Dreier, B., Beerli, R.R., Segal, D.J., Flippin, J.D., and Barbas, C.F., 3rd (2001). Development of zinc finger domains for recognition of the 5'-ANN-3' family of DNA sequences and their use in the construction of artificial transcription factors. J. Biol. Chem. *276*, 29466-29478.

Dreier, B., Fuller, R.P., Segal, D.J., Lund, C.V., Blancafort, P., Huber, A., Koksch, B., and Barbas, C.F., 3rd (2005). Development of zinc finger domains for recognition of the 5'-CNN-3' family DNA sequences and their use in the construction of artificial transcription factors. J. Biol. Chem. *280*, 35588-35597.

Dreier, B., Segal, D.J., and Barbas, C.F., 3rd (2000). Insights into the molecular recognition of the 5'-GNN-3' family of DNA sequences by zinc finger domains. J Mol Biol *303*, 489-502.

Durai, S., Bosley, A., Abulencia, A.B., Chandrasegaran, S., and Ostermeier, M. (2006). A bacterial one-hybrid selection system for interrogating zinc finger-DNA interactions. Comb Chem High Throughput Screen *9*, 301-311.

Durai, S., Mani, M., Kandavelou, K., Wu, J., Porteus, M.H., and Chandrasegaran, S. (2005). Zinc finger nucleases: custom-designed molecular scissors for genome engineering of plant and mammalian cells. Nucleic Acids Res *33*, 5978-5990.

Elrod-Erickson, M., and Pabo, C.O. (1999). Binding studies with mutants of Zif268. Contribution of individual side chains to binding affinity and specificity in the Zif268 zinc finger-DNA complex. J. Biol. Chem. *274*, 19281-19285.

Hacein-Bey-Abina, S., von Kalle, C., Schmidt, M., Le Deist, F., Wulffraat, N., McIntyre, E., Radford, I., Villeval, J.L., Fraser, C.C., Cavazzana-Calvo, M., and Fischer, A. (2003). A serious adverse event after successful gene therapy for X-linked severe combined immunodeficiency. N. Engl. J. Med. *348*, 255-256.

Huang, Y., Paxton, W.A., Wolinsky, S.M., Neumann, A. U., Zhang, L., He, T., Kang, S., Ceradini, D., Jin, Z., Yazdanbakhsh, K., et al. (1996). The role of a mutant CCR5 allele in HIV-1 transmission and disease progression. Nat. Med. 2, 1240-1243.

Hurt, J.A., Thibodeau, S.A., Hirsh, A.S., Pabo, C.O., and Joung, J.K. (2003). Highly specific zinc finger proteins obtained by directed domain shuffling and cell-based selection. Proc. Natl. Acad. Sci. U S A *100*, 12271-12276.

Joung, J.K., Ramm, E.I., and Pabo, C.O. (2000). A bacterial two-hybrid selection system for studying protein-DNA and protein-protein interactions. Proc. Natl. Acad. Sci. U S A *97*, 7382-7387.

Kandavelou, K., Mani, M., Durai, S., and Chandrasegaran, S. (2005). "Magic" scissors for genome surgery. Nat. Biotechnol. *23*, 686-687.

Kim, J.S., and Pabo, C.O. (1998). Getting a handhold on DNA: design of poly-zinc finger proteins with femtomolar dissociation constants. Proc. Natl. Acad. Sci. U S A *95*, 2812-2817.

Kim, Y.G., Cha, J., and Chandrasegaran, S. (1996). Hybrid restriction enzymes: zinc finger fusions to Fok I cleavage domain. Proc. Natl. Acad. Sci. U S A *93*, 1156-1160.

Kim, Y.G., and Chandrasegaran, S. (1994). Chimeric restriction endonuclease. Proc. Natl. Acad. Sci. U S A *91*, 883-887.

Kim, Y. G., Smith, J., Durgesha, M., and Chandrasegaran, S. (1998). Chimeric restriction enzyme: Gal4 fusion to FokI cleavage domain. Biol. Chem. *379*, 489-495.

Li, L., Wu, L.P., and Chandrasegaran, S. (1992). Functional domains in Fok I restriction endonuclease. Proc. Natl. Acad. Sci. U S A *89*, 4275-4279.

Li, S., and Huang, L. (2000). Nonviral gene therapy: promises and challenges. Gene. Ther. *7*, 31-34.

Li, S., and Ma, Z. (2001). Nonviral gene therapy. Curr. Gene. Ther. *1*, 201-226.

Liu, Q., Segal, D.J., Ghiara, J.B., and Barbas, C.F., 3rd (1997). Design of polydactyl zinc-finger proteins for unique addressing within complex genomes. Proc. Natl. Acad. Sci. U S A *94*, 5525-5530.

Liu, Q., Xia, Z., Zhong, X., and Case, C.C. (2002). Validated zinc finger protein designs for all 16 GNN DNA triplet targets. J. Biol. Chem. *277*, 3850-3856.

Lloyd, A., Plaisier, C.L., Carroll, D., and Drews, G.N. (2005). Targeted mutagenesis using zinc-finger nucleases in Arabidopsis. Proc. Natl. Acad. Sci. U S A *102*, 2232-2237.

Lombardo, A., Genovese, P., Beausejour, C. M., Colleoni, S., Lee, Y.L., Kim, K.A., Ando, D., Urnov, F.D., Galli, C., Gregory, P.D., et al. (2007). Gene editing in human stem cells using zinc finger nucleases and integrase-defective lentiviral vector delivery. Nat. Biotechnol. *25*, 1298-1306.

Mandell, J.G., and Barbas, C.F., 3rd (2006). Zinc Finger Tools: custom DNA-binding domains for transcription factors and nucleases. Nucleic. Acids. Res. *34*, W516-523.

Mani, M., Kandavelou, K., Dy, F.J., Durai, S., and Chandrasegaran, S. (2005). Design, engineering, and characterization of zinc finger nucleases. Biochem. Biophys. Res. Commun. *335*, 447-457.

Mani, M., Smith, J., Kandavelou, K., Berg, J.M., and Chandrasegaran, S. (2005). Binding of two zinc finger nuclease monomers to two specific sites is required for effective double-strand DNA cleavage. Biochem. Biophys. Res. Commun. *334*, 1191-1197.

Marshall, E. (1999). Gene therapy death prompts review of adenovirus vector. Science *286*, 2244-2245.

Miller, J.C., Holmes, M.C., Wang, J., Guschin, D.Y., Lee, Y.L., Rupniewski, I., Beausejour, C. M., Waite, A.J., Wang, N.S., Kim, K.A., et al. (2007). An improved zinc-finger nuclease architecture for highly specific genome editing. Nat. Biotechnol. *25*, 778-785.

Moehle, E.A., Rock, J.M., Lee, Y.L., Jouvenot, Y., Dekelver, R.C., Gregory, P.D., Urnov, F.D., and Holmes, M.C. (2007). Targeted gene addition into a specified location in the human genome using designed zinc finger nucleases. Proc. Natl. Acad. Sci. U S A *104*, 3055-3060.

Morton, J., Davis, M.W., Jorgensen, E.M., and Carroll, D. (2006). Induction and repair of zinc-finger nuclease-targeted double-strand breaks in Caenorhabditis elegans somatic cells. Proc. Natl. Acad. Sci. U S A *103*, 16370-16375.

Pabo, C.O., Peisach, E., and Grant, R. A. (2001). Design and selection of novel Cys2His2 zinc finger proteins. Annu. Rev. Biochem. *70*, 313-340.

Pavletich, N.P., and Pabo, C.O. (1991). Zinc finger-DNA recognition: crystal structure of a Zif268-DNA complex at 2.1 A. Science *252*, 809-817.

Porteus, M.H. (2006). Mammalian gene targeting with designed zinc finger nucleases. Mol. Ther. *13*, 438-446.

Porteus, M.H., and Baltimore, D. (2003). Chimeric nucleases stimulate gene targeting in human cells. Science *300*, 763.

Porteus, M.H., Connelly, J.P., and Pruett, S.M. (2006). A look to future directions in gene therapy research for monogenic diseases. PLoS Genet. *2*, e133.

Schnell, M.A., Zhang, Y., Tazelaar, J., Gao, G.P., Yu, Q.C., Qian, R., Chen, S.J., Varnavski, A. N., LeClair, C., Raper, S. E., and Wilson, J.M. (2001). Activation of innate immunity in nonhuman primates following intraportal administration of adenoviral vectors. Mol. Ther. *3*, 708-722.

Schroder, A.R., Shinn, P., Chen, H., Berry, C., Ecker, J.R., and Bushman, F. (2002). HIV-1 integration in the human genome favors active genes and local hotspots. Cell *110*, 521-529.

Shi, Y., and Berg, J.M. (1995). A direct comparison of the properties of natural and designed zinc-finger proteins. Chem. Biol. *2*, 83-89.

Smith, J., Berg, J.M., and Chandrasegaran, S. (1999). A detailed study of the substrate specificity of a chimeric restriction enzyme. Nucleic Acids Res. *27*, 674-681.

Smith, J., Bibikova, M., Whitby, F.G., Reddy, A.R., Chandrasegaran, S., and Carroll, D. (2000). Requirements for double-strand cleavage by chimeric restriction enzymes with zinc finger DNA-recognition domains. Nucleic Acids Res. *28*, 3361-3369.

Szczepek, M., Brondani, V., Buchel, J., Serrano, L., Segal, D.J., and Cathomen, T. (2007). Structure-based redesign of the dimerization interface reduces the toxicity of zinc-finger nucleases. Nat. Biotechnol. *25*, 786-793.

Thomas, C.E., Ehrhardt, A., and Kay, M.A. (2003). Progress and problems with the use of viral vectors for gene therapy. Nat. Rev. Genet. *4*, 346-358.

Urnov, F.D., Miller, J.C., Lee, Y. L., Beausejour, C.M., Rock, J.M., Augustus, S., Jamieson, A. C., Porteus, M.H., Gregory, P.D., and Holmes, M.C. (2005). Highly efficient endogenous human gene correction using designed zinc-finger nucleases. Nature *435*, 646-651.

Van Craenenbroeck, K., Vanhoenacker, P., and Haegeman, G. (2000). Episomal vectors for gene expression in mammalian cells. Eur. J. Biochem. *267*, 5665-5678.

Vasquez, K.M., Marburger, K., Intody, Z., and Wilson, J.H. (2001). Manipulating the mammalian genome by homologous recombination. Proc Natl. Acad. Sci. U S A *98*, 8403-8410.

Wilson, J.H. (2003). Pointing fingers at the limiting step in gene targeting. Nat Biotechnol *21*, 759-760.

Wolfe, S.A., Greisman, H.A., Ramm, E.I., and Pabo, C.O. (1999). Analysis of zinc fingers optimized via phage display: evaluating the utility of a recognition code. J. Mol. Biol. *285*, 1917-1934.

Wright, D.A., Thibodeau-Beganny, S., Sander, J.D., Winfrey, R.J., Hirsh, A.S., Eichtinger, M., Fu, F., Porteus, M.H., Dobbs, D., Voytas, D.F., and Joung, J.K. (2006). Standardized reagents and protocols for engineering zinc finger nucleases by modular assembly. Nat. Protoc. *1*, 1637-1652.

Wright, D.A., Townsend, J.A., Winfrey, R.J., Jr., Irwin, P.A., Rajagopal, J., Lonosky, P.M., Hall, B.D., Jondle, M.D., and Voytas, D.F. (2005). High-frequency homologous recombination in plants mediated by zinc-finger nucleases. Plant J. *44*, 693-705.

Wu, J., Kandavelou, K., and Chandrasegaran, S. (2007). Custom-designed zinc finger nucleases: What is next? Cell Mol. Life Sci. *64*, 2933-2944.

Yew, N.S., Wang, K.X., Przybylska, M., Bagley, R.G., Stedman, M., Marshall, J., Scheule, R.K., and Cheng, S.H. (1999). Contribution of plasmid DNA to inflammation in the lung after administration of cationic lipid:pDNA complexes. Hum. Gene. Ther. *10*, 223-234.

Yew, N.S., Wysokenski, D.M., Wang, K.X., Ziegler, R.J., Marshall, J., McNeilly, D., Cherry, M., Osburn, W., and Cheng, S.H. (1997). Optimization of plasmid vectors for high-level expression in lung epithelial cells. Hum. Gene. Ther. *8*, 575-584.

Chapter 8

Plasmid DNA as Prophylactic and Therapeutic Vaccines for Cancer and Infectious Diseases

Devin B. Lowe, Michael H. Shearer, Cynthia A. Jumper, En-Min Zhou and Ronald C. Kennedy

ABSTRACT

Close to two decades of preclinical studies have validated the concept of plasmid DNA as a protective vaccine strategy for cancer and infectious diseases. However, the crossover application into human studies has been met with poor results based on the DNA vector's inability to provide clinically relevant prophylactic and therapeutic benefit. The overall efficacy of plasmid DNA immunization will no doubt rely on a number of strategies that attempt to increase the vector's immunogenicity while also correcting for factors involved in the specific activation of immune effector cells. This chapter will focus on the difficulties surrounding the use of a DNA vaccine in patients suffering from chronic or infectious diseases with particular interest in resolving the roadblocks for future vaccine success in human cancer cases.

INTRODUCTION

Plasmid DNA represents a highly beneficial vaccine approach to disease that is not easily achieved with other strategies. The DNA vector can be easily engineered and produced to encode a transgene that will express a foreign protein on a large scale in bacterial expression systems (Fig. 1). This transgene can encode a tumor or infectious agent protein that will be expressed as the vaccine. Thus use of the DNA vector and bacterial expression systems extends the benefits of a quick and inexpensive process. Plasmid DNA has also demonstrated its safety in animal and human studies since repeated injections of a DNA construct is well tolerated and neutralizing immune responses to the vaccine are rarely observed. On the contrary, the issue of repeated injections severely limits the applicability of viral vectors based strategies such as adenovirus and vaccinia due to preexisting or the generation of active immunity many individuals have to the viral coat proteins.

A particularly attractive trait for plasmid DNA immunization is that it can elicit broad-based immunity (Weiner and Kennedy, 1999). That is, both antibody and cell mediated immunologic reactions can result from vaccinating with the DNA

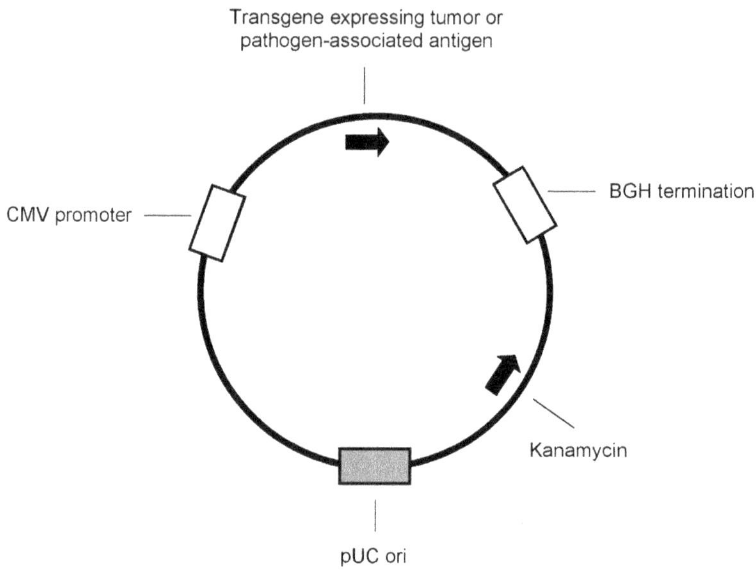

Fig. 1. Representative human clinical trial DNA vector. The plasmid backbone is based on the pcDNA3.1 vector. The encoded transgene of interest is under the control of the CMV promoter. pUC ori denotes the origin of replication while BGH termination represents the polyadenylation site from bovine growth hormone. Antibiotic resistance is conferred to the plasmid by the kanamycin gene. Abbreviations: CMV, cytomegalovirus; BGH, bovine growth hormone.

vector. Recombinant protein and peptide-based strategies serve to skew the response towards one direction. Activating both arms of the immune system has been found to be important in alleviating the progression of certain types of malignancies as well as infectious diseases – specifically in the case of intracellular pathogens. Although it is not completely understood how the DNA vector accomplishes broad based immunity, two models have been suggested. In the first scenario, delivery of the DNA vector to the skin or muscle causes the surrounding cells to become transfected. Antigen presenting cells (APCs) become transfected directly and begin to express the DNA vector's encoded transgenic components. Peptides of the endogenously expressed protein are then presented by the APC through the major histocompatibility complex (MHC) class I receptor, and after the APC travels to the nearest lymph node, CD8+ T cells become activated to the MHC class I displayed peptides. In the second scenario, the APC acquires the DNA vaccine encoded transgenic protein indirectly. Soon after skin or muscle cells become transfected by plasmid DNA, the expressed protein is released by these cells and taken up exogenously by APCs. The uptake of exogenous protein by the APC enters into the classical MHC class II presentation pathway where CD4+ T cells later become activated. However, the APC can also present the protein peptides through MHC class I receptors in a process termed cross-presentation (Jensen, 2007). Therefore, CD8+ T cells also become activated to the processing of an exogenous protein expressed by the DNA vaccine.

The DNA vector has proven fairly effective at providing long lasting protection in rodent models of cancer and infectious diseases. Indeed, the vector's ability to initiate a robust humoral and cell-mediated immune response in mice first provided great optimism for the translational application in humans. This early hope has since waned given the less-than-anticipated results in phase I clinical trials testing the immunogenicity of the vaccine. To push for the success of the next generation of plasmid DNA use, these hurdles faced in the human setting must be overcome by optimizing the conditions needed for successful immune activation to the vector's encoded targets. The aim of this chapter will be to further explore the developments related to using plasmid DNA to vaccinate individuals against cancer and infectious pathogens. Particular interest will be devoted to the achievements and challenges facing the DNA vector in providing immunologic protection against human malignancies.

DNA VACCINES AND MALIGNANT DISEASE

Cancer continues to be an increasing medical issue for the United States, second only to heart disease. According to the American Cancer Society, more than 1.4 million individuals will be diagnosed with some form of malignancy in 2007 while close to 560,000 people will die as a result of the disease (American Cancer Society, 2007). These numbers are problematic, considering that since the "War on Cancer" was declared by President Nixon in 1971, the spread and recurrence of progressive types of cancer have largely been uncontained by the gold standards of care that include surgery, chemotherapy, and radiation. Certainly many types of neoplasms such as those being confined to the breast and prostate have seen increasing prognosis over the years, but these successes can be attributed to the stage of the tumor upon diagnosis and preventative care (*e.g.*, diet, tobacco use, sun exposure) and annual screening tests.

It, therefore, remains imperative that newer strategies be developed and implemented to improve upon existing therapies. Immunotherapy remains one such approach that is capable of harnessing the immune system in some fashion to prevent tumorigenic growth. Unlike surgery, chemotherapy, and radiation, immunotherapeutic strategies attempt to specifically target cancer cells for destruction not only to increase the efficacy of treatment but to also prevent systemic toxicities and overall decreased quality of life.

HISTORICAL PERSPECTIVE OF IMMUNOLOGIC-BASED TUMOR THERAPIES

The development and preventative success of modern day vaccines is firmly rooted in the historical prevention of infectious diseases. Around the early 1700s, the Greek physician, Giacomo Pylarini, first introduced England to the concept of preventing smallpox infection by immunizing individuals with puss taken from the lesions of smallpox patients (a process referred to as variolation). However, the procedure of variolation remained controversial throughout this time period and was avoided by many since its side effects were largely known to include illness and death. It was not until the late 18th century that the method of variolation to prevent human smallpox was improved upon and the age of vaccination ushered in by Jenner's cowpox preparation. The chronology of important vaccine breakthroughs continued with Pasteur's attenuated rabies vaccine in the 1880s, which was soon followed by vaccines for yellow fever in the 1930s and poliovirus in the 1950s. The cumulative work since Jenner's time has resulted in bacterial and viral vaccines that are commonly used for childhood immunizations to prevent pathogens that at one time caused great morbidity and mortality.

Vaccinating against tumorigenic growth is also not a recent scientific development. The methods for inducing an immune response to thwart malignancies can be reliably traced back to the 1700s when patients were deliberately infected with live bacteria in order to achieve tumor regression (Starnes, 1992). These methods followed from the apparent observation that infection could resolve a tumor mass. By the 1800s, a New York surgeon known as William Coley extended these findings and began injecting cancer patients with extracts of killed bacteria (a preparation referred to as "Coley's toxins") so that he could simulate infection in order to induce a fever. Coley's work in this area, particularly in those with inoperable sarcomas, appeared to hold therapeutic benefit for many of his patients and was commonly used in cancer trials by other investigators until the use of the toxins fell out of favor with the medical community by as late as the 1930s (Coley, 1991). It is now realized that Coley's toxins most likely served to induce an innate immune response that prompted non-specific killing of tumor cells.

Moving forward to the modern day, the development of the hybridoma technology in 1975 made it possible to produce monoclonal antibodies for specific therapeutic benefit. Throughout the 1980s and 1990s, work of this kind transcended into clinical trials that tested the efficacy of monoclonal antibodies in treating human tumors. The first of these trials employing hybridoma technology involved the antibody targeting CD20 (*i.e.*, Rituximab [Rituxan]), expressed by B-cell lymphomas (Maloney et al., 1994; McLaughlin et al., 1998), which resulted in its license by the FDA in 1997. The approval of the monoclonal antibody, trastuxumab (Herceptin), used to treat HER-2/neu overexpressing metastatic breast cancers soon followed in 1998. As of this writing, seven additional monoclonal antibody sets have been designed and licensed by the FDA to react with antigenic targets expressed by certain cancer types including acute myeloid leukemia, B-cell chronic lymphocytic leukemia, metastatic colorectal cancer, and head and neck cancer (Carter, 2006). This form of passive immunotherapy with antibody molecules remains the only FDA approved strategies available to cancer patients.

Beginning in the 1980s also, much focus of cancer immunotherapy in human clinical trials centered upon delivering IL-2 activated innate immune effector cells (*i.e.*, lymphokine-activated killer [LAK] cells) to the cancer patient (Yannelli and Wroblewski, 2004). This task eventually proved too burdensome as a high number of LAK cells were required for transfusion purposes and the doses of IL-2 used presented toxicity problems for many patients. The focus of immunotherapy then shifted to the adoptive transfer of tumor infiltrating lymphocytes (TILs) harvested from malignant lesions and cultured *ex vivo*. TILs were found to be more potent at inducing tumor destruction than LAK cells. However, adoptive therapy of TILs was most successful in achieving objective clinical responses in melanoma patients than with other cancer types arising from the breast, colon, and lung, where non-reactive or non-functional TILs were most often obtained.

Altogether, these early adoptive transfer attempts were beset with issues that ranged from the expense and complexities of large-scale culture of immune effector cells to applying the therapy in a universal versus personalized manner to a range of tumors (Yannelli and Wroblewski, 2004). The discovery of the first tumor associated antigen, MAGE-A1, recognized by a melanoma cytotoxic T lymphocyte (CTL) clone reported in 1991, represented a shift in the field of tumor immunotherapy (van der Bruggen et al., 1991; Finn, 2007). The universal approach to cancer treatment could now become realizable if tumor associated antigens were found to be expressed in a sizeable amount of patients with a specific malignancy. Also, rather than delivering a heterogeneous suspension of TILs and avoiding *ex vivo* culture, tumor reactive CTLs could be activated *in vivo* to a specific antigenic tumor target through a vaccine construct such as plasmid DNA. After the MAGE-A1 report, the discovery of other tumor-expressed antigens recognized by T lymphocytes followed, and the current effort has resulted in a long list of cancer target antigens many of which have been incorporated into vaccine-based methods in order to induce an antigen-specific response in the cancer patient (Novellino et al., 2005).

Tumor antigens can be collectively assigned to four categories. The first group is the tumor specific antigen and is represented by common antigens such as CDK-4 and β-catenin. Tumor specific antigens are typically found only in malignancies since the alteration of the antigen leads the cell to a transformed state (Viral oncoproteins are further distinguished by this class). A second group is the cancer/testis antigen. MAGE-A1 and NY-ESO-1 are defined as such since their normal expression is confined to germline tissues, an immunologically privileged site. Overexpression antigens form the third class and is represented by MUC1, HER-2/*neu*, and PSMA. These tumor self antigens are overexpressed on a variety of tissues in comparison to their expression levels on normal cells. This defining property is in contrast to the differentiation group of tumor antigens, which is confined to tumor and normal cells of a specific lineage. For example, Gp100, MART-1/Melan-A, and tyrosinase are expressed by melanomas and melanocytes. CEA, on the other hand, is expressed only in normal cells and cancer types involving the colon.

PLASMID DNA CLINICAL TRIAL APPLICATIONS

A handful of clinical trials have explored the efficacy of DNA vaccines in cancer patients. Of these trials, the types of malignancies under study include melanoma (Rosenberg et al., 2003; Tagawa et al., 2003; Smith et al., 2005; Triozzi et al., 2005a; Cassaday et al., 2007), prostate carcinoma (Pavlenko et al., 2004; Todorova et al., 2005), where a variety of tumor antigens served as targets, B cell lymphoma (Timmerman et al., 2002), where the target antigen is the idiotype associated with the surface expressed B cell receptor, and colorectal cancer (Conry et al., 2002), where CEA was the target antigen (Table 1). The patient population typically consists of individuals who are immunosuppressed as a result of being in late stage disease and having received prior therapies, including surgery, radiation, and chemotherapy. The safety profile of the administered DNA vector is generally well tolerated with common adverse events related to the site of injection (*e.g.*, pain, swelling, redness) that are unrelated to the dosage. However, for the majority of patients in these clinical trials, the plasmid DNA approach is unable to induce an evident vaccine-based immune response and the regression of disease is not achieved (Table 1).

It is worth noting that of the available reported studies, there do remain a few trials particular to prostate cancer that remark on the ability of at least half of the patient population to mount an immune response to vaccination. In a Phase I study, participants with hormone-refractory prostate cancer were vaccinated with a DNA vector encoding PSA alongside injections with recombinant GM-CSF and IL-2. In the cohort given the highest DNA dose after five administrations, 2 of 3 patients responded with a PSA-specific IgG response and T cell-mediated IFN-γ secretion (Pavlenko et al., 2004). A separate Phase I/II study also reported a significant anti-PSMA antibody response in 85.7% of patients as a result of vaccination when assessed by Western blot (Todorova et al., 2005). In this study, individuals diagnosed with prostate cancer received alternating injections of soluble GM-CSF with plasmid DNA encoding PSMA and B7-2 and along with an adenoviral vector expressing PSMA utilizing a prime-boost strategy (see prime-boost strategy section). Yet, considering that the above examples provide basis for plasmid DNA to initiate a targeted immune response in some scenarios, the extent to which disease is alleviated from vaccination cannot be clarified without further study.

To date, the case for the prophylactic and therapeutic success of plasmid DNA is best made through the beneficial results reported from a variety of mouse cancer models. Specific results obtained from these animal studies have been recently reported (Lowe et al., 2006; Yu and Finn, 2006). The overall breakdown in the translational

Table 1. Summary of plasmid DNA use in human cancer clinical trials

Malignancy type	Study treatment	Expressed Tumor Antigen	Vaccine specific response[a]	Objective response[b]	Reference
B-cell lymphoma	Plasmid/antigen + plasmid/GM-CSF	Id	1 of 12 (0.08%)	1 of 12 (0.08%)	Timmerman et al., 2002
Colorectal	Plasmid/antigen	CEA	4 or 17 (23.5%)	0 of 17 (0%)	Conry et al., 2002
Melanoma	Plasmid/antigen	Gp100 Tyrosinase MART-1/Melan-A	0 of 13 (0%) 11 of 26 (42.3%) 0 of 12 (0%)	1 of 22 (4.5%) 0 of 26 (0%) ND	Rosenberg et al., 2003 Tagawa et al., 2003 Triozzi et al., 2005a
	Plasmid/antigen –/- plasmid/GM-CSF	Gp100	0 of 12 (0%)	0 of 12 (0%)	Cassaday et al., 2007
	Plasmid/antigen + vaccinia/antigen	MAGE, MART-1/Melan-A, Tyrosinase, NY-ESO	2 of 7 (28.6%)	ND	Smith et al., 2005
Prostate	Plasmid/antigen + GM-CSF + IL-2	PSA	2 of 3 (66.7%)	ND	Pavlenko et al., 2004
	Plasmid/antigen/B7-2 + adenovirus/antigen + GM-CSF	PSMA	36 of 42 (85.7%)	ND	Todorova et al., 2005

[a] The vaccine specific response does not necessarily correlate with objective response.
[b] Objective response defined as either partial or complete response of the tumor lesion.
Abbreviations: Id, idiotype; ND, not determined

efficacy between murine studies and the clinical response in humans may simply represent a size-to-scale issue. The immunization strategies and plasmid DNA dosages attempted in rodents may not accurately parallel the constraints needed to achieve a human response. A typical human study might employ separate patient cohorts with differing dose levels ranging from 0.2 to 1.0 mg DNA per injection (doses do not usually exceed 2.0 mg). It would, therefore, be appropriate for pre-clinical studies to also be conducted to some degree in non-human primates to further fine tune cancer immunotherapeutic strategies. Although this type of work is largely avoided based on availability and cost, study in animals that are closely related to humans will obviously mimic more accurately the human situation to an immunologic-based vaccine (Kennedy et al., 1997).

The jump from mouse to man is further complicated by animal studies that involve healthy immunocompetent mice inoculated with a malignant cell line to analyze the constraints of an immunologic-based approach. Instead, the large majority of human studies consist of individuals who are highly immunosuppressed with terminal tumors. Certainly the chronic disease characteristics of human cancer would reduce the efficacy of any type of active immunotherapy, but the chronic conditions of cancer have also made it a difficult task to mimic and study in animal models. No doubt, the long-term progression of metastatic disease has proven one great challenge to plasmid DNA immunization therapies.

COMPLEXITIES OF THE TUMOR MICROENVIRONMENT

The clinical approach to vaccinating against malignant growth using plasmid DNA has not only failed in initiating a robust immune response but also in combining strategies to handle the complexities of the tumor microenvironment. It is now generally regarded based on an accumulating body of evidence that tumor cells of various types have the intrinsic ability to evade the immune response and, thus, reduce the efficacy of a particular vaccine strategy. These suppressive mechanisms could be imagined as multiple layers that surround the tumor cell and collectively serve to prevent immune destruction. Tumor-associated mechanisms of immune escape have been previously discussed at length by our group (Lowe et al., 2007a). In order to push for success, cancer immunotherapy will have to involve a multi-pronged approach that not only takes into account the tumor environment at the level of immune attack but also in enhancing the immunogenicity of a DNA vaccine approach as discussed at length in the "improving upon dna vaccine design" section. We briefly summarize below the conceptual vaccine-based difficulties related to the tumor microenvironment.

IMMUNE ACTIVATION AND RESISTANCE TO KILLING

For the greater cancer types, the path to becoming tumorigenic arises from factors that do not impart an obvious immunogenic phenotype to the cell in terms of the host's central tolerance mechanisms. Though the idea of self does not technically qualify as an evasion mechanism employed by neoplasms, vaccinating against tumor-associated self proteins represents a serious obstacle. Not only must the vaccine be able to break self tolerance to generate an appropriate immune response, but the ensuing immune response must not initiate a therapeutically uncontrollable autoimmune reaction.

A tumor cell by definition results from genetic abnormalities obtained either intrinsically or extrinsically (as in the cause of viral infection or carcinogen exposure) (Hanahan and Weinberg, 2000). As such, the tumor cell phenotype can differ from the normal counterpart in a way to favor its continued proliferation and resistance to the checks and balances within the body. An example of this type involves mutational or gene regulation effects that perturb the normal expression of the MHC molecule (designated human leukocyte antigen [HLA] in humans) that displays peptides to activate T cells (Algarra et al., 2004). The malignant cell might also have reduced expression of a tumor associated target, expressed by the MHC/HLA, or complete loss of the peptide from the surface of the cell (Beatty and Paterson, 2000; Sanchez-Perez et al., 2005). Additionally, the amino acid sequence of the antigenic target may be genetically modified in a way to differ from the original peptide (a process referred to as antigenic drift) (Bai et al., 2003).

Tumor cells could also express or shed ligands that bind specific receptors expressed by immune effector cells (Groh et al., 2002; Salih et al., 2002; Danial and Korsmeyer, 2004; Kim et al., 2006). The effects of these inhibitory molecules serve to subvert the killing mechanisms of immune cells that seek out and destroy abnormal cell types such as malignancies.

IMMUNE SUPPRESSION

Established cancers tend to be sustained in a chronically inflamed niche that is maintained by tumor-derived inflammatory substances that include cytokines and reactive oxygen species (Kim et al., 2006). This environment not only confers an advantage to tumor expansion by, for example, stimulating angiogenesis, but the inflamed state is also a highly suppressive environment that prevents the proper activation and effector properties of immune cells.

Many of the soluble factors secreted by malignant cells include VEGF, TGF-β, and IL-10. High levels of these compounds tend to negatively affect the maturation, differentiation, and activity status of APCs such as dendritic cells (DCs) (Zou, 2005). These immunosuppressive DCs can ultimately suppress T cell activation and activate suppressor cells such as T regulatory cells that further downregulate the role of the adaptive immune response.

Although VEGF, TGF-β, and IL-10 represent a collection of the inhibitory compounds tumors secrete, the additive effects of malignant growth ultimately institute immunosuppressive networks that shield the tumor from the actions of immune effector cells. Therefore, it becomes a difficult challenge not only to induce an immune response to non-immunogenic or evasive tumor cells but also to promote an attack on established cancers in light of a highly immunosuppressive environment.

IMMUNOEDITING

The cancer immunoediting hypothesis provides a final collective outlook to the problems associated with tailoring vaccine therapies against malignant disease. Immunoediting is a process described as the selection of tumor variants that are non-responsive to host control (Dunn et al., 2002). Much like Darwinian natural selection, the eventual outgrowth of these types of tumor cells arise as a result of escaping the constraints of the protection mechanisms provided by both the innate and adaptive arms of the immune system. So, with regard to the escape pathways of tumor cells, an active immune response would eventually sculpt a rather impervious non-immunogenic malignant mass.

The concept of immunoediting is particularly unnerving considering that a narrowed vaccine approach to activating the immune response would most likely provide little long term benefit to the cancer patient. That is, vaccinating against one target would eventually select for the outgrowth of a cancer variant that had escaped the vaccine approach through some means. This idea certainly holds true in the case of pursuing antigenic targets that are associated with malignancies. Given that the tumor microenvironment consists of a heterogeneous collection of cells that express antigenic targets differently (or not at all), a targeted immunotherapeutic approach would be effective against many but not all cells. Following this immunoediting logic, a vaccine would have to comprise a collection of tumor associated targets in order to begin to attack the malignancy at various points and reduce the chances of negative selection. However, the construction of a multi-component vaccine would be problematic if multiple targets are unknown for a patient's particular cancer type. The issue of epitope dominance is also a concern in vaccines that encode multiple tumor antigens, and an immunization strategy would have to be developed to ensure that a successful protective polyclonal immune response is initiated.

DNA VACCINES AND INFECTIOUS DISEASES

The majority of modern vaccines serve to prevent infectious diseases through humoral immunity. This approach has led to a successful childhood vaccine campaign that has reduced the incidence of many bacterial and viral pathogens in the United States (Lowe et al., 2007b). Unfortunately, neutralizing antibodies do not effectively combat all types of communicable infections. It is generally regarded that cell-mediated immunity (along with antibody) is needed in cases of intracellular infection to prevent the occurrence and spread of the infection resulting in disease. Many of the problematic intracellular infectious agents such as

HIV-1, *Mycobacterium tuberculosis*, and malaria have proven particularly difficult to control even in the presence of an antibody response (Seder and Hill, 2000). Live-attenuated strategies can induce a CTL response since they simulate natural infection, but there are significant health risks that can result in immunocompromised individuals vaccinated or in cases were the attenuated strain reverts to a pathogenic form. Therefore, plasmid DNA represents a much safer alternative since the vector also elicits broad-based immunity.

Plasmid DNA remains an attractive vaccine approach for the problematic endemic pathogens and emerging infectious diseases that have potential to spread worldwide as well (Lowe et al., 2006). In addition to broad-based immunity, the DNA vector can be quickly engineered with multiple antigenic targets and inexpensively produced on a large scale in comparison to the standard vaccine methods (*i.e.* subunit, killed, live-attenuated). The ability to rapidly inoculate a population with a specific DNA vaccine would certainly aid in the prevention of a pandemic outbreak. For example, influenza infection represents a serious worldwide concern not only in the susceptibility risks associated to the world population but also since the current vaccine methods to grow the virus in chicken eggs and test and distribute the vaccine approximate 1 year (Laddy and Weiner, 2006).

The specific details of DNA vaccine trials as it relates to particular infectious diseases exceed the focus of this chapter. We would, instead, refer the reader to several reviews dedicated to discussing the progress of DNA vaccines as it relates to HIV-1 (Giri et al., 2004) and malaria (Gilbert et al., 2006) where human trials are ongoing (Table 2). Yet, the problems associated with plasmid DNA in human trials involving communicable pathogens closely parallel the work being performed with human cancer. That is, the DNA vector normally suffers from a lack to initiate and maintain an effective CTL and antibody response in the recipient. No doubt, these issues will be better resolved as work in this area progresses to determine the optimal strategies that enhance the immunogenicity of plasmid DNA in humans.

IMPROVING UPON DNA VACCINE DESIGN

Early animal models studying the effects of influenza (Ulmer et al., 1993) and HIV-1 (Wang et al., 1993) infections first demonstrated the potential promise for DNA vaccines to deliver protective immunity in situations of human disease. These early studies along with an accumulating body of evidence for other animal protection models (including cancer) helped propel the first human clinical trail to monitor the safety and efficacy of a DNA vaccine against HIV-1 (MacGregor et al., 1998).

In particular to cancer treatment, the early optimism surrounding plasmid DNA use as a prophylactic and therapeutic tool against human disease has since diminished with the less-than-anticipated results of human clinical trials. However, the issues of low immunogenicity of the vector and lack of efficacy in terms of cancer remission

Table 2. Overview of plasmid DNA application in selected infectious disease trials			
Infectious disease	**Human trials**	**Vaccine specific response**	**Reference**
HIV-1	Phase I/II	Modest antibody/CTL increases	Giri, et al. 2004
Malaria	Phase I/II	Enhanced antibody/ T cell responses	Gilbert et al., 2006

in patient populations cannot be simply thrown out as a failure of the DNA vector strategy as a whole. Instead, the shortfalls of immunotherapies that include plasmid DNA use may instead be the result of compounding issues that have not been fully addressed in the clinical setting. These problems are quite numerous but can essentially be described in terms of immunologic activation from plasmid DNA vaccination as well as the immunologic complexities of the tumor microenvironment previously mentioned.

ENHANCING THE IMMUNE RESPONSE

Certainly a major obstacle with the use of plasmid DNA involves a successful vector that can achieve cellular uptake and eventual localization to the nucleus for transgene expression. Yet considering successful transfection of the vector into a target cell, plasmid DNA vaccination is further confounded by the position that the DNA vector will be lost after successive rounds of cell division without some way of maintaining long-term nuclear persistence in the target cell either through chromosomal integration or as extra-chromosomal units (*i.e.*, episomes) that can replicate autonomously. On the other hand, slow-dividing cells or non-dividing cells can maintain the DNA vector for long periods of time with sustained transgene expression. This was first experimentally shown by Wolff and colleagues with the direct injection of plasmid DNA into mouse skeletal muscle cells (Wolff et al., 1990). Therefore, the division rate and maintenance of the vector in the intended target cell will determine the construction of a safe optimal DNA vector to achieve nuclear persistence and replication with cell division. Additionally, there are clear enhancements over the direct injection method in terms of transgene expression and are represented by techniques that include electroporation (Quaglino et al., 2004) and gene gun delivery (Fynan et al., 1993). These methods and others that attempt to circumvent the issues of vector administration, replication, and nuclear persistence are described at greater length in the "Plasmid-based Expression Systems for Mammalian Cells" and "Plasmids for Gene Therapy" chapters.

The immune response to DNA vaccination might be better fine-tuned to the site and type of response necessary to target the replication site of the infectious agent.

Intranasal and oral delivery of plasmid DNA represents an alternative to the systemic delivery system (*i.e.*, subcutaneous and intramuscular injections) and is achieved through vector carrier methods that include attenuated *Salmonella typhimurium*, liposomes, or microparticles. These mucosa-based injection routes serve to prime mucosal-associated lymphoid tissues (MALT) in order to abrogate the replication and dissemination of the organism in the mucosa. Aside from inducing an immune response at the site of replication, an obvious advantage to MALT priming is that systemic CTL activation can result (Berzofsky et al., 2001; Eo et al., 2001). This would also be advantageous in cancer where targeting the immune response at the site of cancer growth, along with systemic CTL activation, would prevent metastasis to other sites in the body. An additional advantage to mucosal vaccination is that plasmid DNA delivery to mucosal sites can be effective in breaking tolerance to tumor self antigens and eliciting protection to several cancer types (Xiang et al., 2000; Xiang et al., 2001; Niethammer et al., 2002). However, on the whole, mucosal-based activation by plasmid DNA is only scarcely represented in the literature of animal cancer models and has not been extended to study in the clinic. Much of the work instead surrounding MALT activation with plasmid DNA involves models of infectious diseases since many pathogens first initiate infections through gastrointestinal, vaginal, and nasal tracts. But, mucosal-based delivery of DNA vectors are hampered with problems related to inducing a potent and sustained neutralizing antibody and/or CTL response to clear an infection (Eriksson and Holmgren, 2002; Toka et al., 2004).

The inclusion of adjuvants to plasmid DNA design represents one such opportunity to increase the immunogenicity of the DNA vector. Adjuvants in this setting can be collectively characterized by work pertaining to CpG oligodeoxynucleotides (ODNs), cytokines, chemokines, and costimulatory molecules and be delivered concurrently with the DNA vaccine either encoded within the vector or in native form. An improvement on the DNA vaccine design can also include the alternate use of vaccine constructs to enhance the response, represented by prime-boost methods.

CpG ODNs

Plasmid DNA that is derived from bacterial expression systems contains natural non-methylated DNA sequences, termed CpG ODNs. These sequences have affinity for Toll-like receptor-9, which is an intracellular protein expressed by APCs and in humans this includes B cells and DCs. Mice have additional cell types that express Toll-like receptor-9, making the translational jump between murine models and humans again difficult. Upon CpG ODN binding Toll-like receptor-9, APCs secrete cytokines and chemokines that enhance presentation of antigen to downstream effector cells such as CTLs. Several distinct classes of CpG ODNs have been determined that differentially effect APC types and, thus, effect downstream immune mechanisms (Krieg, 2006). It would, therefore, prove advantageous to combine an immunotherapeutic strategy involving plasmid DNA with the optimal CpG ODN sequence type that stimulates the necessary immune effector pathways needed to resolve a targeted disease.

The use of CpG ODNs to successfully alleviate tumor burden and infectious diseases has been reported in many animal model systems. Several human trials have also extended this preclinical work to evaluate the efficacy of enhancing the vaccine response using certain classes of CpG ODNs. The safety profile of CpG ODN use in Phase I clinical trials has shown general tolerance without severe complications or clinical signs of autoimmunity (Krieg et al., 2004; Vicari et al., 2007). Although, to date, no reports have described the additive effect of CpG ODNs and plasmid DNA in humans, several vaccine combination studies, particularly Hepatitis B immunization regimes (Halperin et al., 2003; Cooper et al., 2004; Cooper et al., 2005), provide convincing data that CpG ODNs enhance a protective vaccine-induced immune response. In the Phase I study reported by Spieser and colleagues, the characteristics of CD8+ T cells were assessed in metastatic melanoma patients immunized with a MART-1/Melan-A peptide emulsified with a CpG ODN and Incomplete Freund's adjuvant (Speiser et al., 2005). After only two vaccinations, 6 of 8 CpG ODN-treated patients elicited a greater than 2-fold increase in MART-1/Melan-A reactive CD8+ T cells while no control patients responded at this level. After four vaccinations, CpG ODN patients reached a mean of 43-fold higher MART-1/Melan-A specific CD8+ T cell frequencies than before vaccination (8 of 8 responding), whereas patients vaccinated without CpG ODNs reached a mean of 1.9-fold higher frequencies than before vaccination (4 of 8 responding). Several ongoing Phase I and II clinical trials are in place to test the efficacy of CpG ODNs in combination with other peptide and protein-based vaccines that target melanoma and breast cancers (Krieg, 2007). In the coming years, the results from these studies should provide some idea of CpG ODN applicability towards reducing tumor burden. However, considering the complexities of the tumor microenvironment, the inclusion of adjuvants such as CpG ODNs would provide one of many needed tools to achieve cancer remission.

CYTOKINES AND CHEMOKINES

Cytokines and chemokines represent a further adjuvant approach that can be delivered alongside DNA vaccination. As was described for the importance of Toll-like receptor-9 in immune activation, the status of the environment at the level of the APC determines the degree and type of immune activation. So in cases where cytokines such as IL-4 and IL-10 are dominantly secreted, B cell activation is favored. Pro-inflammatory cytokines on the other hand, represented by TNF-α, IL-12, and IFN-γ influence CTL activity among other effects. The potency of influencing the activation status of an immune effector cell type based on cytokine secretion bias can be appreciated in examples where an infection is resolved by skewing the immune response. One such classical example of the effects of the cytokine balance involves infection with the intracellular bacterium, *Mycobacterium leprae*. If a cytokine dominance favoring B cells existed during infection, growth of the organism would stand unchecked and progress to the lepromatous form of the disease, which is characterized by nodules and skin disfigurations (Lowe et al., 2007b). In contrast, a dominant response favoring CTL activation initiates macrophage function that severely limits bacterial spread and results in the tuberculoid form of leprosy that is characterized by hypopigmented skin macules. Influencing the degree of the immune response has also been realized in cases of malignant disease where skewing the cytokine response in one (Winter et al., 2003; Shah et al., 2005) or both (Lopez et al., 2005; Lowe et al., 2005) directions provides needed results to achieve immunity to specific cancer types. Cytokines and chemokines can be further used in combination to modify the migration and activation status of APCs. GM-CSF is a cytokine molecule that has been commonly employed in many vaccine-based clinical trials to help recruit and activate DCs (Hege et al., 2006; Parmiani et al., 2007). Chemokines also favor recruitment of specific effector populations and increase the antigen presentation properties of APCs (Gajewski, 2007). Delivery of these compounds at the site of plasmid DNA immunization represent potential strategies to elicit T cell homing to and destruction of tumor complexes. Much evidence points to an increased infiltration of T cells and tumor cell destruction with certain neoplasm targets expressing, for example, the CC chemokines, RANTES (Mule et al., 1996; Lavergne et al., 2004) and MIP-1α (Crittenden et al., 2003; Gough et al., 2005). Additional chemokine sets such as the CXC chemokines (MIG and IP-10) have also shown similar effects when tumors were modified to express the molecules (Tominaga et al., 2007). Therefore, modulating the microenvironment at the site of plasmid DNA immunization with cytokines and chemokines not only favors the activation status and antigen presentation of APCs, but these biological adjuvants also serve to recruit effector immune cells such as CTLs for antigen priming and proliferation.

COSTIMULATION

The immune response at the level of T cell activation could be further strengthened with the inclusion of vaccine-encoded costimulatory molecules. Proper T cell interaction with the APC requires the correct peptide:MHC complex, CD4 or CD8 coreceptor, and costimulatory receptors in order to induce successful activation. Multiple costimulation molecules exist that vary in cell-type expression and function, including negative regulation, but the two general T cell activating receptors expressed by activated APCs include B7-1 and B7-2 and bind the constitutively expressed CD28 receptor on T cells.

One approach to modifying the constraints of T cell activation at the level of costimulation involves modifying the infected or tumor cell to initiate priming. For instance, the tumor cell can be resected from the patient, transfected with a DNA vaccine encoding a costimulatory receptor such as the B7 receptor, and reinfused back into the individual. In this sense, the tumor cell is functioning as an APC to directly activate T cells to both known and unknown tumor-associated peptides. Townsend and Allison were among the first to experimentally demonstrate that melanoma cells induced to express the B7 ligand could directly activate CD8+ T cells and potentiate their cytolytic functions without help from CD4+ T cells (Townsend and Allison, 1993). Many other groups have extended upon these first observations and demonstrated similar success with modifying a tumorigenic line with costimulatory ligands to push for the direct priming of T cells and tumor immunity (Yang et al., 1995; Jenkinson et al., 2005; Collins et al., 2006). A similar potential exists for treatment indications with chronically infected individuals.

Several clinical trials have explored supplying cancer patients with autologous or allogeneic tumor cells modified with costimulatory ligands. Autologous transfer represents the ideal situation for a whole cell vaccine since an immune response can be directed to both known and unknown tumor targets. However, the approach has several drawbacks that diminish its feasibility for large scale production and administration. The most important of these technical challenges includes resecting a sufficient number of viable and pure tumor cells for repeated injections (Parney et al., 2006). In the study by Antonia and colleagues, these technical difficulties appeared to have been overcome. Renal cell carcinomas were resected from 15 patients, adapted to tissue culture conditions, and transduced with the B7-1 gene using an adenovirus vector. Each patient then received an autologous subcutaneous injection of the irradiated modified tumor cell along with recombinant IL-2 treatment. Peripheral blood mononuclear cells were further obtained from each patient at time points throughout the therapy, stimulated *in vitro* with autologous tumor cells and IL-2, and analyzed by ELISpot for IFN-γ secretion. Out of 15 patients, 4 individuals were concluded to have increased levels of IFN-γ secretion three weeks post treatment. The overall response rate of 26.7% consisted of 2 patients with a partial response and 2 with stable disease. However, it was not made clear by the authors if the IFN-γ ELISpot data correlated to the partial response or stable disease characteristics of their study population. On the other hand, allogeneic transfer of a stable tumor line requires some knowledge of antigen expression on the patient's lesion (or assumption of shared antigens with the allogeneic vaccine) as well as that patient being matched to MHC antigens on the cell line. It is also important to be able to distinguish between an anti-tumor response and allogeneic response since the latter immune reaction can provide misleading results and also potentially render the tumor vaccine useless (Yannelli and Wroblewski, 2004). In one example using this approach, individuals diagnosed with non-small-cell lung cancer were immunized with an irradiated allogeneic vaccine transfected with the B7-1 ligand (Raez et al., 2004). Approximately 31.6% of study participants were classified with stable disease or a partial response, and anti-tumor CTL activity could be measured in the majority of patients after three immunizations. But, again, ELISpot results did not correlate to tumor response status. A similar reported approach used a B7-1 modified allogeneic breast cancer line that also expressed other costimulatory ligands (Dols et al., 2003). Patients with stage IV breast cancer were immunized with the irradiated allogeneic line and recombinant GM-CSF, but no objective responses were observed.

Intratumoral delivery of targets remains an alternative therapy although the method remains restricted to cases with accessible tumorigenic lesions. One group initiated a Phase I trial with malignant melanoma patients receiving intratumoral injections of a vaccinia virus encoding B7-1 among other costimulatory receptors (Kaufman et al., 2006). Five of thirteen (38.5%) patients experienced an objective response at the lesion receiving vaccinia treatment. This study was in contrast to a trial of metastatic melanoma patients receiving intratumoral delivery with virus vectors encoding the B7-1 receptor and IL-12 gene (Triozzi et al., 2005b). Although, the majority of individuals expressed increased intratumoral levels of B7-1, a greater level of IL-12 mRNA was not evident, and no objective tumor responses were observed in the patients. Interestingly, malignant lesions injected with the viral vectors appeared to correlate to higher intratumoral levels of VEGF and IL-10, two cytokines with immunosuppressive properties.

Overall, clinical approaches that have sought to provide tumor cells with costimulatory ligands have failed to elicit a sufficient anti-tumor response that leads to the regression of disease as preclinical models have suggested. One explanation for this lack of clinical efficacy using genetically modified tumor cells could involve the ligation of inhibitory molecules expressed by activated T cells. Among these include the classical CTLA-4 receptor that binds both B7-1 and B7-2 with much higher affinity than CD28. PD-1 ligation on activated T cells with PD-L1 represents an additional receptor complex that can deliver inhibitor signals to T cells. Interestingly, PD-L1 has been found to be expressed by a number of malignancy types including renal cell carcinoma, non-small cell lung cancer, and breast cancer (Zang and Allison, 2007). An approach to avoiding these effects would involve blocking CTLA-4 and PD-L1 among other molecules that deliver inhibitory signals to activated immune effector cells while delivering plasmid DNA strategies. Indeed, inhibiting CTLA-4 with monoclonal antibodies has been the result of many clinical trials, particularly in melanoma patients receiving peptide-based vaccines (Zang and Allison, 2007). However, the response rate to therapy is normally low and the level of the specific T cell response does not correlate with tumor regression.

The concept of the tumor as an APC is burdened with issues related to the availability of malignant complexes in the patient for resection (*i.e.*, autologous vaccine approaches) as well as properly engaging T cell activation for an array of cancer types at the proper site in the body. For example, MHC II receptors are not typically expressed by tumor cells and this can be problematic in terms of CD4+ T cell priming if CD4 lymphocytes are required effector cells needed to abrogate tumor growth (Thompson et al., 2006). The B7 costimulatory receptors are also one of many receptor complexes expressed by APCs that successfully engage the T cell for immune activation. Therefore, in difficult cases of tumor-derived T cell priming, an optimal ligand set would have to be determined and genetically supplied to the tumor, compounding the complexity of the issue. The spatial location of a genetically modified tumor cell might also play a decisive role in initiating an immune response since priming naïve lymphocytes by the tumor has been suggested to be dependent on occurring in secondary lymphoid organs. Naïve T cells do not typically migrate into peripheral tissues (Jenkinson et al., 2005).

An improvement upon the tumor APC concept involves the genetic manipulation of a true APC type (*e.g.*, DCs) in order to initiate a potent immune response in the cancer patient. The complexities involved with determining an activating set of receptors are reduced since the DC does not have to be supplied with the needed costimulatory ligands to initiate a T cell response. Several human clinical trials are currently underway to study the effects of genetically modified DCs for the treatment of a variety of malignancies. Altogether, the field of DC-based vaccine therapies holds great promise, but the details of this progress are beyond the scope of this chapter. We would refer the reader to other excellent reviews on the subject (Steinman and Banchereau, 2007; Tuyaerts et al., 2007).

PRIME BOOST METHODS

A final strategy discussed here to enhancing the DNA vaccine response entails delivering multiple vaccine approaches in a prime-boost manner. Heterologous prime-boost methods work by focusing the immune response on a targeted antigen that is shared between distinct vaccine constructs (Palmowski et al., 2002; Woodland, 2004). So, in one sense, plasmid DNA could first be delivered as the primary vaccine followed sometime later by booster injections of recombinant protein of the native target antigen or viral vectors encoding the antigen. Although numerous variables are involved with the type of vaccine construct and order of delivery, various animal models have shown success with DNA vectors as priming agents and viral vectors as boosting a targeted response (Woodland, 2004). However, existing or inducing immunity in a patient population to viral products expressed by vectors such as vaccinia and adenovirus can limit the number of booster injections given in a vaccine setting (Ledford, 2008). Or, if possible, the virus vector can be constructed from less common serotype backgrounds.

Human studies have explored the efficacy of heterlogous prime-boosting regimes involving plasmid DNA to initiate a potent CD8+ T cell response. Most examples center upon the prevention of disease that results from HIV-1 (Mwau et al., 2004; Goonetilleke et al., 2006) and malarial infections (McConkey et al., 2003; Epstein et al., 2004) since neutralizing antibodies do little to completely prevent chronic infection with intracellular pathogens of these types. In one example showcasing the promising effects of prime-boosting, volunteers were initially immunized with a DNA vector encoding a malaria antigen and boosted sometime later with a vaccinia vector expressing the same target antigen (McConkey et al., 2003). Individuals immunized with both vaccine types yielded at least a greater than five fold increase in IFN-γ secreting T cells compared to groups immunized with plasmid or viral vector alone. This vaccine-induced response also suggested to confer partial protection from *Plasmodium falciparum* since prime-boosted individuals had a statistically significant delayed time of parasitemia using a sporozoite challenge model.

The efficacy of prime-boost methods have not yet been fully integrated into the clinical setting for cancer patients. In one known example, Smith and colleagues investigated a polyepitope vaccine method on individuals with high risk melanoma recurrence following successful resection (Smith et al., 2005). Overall, the immune response was directed towards one dominant epitope in 46.2% of vaccinated patients with no CTL activity to the other vaccine-encoded antigens. The benefits of prime-boosting were further confused with a specific immune response being elicited in only 2 of 6 patients in the DNA plus viral vector group compared to 4 of 7 patients immunized with viral vector alone. Therefore, much work still remains with optimizing prime-boost vaccine strategies to enhance the plasmid-DNA directed response in individuals stricken with both malignant and infectious diseases. Yet, owing to the success of pre-clinical data and results from selected clinical trials, including the previously described Todorova study (Todorova et al., 2005), prime-boost strategies represent a promising approach to circumventing the low immunogenicity of the DNA vector in humans.

CONCLUSIONS

DNA vaccines provide a number of practical benefits over other forms of vaccine therapies. These factors relate to the ease of genetic construction of the plasmid vector along with the ability to rapidly produce the construct on a large scale using inexpensive methods. Preclinical studies have documented that repeated injections of plasmid DNA is also capable of initiating robust humoral and CTL activity without concerns for safety or neutralizing immune responses that would eventually render the vaccine useless. Although the safety of the DNA vector has been consistently demonstrated in phase I trials of cancer and infectious diseases, this vaccine approach has suffered from an inability to effectively activate immune cells. The plasmid DNA immunization strategy has been somewhat disappointing given its less-than-anticipated results in human trials. The next generation of plasmid DNA use will have to incorporate strategies that seek to enhance the immunogenicity of the vector in cases of cancer and infectious diseases alike. In particular to human malignancies, additional approaches must include targeting the immunosuppressive properties of malignant cells at the level of the tumor microenvironment. It may also prove beneficial to combine active immunization with adoptive cell therapies. In this scenario, autologous DCs could be transfected with plasmid DNA and activated *ex vivo* and infused back into the cancer patient. Although much work still remains, the potential of DNA vaccines, as demonstrated through animal models, provides much hope for future progress to alleviate chronic and infectious diseases through vaccination.

ACKNOWLEDGEMENTS

This work was supported in part by NIH grant RR-12317. D.B.L. is supported by a Dean's Scholarship Award and a scholarship by the ARCS Foundation.

REFERENCES

American Cancer Society. (2007). Cancer Facts and Figures.

Algarra, I., Garcia-Lora, A., Cabrera, T., Ruiz-Cabello, F., and Garrido, F. (2004). The selection of tumor variants with altered expression of classical and nonclassical MHC class I molecules: implications for tumor immune escape. Cancer Immunol. Immunother. 53, 904-910.

Bai, X. F., Liu, J., Li, O., Zheng, P., and Liu, Y. (2003). Antigenic drift as a mechanism for tumor evasion of destruction by cytolytic T lymphocytes. J. Clin. Invest. 111, 1487-1496.

Beatty, G. L., and Paterson, Y. (2000). IFN-gamma can promote tumor evasion of the immune system *in vivo* by down-regulating cellular levels of an endogenous tumor antigen. J. Immunol. 165, 5502-5508.

Berzofsky, J. A., Ahlers, J. D., and Belyakov, I. M. (2001). Strategies for designing and optimizing new generation vaccines. Nat. Rev. Immunol. 1, 209-219.

Carter, P. J. (2006). Potent antibody therapeutics by design. Nat. Rev. Immunol. 6, 343-357.

Cassaday, R. D., Sondel, P. M., King, D. M., Macklin, M. D., Gan, J., Warner, T. F., Zuleger, C. L., Bridges, A. J., Schalch, H. G., Kim, K. M., et al. (2007). A phase I study of immunization using particle-mediated epidermal delivery of genes for gp100 and GM-CSF into uninvolved skin of melanoma patients. Clin. Cancer Res. 13, 540-549.

Coley, W. B. (1991). The treatment of malignant tumors by repeated inoculations of erysipelas. With a report of ten original cases. 1893. Clin. Orthop. Relat. Res., 3-11.

Collins, C. G., Tangney, M., Larkin, J. O., Casey, G., Whelan, M. C., Cashman, J., Murphy, J., Soden, D., Vejda, S., McKenna, S., et al. (2006). Local gene therapy of solid tumors with GM-CSF and B7-1 eradicates both treated and distal tumors. Cancer Gene. Ther. 13, 1061-1071.

Conry, R. M., Curiel, D. T., Strong, T. V., Moore, S. E., Allen, K. O., Barlow, D. L., Shaw, D. R., and LoBuglio, A. F. (2002). Safety and immunogenicity of a DNA vaccine encoding carcinoembryonic antigen and hepatitis B surface antigen in colorectal carcinoma patients. Clin Cancer Res. 8, 2782-2787.

Cooper, C. L., Davis, H. L., Angel, J. B., Morris, M. L., Elfer, S. M., Seguin, I., Krieg, A. M., and Cameron, D. W. (2005). CPG 7909 adjuvant improves hepatitis B virus vaccine seroprotection in antiretroviral-treated HIV-infected adults. Aids 19, 1473-1479.

Cooper, C. L., Davis, H. L., Morris, M. L., Efler, S. M., Adhami, M. A., Krieg, A. M., Cameron, D. W., and Heathcote, J. (2004). CPG 7909, an immunostimulatory TLR9 agonist oligodeoxynucleotide, as adjuvant to Engerix-B HBV vaccine in healthy adults: a double-blind phase I/II study. J. Clin. Immunol. 24, 693-701.

Crittenden, M., Gough, M., Harrington, K., Olivier, K., Thompson, J., and Vile, R. G. (2003). Expression of inflammatory chemokines combined with local tumor destruction enhances tumor regression and long-term immunity. Cancer Res. 63, 5505-5512.

Danial, N. N., and Korsmeyer, S. J. (2004). Cell death: critical control points. Cell 116, 205-219.

Dols, A., Smith, J. W., 2nd, Meijer, S. L., Fox, B. A., Hu, H. M., Walker, E., Rosenheim, S., Moudgil, T., Doran, T., Wood, W., et al. (2003). Vaccination of women with metastatic breast cancer, using a costimulatory gene (CD80)-modified, HLA-A2-matched, allogeneic, breast cancer cell line: clinical and immunological results. Hum. Gene. Ther. 14, 1117-1123.

Dunn, G. P., Bruce, A. T., Ikeda, H., Old, L. J., and Schreiber, R. D. (2002). Cancer immunoediting: from immunosurveillance to tumor escape. Nat. Immunol. 3, 991-998.

Eo, S. K., Gierynska, M., Kamar, A. A., and Rouse, B. T. (2001). Prime-boost immunization with DNA vaccine: mucosal route of administration changes the rules. J. Immunol. 166, 5473-5479.

Eriksson, K., and Holmgren, J. (2002). Recent advances in mucosal vaccines and adjuvants. Curr. Opin. Immunol. 14, 666-672.

Finn, O. J. (2007). Human tumor immunology at the molecular divide. J. Immunol. 178, 2615-2616.

Fynan, E. F., Webster, R. G., Fuller, D. H., Haynes, J. R., Santoro, J. C., and Robinson, H. L. (1993). DNA vaccines: protective immunizations by parenteral, mucosal, and gene-gun inoculations. Proc. Natl. Acad. Sci. U. S. A. 90, 11478-11482.

Gajewski, T. F. (2007). Failure at the effector phase: immune barriers at the level of the melanoma tumor microenvironment. Clin. Cancer Res. 13, 5256-5261.

Gilbert, S. C., Moorthy, V. S., Andrews, L., Pathan, A. A., McConkey, S. J., Vuola, J. M., Keating, S. M., Berthoud, T., Webster, D., McShane, H., and Hill, A. V. (2006). Synergistic DNA-MVA prime-boost vaccination regimes for malaria and tuberculosis. Vaccine 24, 4554-4561.

Giri, M., Ugen, K. E., and Weiner, D. B. (2004). DNA vaccines against human immunodeficiency virus type 1 in the past decade. Clin. Microbiol. Rev. 17, 370-389.

Goonetilleke, N., Moore, S., Dally, L., Winstone, N., Cebere, I., Mahmoud, A., Pinheiro, S., Gillespie, G., Brown, D., Loach, V., et al. (2006). Induction of multifunctional human immunodeficiency virus type 1 (HIV-1)-specific T cells capable of proliferation in healthy subjects by using a prime-boost regimen of DNA- and modified vaccinia virus Ankara-vectored vaccines expressing HIV-1 Gag coupled to CD8+ T-cell epitopes. J. Virol. 80, 4717-4728.

Gough, M., Crittenden, M., Thanarajasingam, U., Sanchez-Perez, L., Thompson, J., Jevremovic, D., and Vile, R. (2005). Gene therapy to manipulate effector T cell trafficking to tumors for immunotherapy. J. Immunol. 174, 5766-5773.

Groh, V., Wu, J., Yee, C., and Spies, T. (2002). Tumour-derived soluble MIC ligands impair expression of NKG2D and T-cell activation. Nature 419, 734-738.

Halperin, S. A., Van Nest, G., Smith, B., Abtahi, S., Whiley, H., and Eiden, J. J. (2003). A phase I study of the safety and immunogenicity of recombinant hepatitis B surface antigen co-administered with an immunostimulatory phosphorothioate oligonucleotide adjuvant. Vaccine 21, 2461-2467.

Hanahan, D., and Weinberg, R. A. (2000). The hallmarks of cancer. Cell 100, 57-70.

Hege, K. M., Jooss, K., and Pardoll, D. (2006). GM-CSF gene-modifed cancer cell immunotherapies: of mice and men. Int. Rev. Immunol. 25, 321-352.

Jenkinson, S. R., Williams, N. A., and Morgan, D. J. (2005). The role of intercellular adhesion molecule-1/LFA-1 interactions in the generation of tumor-specific CD8+ T cell responses. J. Immunol. 174, 3401-3407.

Jensen, P. E. (2007). Recent advances in antigen processing and presentation. Nat. Immunol. 8, 1041-1048.

Kaufman, H. L., Cohen, S., Cheung, K., DeRaffele, G., Mitcham, J., Moroziewicz, D., Schlom, J., and Hesdorffer, C. (2006). Local delivery of vaccinia virus expressing multiple costimulatory molecules for the treatment of established tumors. Hum. Gene. Ther. 17, 239-244.

Kennedy, R. C., Shearer, M. H., and Hildebrand, W. (1997). Nonhuman primate models to evaluate vaccine safety and immunogenicity. Vaccine 15, 903-908.

Kim, R., Emi, M., Tanabe, K., and Arihiro, K. (2006). Tumor-driven evolution of immunosuppressive networks during malignant progression. Cancer Res. 66, 5527-5536.

Krieg, A. M. (2006). Therapeutic potential of Toll-like receptor 9 activation. Nat. Rev. Drug. Discov. 5, 471-484.

Krieg, A. M. (2007). Development of TLR9 agonists for cancer therapy. J. Clin. Invest. 117, 1184-1194.

Krieg, A. M., Efler, S. M., Wittpoth, M., Al Adhami, M. J., and Davis, H. L. (2004). Induction of systemic TH1-like innate immunity in normal volunteers following subcutaneous but not intravenous administration of CPG 7909, a synthetic B-class CpG oligodeoxynucleotide TLR9 agonist. J. Immunother. 27, 460-471.

Laddy, D. J., and Weiner, D. B. (2006). From plasmids to protection: a review of DNA vaccines against infectious diseases. Int. Rev. Immunol. 25, 99-123.

Lavergne, E., Combadiere, C., Iga, M., Boissonnas, A., Bonduelle, O., Maho, M., Debre, P., and Combadiere, B. (2004). Intratumoral CC chemokine ligand 5 overexpression delays tumor growth and increases tumor cell infiltration. J. Immunol. 173, 3755-3762.

Ledford, H. (2008). Merck's HIV vaccine flop brings vectors under closer scrutiny. Nat. Biotechnol. 26, 3-4.

Lopez, M. V., Adris, S. K., Bravo, A. I., Chernajovsky, Y., and Podhajcer, O. L. (2005). IL-12 and IL-10 expression synergize to induce the immune-mediated eradication of established colon and mammary tumors and lung metastasis. J. Immunol. 175, 5885-5894.

Lowe, D. B., Shearer, M. H., Jumper, C. A., and Kennedy, R. C. (2007a). Towards progress on DNA vaccines for cancer. Cell. Mol. Life Sci. 64, 2391-2403.

Lowe, D. B., Shearer, M. H., Jumper, C. A., and Kennedy, R. C. (2007b). Immunity to Bacteria, In Encyclopedia of Life Sciences. (John Wiley and Sons, Ltd: Chichester), http://www.els.net/ [DOI: 10.1002/9780470015902.a000481.pub2].

Lowe, D. B., Shearer, M. H., and Kennedy, R. C. (2006). DNA vaccines: successes and limitations in cancer and infectious disease. J. Cell. Biochem. 98, 235-242.

Lowe, D. B., Shearer, M. H., Tarbox, J. A., Kang, H. S., Jumper, C. A., Bright, R. K., and Kennedy, R. C. (2005). In vitro simian virus 40 large tumor antigen expression correlates with differential immune responses following DNA immunization. Virology 332, 28-37.

MacGregor, R. R., Boyer, J. D., Ugen, K. E., Lacy, K. E., Gluckman, S. J., Bagarazzi, M. L., Chattergoon, M. A., Baine, Y., Higgins, T. J., Ciccarelli, R. B., et al. (1998). First human trial of a DNA-based vaccine for treatment of human immunodeficiency virus type 1 infection: safety and host response. J. Infect. Dis. 178, 92-100.

Maloney, D. G., Liles, T. M., Czerwinski, D. K., Waldichuk, C., Rosenberg, J., Grillo-Lopez, A., and Levy, R. (1994). Phase I clinical trial using escalating single-dose infusion of chimeric anti-CD20 monoclonal antibody (IDEC-C2B8) in patients with recurrent B-cell lymphoma. Blood 84, 2457-2466.

McConkey, S. J., Reece, W. H., Moorthy, V. S., Webster, D., Dunachie, S., Butcher, G., Vuola, J. M., Blanchard, T. J., Gothard, P., Watkins, K., et al. (2003). Enhanced T-cell immunogenicity of plasmid DNA vaccines boosted by recombinant modified vaccinia virus Ankara in humans. Nat. Med. 9, 729-735.

McLaughlin, P., Grillo-Lopez, A. J., Link, B. K., Levy, R., Czuczman, M. S., Williams, M. E., Heyman, M. R., Bence-Bruckler, I., White, C. A., Cabanillas, F., et al. (1998). Rituximab chimeric anti-CD20 monoclonal antibody therapy for relapsed indolent lymphoma: half of patients respond to a four-dose treatment program. J. Clin. Oncol. 16, 2825-2833.

Mule, J. J., Custer, M., Averbook, B., Yang, J. C., Weber, J. S., Goeddel, D. V., Rosenberg, S. A., and Schall, T. J. (1996). RANTES secretion by gene-modified tumor cells results in loss of tumorigenicity *in vivo*: role of immune cell subpopulations. Hum. Gene. Ther. 7, 1545-1553.

Mwau, M., Cebere, I., Sutton, J., Chikoti, P., Winstone, N., Wee, E. G., Beattie, T., Chen, Y. H., Dorrell, L., McShane, H., et al. (2004). A human immunodeficiency virus 1 (HIV-1) clade A vaccine in clinical trials: stimulation of HIV-specific T-cell responses by DNA and recombinant modified vaccinia virus Ankara (MVA) vaccines in humans. J. Gen. Virol. 85, 911-919.

Niethammer, A. G., Xiang, R., Becker, J. C., Wodrich, H., Pertl, U., Karsten, G., Eliceiri, B. P., and Reisfeld, R. A. (2002). A DNA vaccine against VEGF receptor 2 prevents effective angiogenesis and inhibits tumor growth. Nat. Med. 8, 1369-1375.

Novellino, L., Castelli, C., and Parmiani, G. (2005). A listing of human tumor antigens recognized by T cells: March 2004 update. Cancer Immunol. Immunother. 54, 187-207.

Palmowski, M. J., Choi, E. M., Hermans, I. F., Gilbert, S. C., Chen, J. L., Gileadi, U., Salio, M., Van Pel, A., Man, S., Bonin, E., et al. (2002). Competition between CTL narrows the immune response induced by prime-boost vaccination protocols. J. Immunol. 168, 4391-4398.

Parmiani, G., Castelli, C., Pilla, L., Santinami, M., Colombo, M. P., and Rivoltini, L. (2007). Opposite immune functions of GM-CSF administered as vaccine adjuvant in cancer patients. Ann. Oncol. 18, 226-232.

Parney, I. F., Chang, L. J., Farr-Jones, M. A., Hao, C., Smylie, M., and Petruk, K. C. (2006). Technical hurdles in a pilot clinical trial of combined B7-2 and GM-CSF immunogene therapy for glioblastomas and melanomas. J. Neurooncol. 78, 71-80.

Pavlenko, M., Roos, A. K., Lundqvist, A., Palmborg, A., Miller, A. M., Ozenci, V., Bergman, B., Egevad, L., Hellstrom, M., Kiessling, R., et al. (2004). A phase I trial of DNA vaccination with a plasmid expressing prostate-specific antigen in patients with hormone-refractory prostate cancer. Br. J. Cancer 91, 688-694.

Quaglino, E., Iezzi, M., Mastini, C., Amici, A., Pericle, F., Di Carlo, E., Pupa, S. M., De Giovanni, C., Spadaro, M., Curcio, C., et al. (2004). Electroporated DNA vaccine clears away multifocal mammary carcinomas in her-2/neu transgenic mice. Cancer Res. 64, 2858-2864.

Raez, L. E., Cassileth, P. A., Schlesselman, J. J., Sridhar, K., Padmanabhan, S., Fisher, E. Z., Baldie, P. A., and Podack, E. R. (2004). Allogeneic vaccination with a B7.1 HLA-A gene-modified adenocarcinoma cell line in patients with advanced non-small-cell lung cancer. J. Clin. Oncol. 22, 2800-2807.

Rosenberg, S. A., Yang, J. C., Sherry, R. M., Hwu, P., Topalian, S. L., Schwartzentruber, D. J., Restifo, N. P., Haworth, L. R., Seipp, C. A., Freezer, L. J., et al. (2003). Inability to immunize patients with metastatic melanoma using plasmid DNA encoding the gp100 melanoma-melanocyte antigen. Hum. Gene. Ther. 14, 709-714.

Salih, H. R., Rammensee, H. G., and Steinle, A. (2002). Cutting edge: down-regulation of MICA on human tumors by proteolytic shedding. J. Immunol. 169, 4098-4102.

Sanchez-Perez, L., Kottke, T., Diaz, R. M., Ahmed, A., Thompson, J., Chong, H., Melcher, A., Holmen, S., Daniels, G., and Vile, R. G. (2005). Potent selection of antigen loss variants of B16 melanoma following inflammatory killing of melanocytes *in vivo*. Cancer Res. 65, 2009-2017.

Seder, R. A., and Hill, A. V. (2000). Vaccines against intracellular infections requiring cellular immunity. Nature 406, 793-798.

Shah, S., Divekar, A. A., Hilchey, S. P., Cho, H. M., Newman, C. L., Shin, S. U., Nechustan, H., Challita-Eid, P. M., Segal, B. M., Yi, K. H., and Rosenblatt, J. D. (2005). Increased rejection of primary tumors in mice lacking B cells: inhibition of anti-tumor CTL and TH1 cytokine responses by B cells. Int. J. Cancer 117, 574-586.

Smith, C. L., Dunbar, P. R., Mirza, F., Palmowski, M. J., Shepherd, D., Gilbert, S. C., Coulie, P., Schneider, J., Hoffman, E., Hawkins, R., et al. (2005). Recombinant modified vaccinia Ankara primes functionally activated CTL specific for a melanoma tumor antigen epitope in melanoma patients with a high risk of disease recurrence. Int. J. Cancer 113, 259-266.

Speiser, D. E., Lienard, D., Rufer, N., Rubio-Godoy, V., Rimoldi, D., Lejeune, F., Krieg, A. M., Cerottini, J. C., and Romero, P. (2005). Rapid and strong human CD8+ T cell responses to vaccination with peptide, IFA, and CpG oligodeoxynucleotide 7909. J. Clin. Invest. 115, 739-746.

Starnes, C. O. (1992). Coley's toxins. Nature 360, 23.

Steinman, R. M., and Banchereau, J. (2007). Taking dendritic cells into medicine. Nature 449, 419-426.

Tagawa, S. T., Lee, P., Snively, J., Boswell, W., Ounpraseuth, S., Lee, S., Hickingbottom, B., Smith, J., Johnson, D., and Weber, J. S. (2003). Phase I study of intranodal delivery of a plasmid DNA vaccine for patients with Stage IV melanoma. Cancer 98, 144-154.

Thompson, J. A., Dissanayake, S. K., Ksander, B. R., Knutson, K. L., Disis, M. L., and Ostrand-Rosenberg, S. (2006). Tumor cells transduced with the MHC class II Transactivator and CD80 activate tumor-specific CD4+ T cells whether or not they are silenced for invariant chain. Cancer Res. 66, 1147-1154.

Timmerman, J. M., Singh, G., Hermanson, G., Hobart, P., Czerwinski, D. K., Taidi, B., Rajapaksa, R., Caspar, C. B., Van Beckhoven, A., and Levy, R. (2002). Immunogenicity of a plasmid DNA vaccine encoding chimeric idiotype in patients with B-cell lymphoma. Cancer Res. 62, 5845-5852.

Todorova, K., Ignatova, I., Tchakarov, S., Altankova, I., Zoubak, S., Kyurkchiev, S., and Mincheff, M. (2005). Humoral immune response in prostate cancer patients after immunization with gene-based vaccines that encode for a protein that is proteasomally degraded. Cancer Immun. 5, 1.

Toka, F. N., Pack, C. D., and Rouse, B. T. (2004). Molecular adjuvants for mucosal immunity. Immunol. Rev. 199, 100-112.

Tominaga, M., Iwashita, Y., Ohta, M., Shibata, K., Ishio, T., Ohmori, N., Goto, T., Sato, S., and Kitano, S. (2007). Antitumor effects of the MIG and IP-10 genes transferred with poly [D,L-2,4-diaminobutyric acid] on murine neuroblastoma. Cancer. Gene. Ther. 14, 696-705.

Townsend, S. E., and Allison, J. P. (1993). Tumor rejection after direct costimulation of CD8+ T cells by B7-transfected melanoma cells. Science 259, 368-370.

Triozzi, P. L., Aldrich, W., Allen, K. O., Carlisle, R. R., LoBuglio, A. F., and Conry, R. M. (2005a). Phase I study of a plasmid DNA vaccine encoding MART-1 in patients with resected melanoma at risk for relapse. J. Immunother. 28, 382-388.

Triozzi, P. L., Allen, K. O., Carlisle, R. R., Craig, M., LoBuglio, A. F., and Conry, R. M. (2005b). Phase I study of the intratumoral administration of recombinant canarypox viruses expressing B7.1 and interleukin 12 in patients with metastatic melanoma. Clin. Cancer Res. 11, 4168-4175.

Tuyaerts, S., Aerts, J. L., Corthals, J., Neyns, B., Heirman, C., Breckpot, K., Thielemans, K., and Bonehill, A. (2007). Current approaches in dendritic cell generation and future implications for cancer immunotherapy. Cancer Immunol. Immunother. 56, 1513-1537.

Ulmer, J. B., Donnelly, J. J., Parker, S. E., Rhodes, G. H., Felgner, P. L., Dwarki, V. J., Gromkowski, S. H., Deck, R. R., DeWitt, C. M., Friedman, A., et al. (1993). Heterologous protection against influenza by injection of DNA encoding a viral protein. Science 259, 1745-1749.

van der Bruggen, P., Traversari, C., Chomez, P., Lurquin, C., De Plaen, E., Van den Eynde, B., Knuth, A., and Boon, T. (1991). A gene encoding an antigen recognized by cytolytic T lymphocytes on a human melanoma. Science 254, 1643-1647.

Vicari, A. P., Schmalbach, T., Lekstrom-Himes, J., Morris, M. L., Al-Adhami, M. J., Laframboise, C., Leese, P., Krieg, A. M., Efler, S. M., and Davis, H. L. (2007). Safety, pharmacokinetics and immune effects in normal volunteers of CPG 10101 (ACTILON), an investigational synthetic toll-like receptor 9 agonist. Antivir. Ther. 12, 741-751.

Wang, B., Ugen, K. E., Srikantan, V., Agadjanyan, M. G., Dang, K., Refaeli, Y., Sato, A. I., Boyer, J., Williams, W. V., and Weiner, D. B. (1993). Gene inoculation generates immune responses against human immunodeficiency virus type 1. Proc. Natl. Acad. Sci. U. S. A. 90, 4156-4160.

Weiner, D. B., and Kennedy, R. C. (1999). Genetic vaccines. Sci. Am. 281, 50-57.

Winter, H., Hu, H. M., Poehlein, C. H., Huntzicker, E., Osterholzer, J. J., Bashy, J., Lashley, D., Lowe, B., Yamada, J., Alvord, G., et al. (2003). Tumour-induced polarization of tumour vaccine-draining lymph node T cells to a type 1 cytokine profile predicts inherent strong immunogenicity of the tumour and correlates with therapeutic efficacy in adoptive transfer studies. Immunology 108, 409-419.

Wolff, J. A., Malone, R. W., Williams, P., Chong, W., Acsadi, G., Jani, A., and Felgner, P. L. (1990). Direct gene transfer into mouse muscle *in vivo*. Science 247, 1465-1468.

Woodland, D. L. (2004). Jump-starting the immune system: prime-boosting comes of age. Trends. Immunol. 25, 98-104.

Xiang, R., Lode, H. N., Chao, T. H., Ruehlmann, J. M., Dolman, C. S., Rodriguez, F., Whitton, J. L., Overwijk, W. W., Restifo, N. P., and Reisfeld, R. A. (2000). An autologous oral DNA vaccine protects against murine melanoma. Proc. Natl. Acad. Sci. U. S. A. 97, 5492-5497.

Xiang, R., Silletti, S., Lode, H. N., Dolman, C. S., Ruehlmann, J. M., Niethammer, A. G., Pertl, U., Gillies, S. D., Primus, F. J., and Reisfeld, R. A. (2001). Protective immunity against human carcinoembryonic antigen (CEA) induced by an oral DNA vaccine in CEA-transgenic mice. Clin. Cancer Res. 7, 856s-864s.

Yang, G., Hellstrom, K. E., Hellstrom, I., and Chen, L. (1995). Antitumor immunity elicited by tumor cells transfected with B7-2, a second ligand for CD28/CTLA-4 costimulatory molecules. J. Immunol. 154, 2794-2800.

Yannelli, J. R., and Wroblewski, J. M. (2004). On the road to a tumor cell vaccine: 20 years of cellular immunotherapy. Vaccine 23, 97-113.

Yu, M., and Finn, O. J. (2006). DNA vaccines for cancer too. Cancer Immunol. Immunother. 55, 119-130.

Zang, X., and Allison, J. P. (2007). The B7 family and cancer therapy: costimulation and coinhibition. Clin. Cancer Res. 13, 5271-5279.

Zou, W. (2005). Immunosuppressive networks in the tumour environment and their therapeutic relevance. Nat. Rev. Cancer 5, 263-274.

Index

M

N

O

P

R

S

T

U

V

X

Y

Z

Other Books of Interest

www.ingramcontent.com/pod-product-compliance
Lightning Source LLC
LaVergne TN
LVHW011753200926
848484LV00003B/697

* 9 7 8 1 9 0 4 4 5 5 3 5 6 *